MANAGEMENT OF TECHNOLOGY

MANAGEMENT OF TECHNOLOGY

Managing Effectively in Technology-Intensive Organizations

Hans J. Thamhain

WILEY

John Wiley & Sons, Inc.

Published by John Wiley & Sons, Inc., Hoboken, New Jersey
Published simultaneously in Canada

For general information on our other products and services or for technical support, please contact our Customer Care Department within the United States at (800) 762-2974, outside the United States at (317) 572-3993 or fax (317) 572-4002.

Wiley also publishes its books in a variety of electronic formats. Some content that appears in print may not be available in electronic books. For more information about Wiley products, visit our web site at www.wiley.com.

Library of Congress Cataloging-in-Publication Data:

Thamhain, Hans J. (Hans Jurgen), 1936–
    Management of Technology: managing effectively in technology-intensive organizations / Hans Thamhain.
        p. cm.
    ISBN-13: 978-0-471-41551-0 (cloth)
    ISBN-10: 0-471-41551-0 (cloth)
    1. High technology industries—Management. 2. Information technology—Management. 3. Technological innovations—Management. I. Title.
    HD62.37.T49 2005
658—dc22

2005004374

Printed in the United States of America

10 9 8 7 6 5 4 3 2 1

CONTENTS

PREFACE

The magnitude and speed of technological advances over the past decades are stunning, reshaping our world and influencing virtually every aspect of life. This current rate of change is not likely to slow down any time soon. From a business perspective, technology is the catalyst that creates wealth. It also lowers the barriers to market entry and levels the competitive playing field for virtually any business, from technical products to medical and financial services to transportation and retailing.

As the important role of technology is being realized, business leaders are under great pressure to leverage technology to obtain better, faster, and cheaper results, gaining sustainable competitive advantages.

Time after time, managers have told me that the biggest challenge they face is not so much in understanding and applying technology, but in integrating and transferring technologies into marketable products, services, and processes that add value to the enterprise. This requires specialized skills in planning, organizing, and guiding multidisciplinary activities. It also requires a great deal of people skills in building cross-functional teams and leading them toward desired results. This involves effective motivation, power and resource sharing, communication both vertically and horizontally, and conflict management. To get results, technology managers must be social architects who understand the culture and value system of the enterprise, and can relate socially as well as technically. The days of managers who get by with only technical expertise or pure administrative skills are gone.

This book is written from the technical manager's perspective, for managers and professionals who must function effectively in technology-oriented business environments, such as R&D, product development, operations, marketing, and field services. As technology crosses virtually all levels and all disciplines of an enterprise, the principles of managing in technology are relevant not only in engineering and R&D, but also in any organization or business that must effectively deal with the application, integration, and transfer of technology. Financial institutions, hospitals, and law enforcement are just a few examples of the vast array of organizations outside the traditional engineering-scientific community that see themselves as high-technology enterprises.

In addition to the objective of professional reference, this book also is designed as a text for college courses in technology management. It integrates today's contemporary management practices with the emerging body of knowledge on management of technology, which is further linked with the contemporary concepts of organizational behavior. Managers and technology-oriented professionals at all levels should find this text useful in gaining an understanding of the organizational process, organizational dynamics, and critical success factors that drive technology-based business performance. Such insight can help in fine-tuning leadership style, resource alloca-tion, and organizational developments, hence continuously improving the enterprise's ability to compete effectively in today's complex global markets.

The book also includes my findings of 10 years of formal field research in the area of engineering and technology management. It further integrates the observations and experiences of my 20 years of R&D and technology management with ITT, Westinghouse, General Electric, and GTE/Verizon prior to my current teaching and research career.

I would like to express my appreciation to the many colleagues who encouraged and nurtured the development of this book. Special thanks go to the large number of professionals who contributed valuable information via the formal field research which is summarized in this book.

Waltham, Massachusetts

Hans J. Thamhain
January 2005

1

CHALLENGES OF MANAGING IN HIGH TECHNOLOGY

HUGHES AND THE DIRECT BROADCASTING BUSINESS

Today, Hughes Electronics, a unit of General Motors, is a successful part of the communications revolution. The company designs, manufactures, and markets advanced electronic systems, including telecommunications equipment, offering digital television entertainment and information programming via satellite Hughes operates a network of satellites, including DIRECTV, the largest US direct broadcast satellite (DBS) system, marketed via PanAmSat Corporation, offering 150 channels of movies, cable TV programs, and sporting events directly to anyone in the United States, Canada, and part of South America. Hughes had secured part of this market niche already in 1997 by forming an alliance between its PanAmSat subsidiary and SatMex (Satellites Mexicanos) to bid for an additional strategic satellite position. Other bidders included Primestar and EchoStar. At 77 degrees west longitude the Mexican slot was the last to allow a satellite to beam full conus coverage of the United States and Mexico.

The direct broadcast satellite (DBS) ventures are part of the world's most complex, but also fastest growing businesses. Today, DIRECTV/Hughes markets its services through PanAmSat to more than 10 million subscribers with annual revenues of nearly $7 billion. For Hughes Electronics, a subsidiary of General Motors, this new business started 15 years ago, when the Federal Communications Commission (FCC) set aside part of the radio spectrum for

TV programs. But digital compression had not been invented then, and satellites were more primitive. Yet, competitors such as Rupert Murdoch's Sky Broadcasting in Europe and Hutchison Whampoa of Hong Kong broadcasting into Asia started their ventures as early as 1990. DIRECTV/Hughes' satellite factory, "High Bay" in El Segundo near Los Angeles International Airport, appears unusual by traditional norms of manufacturing. There are no assembly lines, no conveyor belts, and no grinding machine tools. Workers gather around a half-dozen shiny objects, which create a *Star Wars* ambiance. High Bay is the plant where Hughes finishes satellites before launching them into space. Many of the satellites are more conventional telecommunications relays that will handle international calls. But the stars of High Bay are the body-stabilized DBS models, which are crammed with stuff: power amplifiers, radio-wave propagators, titanium fuel tanks, navigational gear, explosive charges for deploying solar panels, azure-blue glass solar cells, thruster jets, antennas, solar panels. Hughes hopes that this product will transform the television industry, and the anticipation in this plant is palpable. The combination of great stakes, exotic technology, and painstaking team efforts by many people, culminating in a single event that may bring instantaneous success or the total disaster of launch failure, makes satellite building the most tension-laden business in the world.

At the beginning of 2004, Hughes Electronics planned to sell its DIRECTV business for $6.6 billion to News Corporation, while GM intended to sell the remaining Hughes Electronics businesses, including satellite network operator PanAmSat and satellite equipment maker Hughes Network Systems, to EchoStar (DIRECTV's main rival) for $18 billion. However, the GM-EchoStar deal was rejected by the FCC.

Source: Hughes Electronics Corporation at *www.hughes.com*.

1.1   MANAGING IN TODAY'S HIGH-TECH BUSINESS ENVIRONMENT

The complexities and challenges faced by Hughes Electronics and its DIRECTV business are quite common in today's technology-based business environment.[1] Activities often cluster around projects with team efforts that span organizational lines involving a broad spectrum of personnel, support groups, subcontractors, vendors, partners, government agencies, and customer organizations. Effective linkages, cooperation, and alliances among various organizational functions are critical for proper communication, decision making, and control. This requires sophisticated

---

[1] For additional discussion on these challenges and issues of high-tech management see Armstrong (2000), Barkema (2002), Barner (1997), Dillon (2001), Gray and Larson (2000), and Thamhain (2002).

teamwork and, as is typical for many high-tech organizations, the ability to manage across functional lines with little or no formal authority, dealing effectively with resource sharing and multiple reporting relationships and accountabilities.

Yet, to be sure, technology and its management are not new phenomena. It has been around for a long time. From Noah's Ark and the Egyptian pyramids to railroads and steel mills, people have used and managed technology, and developed special skills to deal with the challenges, risks, and uncertainties. However, today's technologies have created a new business environment that has pushed conventional boundaries ever farther, with high risks and great opportunities for big gains. New technologies, especially those related to computers and communication, as shown in the Hughes DBS situation, have radically changed the workplace and transformed our global economy, with a focus on effectiveness, value, and speed. Every organization is under pressure to do more things faster, better, and with fewer resources. Speed especially has become one of the great equalizers of competitive performance. As for the Hughes broadcasting satellite, a system that requires five years of development time, it may be obsolete in a couple of years, unless provisions for continuous upgrading and enhancement have been built into the system and are implemented according to evolving market needs. The impact of accelerating technology is even more visible in consumer markets. A new computer product introduced today will be obsolete in four months, while the firm's concept-to-market development cycle may take a year or more! Hence, the new breed of business leaders must deal effectively with a broad spectrum of contemporary challenges that focus on time-to-market pressures, accelerating technologies, innovation, resource limitations, technical complexities, social and ethical issues, operational dynamics, cost, risks, and technology itself, as summarized in Table 1.1. Traditional linear work processes and top-down controls are no longer sufficient, but are gradually being replaced with alternate organizational designs and new management techniques and business processes, such as concurrent engineering, design-build, and Stage-Gate protocols (Thamhain 2003). These techniques offer more sophisticated capabilities for cross-functional integration, resources mobility, effectiveness, and marker responsiveness, but they also require more sophisticated management skills and leadership.

Taken together, the business environment is quite different from what it used to be. New technologies and changing global markets have transformed our business communities. Companies that survive and prosper in this environment have the ability to deal with a broad spectrum of contemporary challenges that focus on speed, cost, and quality. They have also shifted their focus from managing specific functions efficiently to an integrated approach of business management with particular attention being paid to organizational interfaces, human factors, and the business process.

1.2 MOT SCOPE AND FOCUS

Both the scope and the definition of *management of technology (MoT)* have been the subject of intense debate, controversy, and confusion. The words "management" and

Table 1.1   Characteristics and Challenges of Today's Technology-Based Businesses

Characteristics and challenges:

- High task complexities, risks, and uncertainties
- Fast-changing markets, technology, regulations
- Intense competition, open global markets
- Resource constraint, tough performance requirements
- Tight, end-date-driven schedules
- Total project life-cycle considerations
- Complex organizations and cross-functional linkages
- Joint ventures, alliances and partnerships; need for dealing with different organizational cultures and values
- Complex business processes and stakeholder communities
- Need for continuous improvements, upgrades, and enhancements
- Need for sophisticated people skills, ability to deal with organizational conflict, power, and politics
- Virtual organizations, markets, and support systems
- Increasing impact of IT and e-business

Resulting in demand for:

- ➤ High market responsiveness
- ➤ Fast developments
- ➤ Low cost
- ➤ High levels of creativity, innovation, and efficiency

"technology" each carry different meanings and boundaries, and in combination, they stand for a wide array of actions, methods, tools, and techniques. For some people, MoT relates to scientific research and the development of new concepts. To others, MoT means engineering design and development, manufacturing, or operations management, while yet others relate MoT to managing hospitals, financial businesses, the Olympic Games, or eBay. Indeed the scope of MoT is very broad and diverse. Its boundaries also overlap considerably with those of the major disciplines of science, engineering, and management. Furthermore, with the increasing complexity of our business environment, MoT focuses more strongly on "managing" the organizational processes and the people affiliated with them.

Over the last 20 to 30 years, management literature has shifted its technology focus from R&D to new product development and then to product enhancement. In more current times, its emphasis has been gravitating toward market development and e-commerce. However, the scope of technology management has not changed. *In its fundamental form, the scope of technology management* runs parallel to the general field of management. It includes the planning, organizing coordinating, and integrating of all the resources needed to achieve the enterprise-specific goals and objectives. However, what makes MoT unique and differentiates it from other established fields of organization and management are the special knowledge and skill requirements for applying the technology, including organizing and coordinating technology resources, and directing the people involved with it. In this context, *management of technology* can be broadly defined as the art and science of *creating value* by using technology together with the other resources of an organization. This definition points toward the well-supported position that management of technology is not be confined to R&D, engineering, or scientific work, but includes many other facets of the enterprise and its environment. This is the context in which the definition of MoT will be further developed and discussed in the next section.

1.3 DEVELOPING A FORMAL DEFINITION

By its very nature, *management of technology* is multidisciplinary. It involves at least two established areas of knowledge: (1) organization and management and (2) natural sciences and engineering, plus a broad spectrum of disciplines from other fields, in support of either management or technology, such as social sciences, information technology, and industrial engineering.

In 1987, the National Research Council released the following definition of management of technology:

> *Management of Technology links engineering, science and management disciplines to plan, develop and implement technological capabilities to shape and accomplish the strategic and operational goals of an organization.*

The specific dimensions of this definition are worth emphasizing. They are also graphically summarized in Figure 1.1.

1. MoT involves the *management of engineering, natural science, and social science.*
2. MoT involves *administrative science* in the planning, decision making, development, and implementation of technology.
3. MoT focuses on the *development of operational capabilities* such as manufacturing, distribution, and field services.
4. MoT involves *operational processes, tools and techniques, and people* who make it all happen.
5. MoT involves *guidance and leadership* aimed toward the development of products and services.
6. MoT is influenced by business strategy, organizational culture, and the business environment, and vice versa.
7. MoT involves managing many *interdisciplinary* components and managing their *integration* into a whole system. It also involves managing the system.

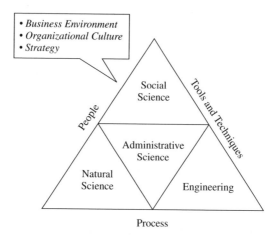

Figure 1.1 Dimensions of MoT.

Beginning with the 1990s a body of specialized knowledge emerged linking MoT with the broader enterprise operation and its strategy. Technology became broadly recognized as a resource for conducting business more effectively throughout all functions of the enterprise, from concept development to distribution, and for sustaining economic growth.[2] Moreover, technology became a strategic weapon for positioning products and services uniquely in the marketplace, and the *management of technology* emerged as a managerial specialty and professional field.

Today, businesses must function in an environment of increasing global competition, rapid technological changes, time compression, and ultra-efficiency. Virtually every organization aims to make things better, faster, and cheaper. Success is driven to a significant degree by an organization's ability to integrate and utilize technology effectively. *Management of technology* is recognized as a core competency, critical to the survival and growth of virtually every enterprise. It includes to some extent the creation of new technology, but in most situations it focuses on the *application* of technology toward the creation and improvement of products and services, and the enhancement of business processes toward more effective, faster, more agile, and more socially acceptable operations. In this context, management of technology can be broadly defined as follows:

> *Management of technology (MoT)* is the art and science of *creating value* by using technology together with other resources of an organization.

---

[2] The National Science and Technology Council (NSTC) concluded in 1996 (Report: *Technology in the National Interest*) that technological progress is the single most important determining factor in a nation's sustained economic growth, crediting one half of the U.S. economic growth during the past 50 years to advances in technology.

From my vantage point, this is a good working definition which focuses on the managerial processes, skills, tools, and techniques needed to achieve business results. It also establishes the platform for the discussions in this book. However, the term *technology management* can be defined in many ways depending on the emphasis and context in which technology and its management are seen as part of the enterprise.[3] Within today's enterprise, technology management crosses all disciplines, functions, and levels of the organization. It seems less important to arrive at a perfect definition than it is to understand the role of technology for the enterprise and the way it can be leveraged toward business performance. A more advanced discussion on the unique characteristics of technology management is provided in Chapter 2.

1.4 THE SPECIAL ROLE OF ENGINEERING MANAGEMENT

Engineering management (EM) focuses more specifically on managing in an *engineering* environment. It deals with the application of engineering principles and organizational and people skills to the delivery of engineering-based results. Engineering management is often conducted within a project context. There is a *great deal of overlap* between *engineering management (EM)* and *technology management (TM)*. Yet, a persuasive argument can be made that EM, with its somewhat better established theoretical framework,[4] is a subset of TM, regarding its body of knowledge and managerial practice. Engineering a special area of the broader field of technology, but the management of engineering focuses more exclusively on the operational aspects of the firm, while technology management covers virtually all facets of the enterprise.

When analyzing the two areas of management from a *career path perspective*, technology management often becomes the umbrella for engineering management. Engineers, scientists, and other technical personnel with managerial aspirations

---

[3] There are a large number of definitions for MoT. Often each subject expert creates his or her own definition with focus on a specific area of managerial focus. Here is a sample of MoT definitions from well-known scholars of management: Michael Badawy of Virginia Polytechnic Institute states "technology management . . . is the architecture or configuration of management systems, policies and procedures, governing the strategic and operational functioning of the enterprise in order to achieve its goals and objectives." According to Steven Wheelwright of Harvard University, MoT is based on "the theoretical and practical knowledge, skills and artifacts, useful to the development of products and services as well as their delivery systems." Peter Drucker focuses more on the human side, referring to technology management as an "interrelated system of intercommunicating units and activities." Ray Gehani of the University of Akron, in his book *Management of Technology and Operations* (1998), describes technology management as "the transformation of proprietary know-how to commercialization."

[4] The beginnings of engineering management as a formal discipline can be traced back to the 1940s. One of the early recognitions of EM as a profession was the formation of the Engineering Management Society, as part the newly organized IEEE, in 1963. The body of knowledge of engineering management has been developed for a long time with the support of many government institutions, universities, and professional organizations. Many publications helped in documenting the building blocks of EM. One of the oldest and highly respected journals reporting EM research today is the *IEEE Transactions on Engineering Management*, now in its fiftieth year of publication.

might grow into engineering management positions, often via project team leader-ship assignments. Depending on the company's size, these engineering management positions can be connected to long career ladders, leading to senior management positions such as director of engineering, director of manufacturing, or VP of mar-keting. However, with increasing responsibilities, the engineering manager becomes involved with the broader issues of the enterprise, requiring more interdisciplinary, integrative, and strategic business skills, which span the wide spectrum of technol-ogy areas within the company and across its boundaries. At this point the engineer-ing manager's responsibilities strongly overlap with the responsibilities for managing the technology of the business; essentially the *engineering manager* becomes a *manager of technology*.

1.4.1   Educational Impact

This observation also has educational implications for both students and educators. Choosing an EM versus MoT curriculum is a difficult decision, especially for engi-neering professionals in their early career stages. William Lannes (2001), director of engineering management at the University of New Orleans, provides a very good perspective with the following advice:

> *We differentiate between the two options of EM and TM by telling students that if they want to remain in their technical area of expertise, but see a need for improving their managerial skills and understanding of business, then the* Engineering Management Program *is right for them. On the other side, if they want to use their engineering knowledge and expertise as a stepping stone to grow into a stage in their careers in which they have to make executive decisions on the use of technology in their corpora-tions, then the* Master's Program in Management of Technology *should be considered.*

As a final note to the EM versus MoT discussion, it should be emphasized that nei-ther field should be seen as "superior" to the other. I am saying this as an engineer who has worked in and studied both the EM and TM areas. Clearly, both fields over-lap a great deal, in concept and in practice. While one could argue that *engineering management* has a more established body of knowledge with better-defined bound-aries and career paths, it is the area of *technology management* that crosses virtually all organizational areas and levels of the enterprise, regardless of its industry, mar-kets, or business specialization. Many of the managerial issues, tools, and processes apply to both concepts and are discussed in this book under the label of *management of technology*. Hence the focus of the following chapters is on *managing* effectively in technology-based organizations.

1.5   GLOBAL DIMENSIONS

Globalization and technology are perhaps the two most significant drivers of business performance. They are interrelated. Companies benefit from economies of scale and scope by exploiting technology globally, but they also need technology to operate globally. Therefore, technology is both an enabler of globalization and vice versa. Yet, multinational commerce flourished long before the computer age. Motivated by opportunities for economic gain, companies have ventured into international operations since the dawn of civilization. However, in the past, the ability to gain access to supporting resources, such as labor, raw material, energy, and knowledge, was limited. In addition, coordination and integration of activities across international borders was difficult and costly. While these are still major challenges, new multinational trade agreements and new technologies, especially computers and information technology, have radically changed the way an enterprise can operate and have transformed the business environment into a global economy that is reoriented toward service and knowledge work, with a greater mobility of resources, skills, processes, and technology itself. Driven by the promise of new markets and cheaper resources, and in many cases access to new knowledge and talent, today's companies are trying to take advantage of the new global marketplace. As demonstrated by Nokia, Harley Davidson, Volvo, Wendy's, and Briggs & Stratton, just to mention a few well-recognized names, not only the giants in their industry but any company in any market can operate multinationally with potential benefits in the following areas:

- New market penetration
- Accessing new and complementary knowledge and talents
- Accessing resources more economically
- Economies of scale and scope
- Technology and resource sharing
- Capacity enhancement
- Joint ventures and partnering

Taken together, these benefits translate into "faster-cheaper-better" business results, specifically (1) faster market response, (2) lower costs, and (3) more innovative, higher-quality products and services, all aimed at a favorable impact on overall business performance. Thanks to standardization, digitization, favorable weight-size-value ratios, and the Internet, high-technology firms are especially well positioned to take advantage of these opportunities and have to reevaluate their own business models to take advantage of the new e-commerce environment.

Yet, despite the many benefits, there are great challenges in dealing with the enormous dynamics, complexities, and risks of global enterprising. As a result, many companies have opted for alliances, partnerships, joint ventures, or acquisitions, rather than in-house expansion, for reaping the benefits of global business. This introduces important strategic business dimensions. Companies that fund and manage multinational activities internally (from R&D to manufacturing and marketing) instead of outsourcing them do so under the assumption that long-term economies of scale or scope or a monopolistic advantage can be gained. Actual success will often depend on how well these assumptions hold in our ever-changing world of business. In addition to the sourcing decision, companies must choose the proper organizational design for optimizing managerial control. As researched by West (2004), firms that carry out only a small part of their activities abroad try to exercise relatively tight cross-border control from their home base. However, as the importance and scope of these activities grow, the organizations abroad need more autonomy. The greater the cultural and geographic distances, the more likely that they are going to get the additional power. This will increase the quality and effectiveness of local decision making, resulting in greater organizational flexibility, speed, and adaptability. The potential *downside* is the deterioration of the parent company's network structure, linkages, interoperability, and integration processes, precisely the mechanisms that drive the economics of scope and scale within the enterprise. This dilemma and managerial complexity grow with the scope of multinational operations. That is, the more subsidiaries, R&D laboratories, or field operations a company runs, the more complex becomes integration with its total enterprise, a condition necessary for wringing out operational synergism. Especially from the technology side, the creation, transfer, and coordination of knowledge, key to successful R&D, is one of the most difficult areas in multinational management. Flexible and innovative organizational designs and policies have been advocated by business leaders and researchers to cope with this challenge. Concepts such as the *transnational management model* have been proposed for some time (Bartlett and Ghoshal 1989) to deal with the polarized issues of local authority, integrated decision making, and central control. Management in our global environment requires more than a simple theory or a set of business policies—it requires a well-defined, yet flexible business process that has been worked out and agreed to by all key stakeholders, an approach that is being advocated for managing technology throughout this book.

1.6   IMPACT OF INTERNET AND E-COMMERCE

Modern computers and information technology have created vast opportunities for managers and their organizations for leveraging their resources. With over 500 million end users worldwide and increasing exponentially, the Internet and all its global subnetworks has become one of the most dominant enablers toward better-faster-cheaper business operations. From in-house R&D to joint developments, shareholder services, medical surgery, and technical field support, the Internet opens up new and more effective ways of operating the enterprise and conducting its business with customers, suppliers, regulators, and other partners, and has become the operational backbone for many companies. Distance and time no longer matter. The cost, speed, and capacity of data handling, together with its 24/7 access, define the advantages of the Internet and its value for e-commerce. Business activities from design to manufacture, and from accounting to human resources, consist to a large degree of information processing, transferring, storing, and retrieving. Hence, Internet technology enables the enterprise to conduct more business over geographically dispersed areas with great speed, flexibility, and economics. In addition, sophisticated systems interoperability and data processing allow the enterprise to conduct vastly more complex ventures than was ever possible in the past, such as the recent Mars explorations and the development of nanosecond-speed microprocessors. While these technologies have created great opportunities for running the enterprise more efficiently and pursuing new business ventures, they have also created enormous pressures for delivering cutting-edge products of superior quality and service, and produced a business environment with fast changes, high risk, uncertainty, and virtually no barriers to competitive entry. This has profound *implications especially on the practice of technology management*. The value of technology-based companies lies increasingly in the creation and transfer of *knowledge*, rather than in the brick-and-mortar assets that appear on the balance sheet. In fact, it is estimated that intellectual property, including R&D, franchises, patents, brands, ideas, and experience, accounts for 85 percent of the market value of technology companies (Cairncross 2002). To run these businesses calls for different leadership and managerial skill sets that focus especially on knowledge generators: the human side of the enterprise. This includes virtually all the people of a technology-based organization, plus many people at the supplier and customer interface. In addition, managers must pay attention to the decision-making processes and build collaborative communities and learning organizations. Knowledge creation relies on information access. Managers must provide the people in the organization with systems for sharing intellectual capital and new ideas that emerge. They must continuously develop the infrastructure needed for effectively handling the increasingly vast amount of information that flows across the enterprise and all of its partner communities. Using Internet technology, *intranets* and *extranets* have become important electronic data interchange (EDI) systems of choice for moving large sets of data, centrally managed and controlled, among validated user communities, anywhere, anytime.

1.7   TECHNOLOGY AND SOCIETY

Technological progress is the single most important factor driving a nation's sustained economic growth.[5] In the United States, as much as half of the economic growth gained over the past 50 years is being credited to technological advances (Khalil 2000). Yet, the issues underlying economic growth are highly complex and interrelated with a vast array of ethical values and social responsibilities. For a long time scholars have argued over which processes drive technological advancement. Are these drivers derived largely from scientific or social sources? Two primary theories evolved.

Rational-Objective View. According to the oldest and still prevailing theory, technological advances result from purposeful, empirical/scientific discovery. That is, new technology is, by and large, the result of rational thought, independent of social influences (Volti 1995). In its purest interpretation, this theory argues that *science leads and society follows*, a view that is labeled by Pacey (1983) as *"technological determinism."*

Social Constructivism. This theory makes the compelling argument that technological advances are, by and large, the result of social influences. Thomas

[5]Cf. *Technology in the National Interest,* Special Report by the U.S. National Science and Technology Council, Office of Technology Policy, U.S. Department of Commerce, Washington, DC, 1996.

Kuhn presents one of the classic discussions of this concept in his widely quoted book *The Structure of Scientific Revolutions* (1962), which provides the basis for much of today's theory on social constructivism (sometimes also referred to as *social constructionism*). In its extreme form, social constructivism argues that there is *no basis* for the belief that technological advances result strictly from logical thought processes, because all scientific processes are influenced by social value systems (Pool 1997). This is the *cultural dimention* of technology development and its penetration into our society (Pacey 1983).

So much for the two poles of the theoretical spectrum. The applied literature and practice of MoT take more of a "middle-of-the-road" approach, recognizing both the power of focused scientific discovery *and* the social/cultural influences (Haddad 2002). In addition technological advances are potentially influenced by other factors, such as the organizational culture of the enterprise, personal objectives of powerful leaders, and special events, such as September 11, that create a sudden paradigm shift of culture and values within the society.

Social Responsibility. Social responsibilities and business ethics add yet another dimension to the management of technology. Many of our social problems, from health to environmental pollution and unethical business practices, are caused or promoted by technology. Yet, it requires more technology to solve these problems. This puts pressure on the leadership of technology-based companies to run the enterprise in a socially responsible way. From maintaining environmental quality to performing community service, companies are expected to contribute their share to society as a whole. The more technology-oriented an enterprise is seen as being, the more technological solutions are expected from it by society. Business performance is no longer a simple financial measure, but has been redefined by consumers, competitors, and regulators into a complex set of parameters that translate into market capitalization. Government and public pressures have pushed companies toward treating social issues more realistically, focusing on longer-range cost-benefit analyses and integrating environmental challenges in particular into the total business process (Singh 2000). The challenges are obvious for companies that sell cigarettes, breast implants, or telephone services; although they are more subtle, they are equally serious for technology businesses in automotives, food, pharmaceuticals, or computers, requiring sophisticated business strategies and management skills and great leadership.

1.8 FUTURE TRENDS

The magnitude and speed at which technological advances and changes in management practices have occurred over the past 50 years are stunning. "Technology is reshaping our world and has influenced our life at a speed unimaginable just a few years ago." This conclusion, reached by the U.S. National Science and Technology Council in its 2000 annual report, is typical for the changing environment we are experiencing. There is no reason to believe that the current rate of change will slow down anytime soon. The great challenge is to harness this newly emerging technology for the benefit of society. This is where *management of technology* can make a difference. Great potential exists for further advances in the area of technology integration toward larger multidisciplinary applications, such as those needed for electric cars and seamless product development from R&D to manufacturing and distribution. To benefit from these broadband technology applications, our business processes, management techniques, educational methods, and lifestyles have to change further. The current trend toward flatter, leaner organizations, collaboratively networked, with higher levels of shared authority, operating-level autonomy, and automated work processes, is expected to continue. Economically, these trends will lead to a steady increase of gross domestic product (GDP) for those who can participate in the exploitation of technology. More difficult to determine is the true benefit of these technological advances to society. While the quality of life is indeed difficult to quantify or to assess uniformly across all people, scholars agree that technology advances, by and large, have had a positive effect on our quality of life, and this trend is expected to continue.

1.9 SUMMARY OF KEY POINTS AND CONCLUSIONS

The key points that have been made in this chapter include:

- *Management of technology* is the art and science of *creating value* by using technology together with other resources of an organization.
- Technology is reshaping our world and influencing our lives at a speed unimaginable just a few years ago. In particular, computers and communications have created a new business environment focused on effectiveness, value, and speed.
- Companies that survive and prosper in today's environment have the ability to deal with a broad spectrum of contemporary challenges that focus on speed, cost, and quality.

   Traditional linear work processes and top-down controls are no longer sufficient, but are gradually being replaced with alternate organizational designs and new management techniques and business processes.

- Engineering and technology managers are involved with the broad issues of the enterprise, requiring more interdisciplinary and strategic business skills that span a wide spectrum of technology areas within the company and across its boundaries.
- Engineering managers are to a large extent responsible for managing technology; essentially, the *engineering manager* has become a *manager of technology*.
- As a result of the complex dynamics and enormous risks of global enterprising, many companies have opted for alliances, partnerships, joint ventures, or acquisitions rather than in-house expansion.

- Modern computers and information technology have created vast opportunities for leveraging their resources. Distance and time no longer matter. The cost, speed, and capacity of data handling, together with 24/7 access, define the advantages of the Internet and its value for e-commerce.
- The value of technology-based companies lies increasingly in the creation and transfer of knowledge.
- The current trend toward flatter, leaner organizations, collaboratively networked, with higher levels of shared authority, operating-level autonomy, and automated work processes, is expected to continue.
- To run technology-based businesses requires new leadership and managerial skills that focus on knowledge generators, the human side of the enterprise.
- Social responsibility and business ethics are important dimensions of technology management. Many of our social problems, from health to the environment, are caused by technology. Yet, it requires more technology to solve these problems.

1.10 CRITICAL THiNKING: QUESTIONS FOR DISCUSSION

1. What is the role of technology in today's enterprises?
2. What characteristics differentiate high- and low-technology companies?
3. What are the benefits of establishing a chief technology officer (CTO) position?
4. How does technology affect the "globalization of business"?
5. Should the U.S. government develop a strong national technology policy and influence "desirable" technology developments?
6. Why are traditional management styles with emphasis on central control apparently ineffective in high-technology organizations?
7. Analyze your company in terms of its technology-supporting infrastructure and resource functions. What changes would you recommend?
8. Analyze your company in terms of organizational structure and the effectiveness of its management style. What changes would you recommend?
9. What kind of changes in organizational structure and leadership style do you see for companies 10 years (or 20 years) from now?

1.11 REFERENCES AND ADDITIONAL READINGS

Armstrong, D. (2000) "Building teams across borders," *Executive Excellence,* Vol. 17, No. 3 (March), p. 10.

Barkema, Harry G., Baum, Joel A., and Manix, Elizabeth A. (2002) "Managing challenges in a new time," *Academy of Management Journal,* Vol. 45, No. 5 (October), pp. 916–930.

Barner, R. (1997) "The new millennium workplace," *Engineering Management Review* (IEEE), Vol. 25, No. 3 (Fall), pp. 114–119.

Bartlett, C. and Ghoshal, S. (1989) *Managing Across Borders: The Transnational Solution,* Boston: Harvard Business School Press.

Cairncross, F. (2002) *The Company of the Future*, Cambridge, MA: Harvard Business School Press.

Dillon, P. (2001) "A global challenge," *Forbes Magazine*, Vol. 168 (September 10), pp. 73+.

Gaynor, G. H., editor (1986) *Handbook of Technology Managing*, New York: McGraw-Hill.

Gehani, R. R. (1998) *Management of Technology and Operations*, New York: Wiley.

Gray, C. and Larson, E. (2000) *Project Management*, New York: Irwin McGraw-Hill.

Haddad, C. J. (2002) *Managing Technological Change*, Thousand Oaks, CA: Sage Publications.

Khalil, T. (2000) *Management of Technology*, New York: McGraw-Hill.

Kruglianskas, I. and Thamhain, H. J. (2000) "Managing technology-based projects in multi-national environments," *IEEE Transactions on Engineering Management*, Vol. 47, No. 1 (February), pp. 55–64.

Kuhn, T. S. (1962) *The Structure of Scientific Revolutions*, Chicago, IL: University of Chicago Press.

Lannes, W. J. (2001) "What is engineering management?" *IEEE Transactions on Engineering Management*, Vol. 48, No. 1 (February), pp. 107–110.

Marshall, Edward (1995) *Transforming the Way We Work*, New York: AMACOM.

National Research Council (1987) *Management of Technology: The Hidden Competitive Advantage,* Washington, DC: National Academy Press, Report No. CETS-CROSS-6.

Oakey, R. (2003) "Technical entrepreneurship in high technology small firms: some observations on the implications for management," *Technovation,* Vol. 23, No. 8. (August), p. 679.

Pacey, A. (1983) *The Culture of Technology*, Cambridge, MA: MIT Press.

Pool, R. (1997) *Beyond Engineering: How Society Shapes Technology*, Cambridge, MA: MIT Press.

Singh, J. (2000) "Making business sense of environmental compliance," *Sloan Management Review*, Vol. 41, No. 3 (Fall), pp. 91–99.

Thamhain, H. J. (2002) "Criteria for effective leadership in technology-oriented project teams," Chapter 16 in *The Frontiers of Project Management Research* (Slevin, Cleland, and Pinto, eds.), Newton Square, PA: Project Management Institute, pp. 259–270.

Thamhain, H. J. (2003) "Managing innovative R&D teams," *R&D Management*, Vol. 33, No. 3 (June), pp. 297–312.

Thamhain, H. J. and Wilemon, D. L. (1999) "Building effective teams in complex project environments," *Technology Management*, Vol. 5, No. 2 (May), 203–212.

U.S. National Science and Technology Council, Office of Technology Policy (1996) *Technology in the National Interest,* Special Report, Washington, DC: U.S. Department of Commerce.

U.S. National Science and Technology Council, (2002), Annual Report, Washington, DC.

Volti, R. (1995) *Society and Technological Change*, New York: St. Martin's Press.

West, D. C. (2004) "Global marketing," Chapter 22.1 in *Technology Management Handbook* (R. Dorf, ed.), Boca Raton, FL: CRC Press.

2

MANAGING IN AN
E-BUSINESS WORLD

REPROGRAMMING AMAZON

At its nine massive distribution centers from Fernley, Nevada, to Bad Hersfeld, Germany, workers scurry around the clock to fill up to 1.7 million orders a day—picking and packing merchandise, routing it onto conveyors, and shipping the boxes to every corner of the world. It's an impressive display, but utterly misleading. The kind of work that will truly determine Amazon's fate is happening in places like the tiny, darkened meeting room at its Seattle headquarters.

Jeffrey A. Wilke, a compact, intense senior vice president who runs Amazon's worldwide operations and customer service, and an engineering team are trying out a "beta" test version of new software they wrote. When the buying automation program is ready for prime time in mid-2004, Amazon's merchandise buyers will be able to quickly and accurately forecast product demand, find the best suppliers, and more. The effort is one of scores of technology projects under way at Amazon that ultimately may change the entire experience of shopping online—and Amazon itself.

Just as most folks have come to view Amazon as a retailer that happens to sells online, guess what? It's morphing into something new. In ways few people realize, Amazon is becoming more of a technology company—much like Microsoft and Wal-Mart. "What gets us up in the morning and keeps us here late at night is technology," says founder and chief executive officer Jeffrey P. Bezos. "From where we sit, advanced technology is everything."

Building on a raft of tech initiatives, from ever-richer Web sites to new search technology, Bezos aims to reprogram the company into something even more potent. The notion is to create a technology-driven nexus for e-commerce that's as pervasive and powerful as Microsoft's Windows operating software is in computing. That's right: Bezos hopes to create a Windows for e-commerce. Since last year, Amazon has been steadily tuning innovations it developed for its own retail site into so-called Amazon Web Services that make it easy for other merchants to list their products on Amazon's 37 million customer base. When a product sells, Amazon takes a commission of about 15 percent and these revenues have much higher margins than Amazon's own retail business.

While Amazon isn't exiting retailing, its tech initiatives could help Amazon to bust out of the conceptual prison of stores and virtual confines of a single Web site, and become an online mall—a piece of business that generates gross margins about double its 25 percent retail margins. By plugging into its massive e-commerce system, thousands of retailers from mom-and-pop shops to Lands' End, Circuit City, Target, and Toys "R" Us, which are featured on Amazon.com.

If Bezo's new plans work, Amazon could become a service that would allow anyone, anywhere to find whatever they want to buy—and to sell whatever they want almost as easily. However, Bezos' vision won't be easy to fulfill. Trying to be a world-class retailer, a leading software developer, and a service provider simultaneously strikes some observers as a nearly impossible endeavor. As both retailer and mall operator, Amazon has divided loyalties. Yet, Web services may offer the most expansive potential of all Amazon's tech initiatives. It's still early, but it's possible that Amazon has latched onto one of tech's juiciest dynamics—a self-reinforcing community of supporters. Indeed, it seems to be harnessing the same "viral" nature of the open-source movement that made Linux a contender to Windows.

Amazon, the next Microsoft? Not so fast. For all the promise, building a broad platform is about as tough as a goal gets in the tech business—as nearly every competitor to Microsoft can attest. And unlike most tech companies, Amazon has to contend not just with bits and bytes, but also with the bricks and mortar of warehouses and the fickle fingers of Web shoppers. But for now, at least, Amazon's pursuit of cutting-edge technology has given it time to figure out what comes next.

A more detailed description of this business scenario is provided in Case 2.1, at the end of this chapter.

Source: Excerpted with permission from *Business Week*, December 22, 2003, pp. 82–86, "Reprogramming Amazon," by Robert D. Hof.

2.1   A CHANGING ENVIRONMENT

Amazon is hardly alone in facing both challenges and opportunities in a steadily changing environment. A company in any industry from computers to retailing and medical services is confronted with the need to adapt to the changing business climate that seems to demand better, faster, and cheaper solutions. Many factors, such as globalization and the transparency created by the Internet, drive these demands. Yet, the most powerful catalysts are technological advances. Creating new products and services, improving existing ones, responding to customer and market needs faster, and taking advantage of economies of scale are all affected by technology. Utilizing existing technologies or developing new ones is expensive and risky. But companies have no choice. They either take advantage of the new business potential or someone else will do it, penetrating into their market and customer base. Many of the new and fast-changing technologies, such as communications and computers, have forced enterprises to improve their existing products on a much faster scale than ever before. In addition, the new electronic environment has pushed many companies to use e-commerce technologies as their basis for conducting business. It is interesting to note that small businesses have a distinct advantage in this new digital world. They usually carry less financial overhead and, by the nature of their smaller size, operate more flexibly and can respond more quickly to changes in the business environment. At the same time, many of the advantages larger companies enjoyed in the past, such as customer loyalty and unique business infrastructure, can be siphoned off by others with advanced communication and computer technology. As a result, barriers to entry are diminishing for any business, and companies have to rethink their business strategy to remain competitive and profitable.

One of the biggest catalysts for change in our environment is the shift from an industrial society to an information society, a development that Alvin Tofler had already identified in his book *The Third Wave* in 1980. A similar conclusion was reached by John Naisbitt in *Megatrends* (1982). He suggested that a company's value is derived from knowledge, which in turn is based on information. Indeed, in today's world information technology supports and affects virtually every business operation. Large amounts of capital are being invested in information-processing equipment, software, and skill development. Yet, the real challenge is to wring value out of these investments.

Probably no other society has exploited the opportunities of the information age better than Japan. With very limited natural resources, Japan has created enormous value, using knowledge in general and technology in particular to create superior goods and services. In addition, virtually every company realizes that an increasingly large portion of the value of goods and services is derived from utility to the user. As pointed out by Sakaiya and March (1991), the material content of a microprocessor chip does not change over time, but its knowledge (or utility) value does. The same is true for washing machines and online banking. However, the customer-perceived value of a given product changes with the changing application environment and with the introduction of new products. A good example is the computer. The value of a laptop computer is determined not so much by its components and functional specs, but by its abilities to perform certain tasks, its portability and inter-connectability, and so on. This perceived value, when aggregated across the market, eventually determines the supply-demand price equilibrium. However, changes on either the demand or the supply side can quickly drive the value and price for this product down. Assuming that wireless Internet connectability or longer offline power operability becomes available, and assuming that the market (user) wants those features, many people would prefer the new computer and be willing to pay a higher price for the added value, while the value and market price for the old computer would drop. Similar arguments can also be made from the demand side. Imagine how the perceived value of laptops would be affected if air travel regulations no longer allowed computers to be carried or checked on airplanes.

This value perception has broad implications for the product life cycle. As pointed out by many scholars (Cairncross 2002, Fellenstein and Wood 2000, Gehani 1998, Haddad 2002), the more knowledge- and information-based that goods and services are, the faster they lose their value over time. An extreme example is the newspaper! But, computer chips and bank/investment products with life cycles of 3 to 5 months represent equally tough management challenges, especially when considering the development time and cost of these products, which often exceed by far the market life-cycle time and revenue. Hence, organizational agility and responsiveness to market conditions are critical characteristics necessary for technology-based companies to survive and prosper in today's business environment of the information age.

2.2 THE UNIQUE NATURE OF MANAGING TECHNOLOGY-BASED BUSINESSES

Managers in technology-based companies argue that their business environment is different, requiring unique organizational structures, policies, interaction of people, and economic behavior. Responding to this claim, scholars have searched for patterns of managerial behavior and broader organizational characteristics associated with different levels of technology-based businesses. They also have tried to determine specific organizational structures, leadership styles, and other enterprise systems most advantageous in high-tech business environments. This research has been published in the literature and summarized in two major handbooks (Gaynor 1996, Dorf 2004), providing useful insight into the functioning and dynamics of technology-based enterprises. Yet, no single theory or body of knowledge has emerged that could point toward the uniqueness of managing in technology, nor have these studies reached any unified conclusions on a best way to organize or manage in technology-based environments. Still, managers in technology know that their environment is unique!

In the frequently cited study *Made in America: Regaining the Productive Edge*, the MIT Commission on Industrial Productivity concluded as early as 1990 that:

> *For too long business schools have taken the position that a good manager could manage anything, regardless of its technological base. It is now clear that this view is wrong. While it is not necessary for every manager to have a science or engineering*

degree, every manager does need to understand how technology relates to the strategic positioning of the firm, how to evaluate alternative technologies and investment choices, and how to shepherd scientific and technical concepts through the innovation and production processes to the marketplace (Dertouzos et al 1990).

Increasingly, both corporate executives and academicians have debated how organizations can best prepare leaders for the unique challenges of managing in a technology-based enterprise, which again raises the question, "What is unique about managing in technology?"

To find a way out of these quandaries, some researchers have used the paradigm approach. Similar to the experiments conducted by Thomas Kuhn (1970), field researchers have observed managers and organizational interactions in various technology settings. They have studied the phenomena gleaned from these observations to determine the patterns of organizational behavior. These patterns, so-called paradigms, provide building blocks for a body of knowledge.

Based on these field observations, *technology-intensive enterprises exhibit certain unique characteristics*, which are summarized here in 16 distinct categories:

1. *Value Creation by Applying Technology.* Technology-based companies, focus on technology as a primary factor for creating wealth. They exploit or commercialize technology. Examples range from plastics and fiber optics to financial services and e-commerce. The technology-based enterprise competes through technological innovation. It creates new products and services by applying technology in form of unique processes, systems, equipment, or advanced materials. This value addition is part of the innovation process, where the final product, such as the Internet service or computer chip, is worth a lot more than its ingredients. This is also supported by macroeconomic research. According to the U.S. National Science and Technology Council, NSTC, (1996), "technical progress is the single most important factor in determining a nation's sustained economic growth."

2. *Use of Advanced Technology within the Firm's Operations.* High-tech organizations use computer-aided methods, advanced automation, sophisticated communication systems, the latest materials and software, and other advanced technology extensively in their business processes to gain operational advantages that lead to faster, better, and more economical products and services.

3. *Highly Educated, Skilled Workers.* The use of high technology in components, equipment, systems, and processes requires special skills and knowledge, as well as sophisticated methods of decision making, teamwork, and project integration.

4. *Replacement of Manual Labor with Technology.* Technology-based companies utilize extensive automation within their operation and "plug-and-play" subsystems, such as microprocessors, to add value to a product while reducing its labor content. In addition to economic benefits, this often results in higher speed and greater quality and reliability, which explains in part the

fact that technology-based organizations are mostly *capital* (rather than labor) *intensive*.

5. *Infrastructure.* High-tech organizations provide special state-of-the-art equipment, facilities, software tools, and education/training in support of the complex work to be performed.

6. *High R&D and Product Development Expenditures.* The ratio of R&D and new product development expenditures to sales or operations is often higher for technology-based organizations. Typical R&D expenditures of high-tech companies run to 10 percent of sales, double the average across industries. Competing through technology is expensive.

7. *High Risks.* Investing in technology ventures is very risky. High-tech organizations are exposed to a large number of risks due to market uncertainties, emerging and changing technologies, and regulatory ambiguities. In order to survive and prosper, these companies must have unique business processes, decision-making tools, and leadership styles for processing these risk factors effectively.

8. *Continuing Changes.* Technology-based business environments are fast changing. As with risks, high-tech organizations must have business processes, decision-making tools, and leadership styles conducive to dealing effectively with the challenges of their constantly changing environment.

9. *Complex Decision-Making Processes.* As a result of the high risks, great uncertainties, and constant changes that high-tech companies are exposed to, top-down, or centralized, management and decision making is usually ineffective. High-tech companies rely to a large extent on *distributed decision making,* which helps in risk sharing and commitment building at the project team level.

10. *Short Product Life Cycles.* Continuously changing market conditions and emerging technologies drive shorter product life cycles for technology companies. As a result we see these companies reducing the time to market of new product developments.

11. *Quick Market Response.* High-tech companies are fast, agile, and flexible in responding to business opportunities and threats. Often at the expense of central control and unified strategies, they have flatter organizational structures and more autonomous business processes that enable quick reaction to their constantly changing environment.

12. *Fast Growth Potential.* Technology-based enterprises have the potential for fast growth of their businesses, especially in terms of sales and market capitalization, as demonstrated by many companies in the areas of computers, pharmaceuticals, and e-commerce, to name a few.

13. *Low Barriers to Entry.* Traditional barriers to entry, such as infrastructure, brand loyalty, and established supply chains, are virtually nonexistent for high-tech businesses. In particular "new and emerging technologies" can reset the competitive field to "ground zero," wiping out any competitive advantage of established products and services (Andrew and Sirkin 2003).

14. *Low Profitability in Spite of Strong Cash Flow.* Investing in technology is very costly. High-tech businesses need large amounts of cash, often more than they can generate. They are the classical "stars" in BCG's Growth-Share Matrix.[1] As a result, strong financial leveraging and low profitability are quite common, even for well established high-tech giants such as Amazon, Intel, Maxtor, Microsoft, and Pfizer.

15. *Threats to Survival and High Failure Rate.* The high cost of operations, combined with the uncertainties of the business environment, in terms of emerging technologies, changing markets, and complex regulations, makes technology-based companies highly vulnerable to failure.

16. *Many Alliances, Joint Ventures, and Partnerships.* Because of the high costs, risks, and complexities of high tech, virtually no company has the resources to handle all the facets of a technology development, its rollout, and its field support single-handedly. Resource pooling from cooperative agreements to acquisitions is quite common among high-tech enterprises to raise the resources for implementing the new venture in a timely fashion.

To summarize the areas that are unique and different in managing technology, let us focus on six selected business subsystems, as graphically shown in Figure 2.1.

➤ *Work.* Technology-oriented work is by and large more complex, requiring special skills, equipment, tools, processes, and support systems. The unit of work is often a project organized and executed by a multidisciplinary team. Cross-functional integration, progress measurements, and controlling the work toward desired results are usually more challenging with increasing technology orientation, involving creativity, risks, and uncertainties. Work processes are often "nonlinear," with solutions evolving incrementally and iteratively with many interfunctional dependencies.

Impact Areas: Organizational structure, work planning, work processes (e.g., project management), personnel recruiting and advancement, skill development, management style, organizational culture, and business strategy.

➤ *People.* Because of the type of work and its challenges, technology-oriented environments attract different people. On average, these people have highly specialized skills and are very good at applying them. They are better educated, are self-motivated/directed, require a minimum of supervision, and enjoy the autonomy and freedom of decision making, while being willing to take on responsibilities. They enjoy problem solving and find technical challenges and resource and schedule constraints to some degree motivating and intellectually stimulating. People in technology-oriented work environments often enjoy a sense of community and team spirit, while having little tolerance for personal conflict, anxieties, and organizational politics.

---

[1] The Growth-Share Matrix, developed by the Boston Consulting Group (BCG), is a tool for explaining cash flaw of products through their life cycle. For further information see *www.bcg.com.*

Figure 2.1   Business subsystems unique to technology-intensive organizations.

Impact Areas: Because of the relationship between people and work issues, the two impact areas are similar. Organizational structure, work plans and processes, personnel recruiting and advancement, skill development, management style, and organizational culture are the primary areas affected by people in technology.

➤ *Work Process.* The nature of high-tech work and its business environment requires the ability to deal effectively with complexities, uncertainties, speed, and innovation. This influenced the evolution of work processes that are less sequential and centrally administered, but more team based, self-directed, and agile, structured for parallel, concurrent execution of the work. New organizational models and management methods, such as the Stage-Gate, concurrent engineering, and design-build processes, evolved together with the refinement of long-held concepts such as the matrix, project management, and product management.

Impact Areas: Because of the effect on the people and their work, the work process design affects primarily people issues, management style, and organizational culture. In addition, the work process affects management tools, such as scheduling, budgeting, and project performance analysis, as well as operational effectiveness, such as time to market, cost, and flexibility.

➤ *Managerial Tools and Techniques.* The unique nature of the work and its business environment creates the need for a special set of tools and techniques for effective administration and management in technology-based organizations. Virtually all of the tools and techniques are being used in both high-tech and low-tech organizations, and in many cases, the tools were around long before the high-tech era. However, in their specific application and integration with the enterprise, these tools and techniques often fulfill a unique function and play a unique role. Examples are project schedules and integrated product development methods that have been tailored to respond to the pressures of a faster, more competitive, and more team-directed work environment. The wide spectrum of tools and techniques can be grouped into five major categories according to their application: (1) project management, (2) product management,

(3) quality control, (4) general management, including legal, human resources (HR), accounting, and training, and (5) strategic.

Impact Areas: The effectiveness of tools and techniques in the enterprise is strongly influenced by the people who use them. Therefore, stakeholder involvement during the tool selection, development, and implementation is critical. The application of many tools involves trade-offs, such as efficiency versus speed, control versus flexibility, or optimization versus risk. All of these factors need to be carefully considered and make the implementation of managerial tools and techniques a great challenge.

➤ *Organizational Culture.* The challenges of technology-driven environments create a unique organizational culture with its own norms, values, and work ethics. These cultures are more team oriented in terms of decision making, work flow, performance evaluation, and workgroup management. Authority must often be earned and emerges within the workgroup as a result of credibility, trust, and respect, rather than organizational status and position. Rewards come to a considerable degree from satisfaction with the work and its activities. Recognition of accomplishments becomes an important motivational factor for stimulating enthusiasm, cooperation, and innovation. It is also a critical catalyst for unifying project teams, encouraging risk taking, and dealing with conflict.

Impact Areas: Organizational culture has a strong influence on people and the work process. It affects organizational systems from hiring practices to performance evaluation and from reward systems to organizational structure and management style.

➤ *Business Environment.* Technology-oriented businesses operate in an environment that is fast changing in terms of market structure, suppliers, and regulations. Short product life cycles, intense global competition, and strong dependency on other technologies and support systems are typical for these businesses, which operate in markets with low brand loyalty, low barriers to entry, and fast and continuously improving price-performance ratios.

Impact Areas: The need for speed, agility, and efficiency affects not only the work process design, the organization and execution of work, and the management methods, tools, and techniques, but also business strategy and competitive behavior, which often focus on cooperation and resource leveling via alliances, mergers, acquisitions, consortia, and joint ventures.

Looking at these complex characteristics, with their embedded challenges and opportunities, it is hard to imagine a job more complex than managing a technology-based business. However, this is just the beginning of an exploration of MoT, leading to questions of "what drives these unique characteristics? how are the organizational structures, policies, interactions of people, and economics of these high-tech companies affected? and how do all of these factors shape the enterprise and our business environment?"

Working with technology has always created specialized work environments with particularly skilled people, distinct functional capabilities, and specialized tools and techniques. However, the unique nature of technology and its management did not emerge until the 1900s. The astounding advances that brought us from the pony express to interplanetary spacecraft, from the first radio transmission to cellular telephony, and from the basic understanding of the human body to human genome mapping all happened over the past 100 years. In particular the events of the last 30 years have transformed technology management into a unique multidisciplinary field with its own terminology, standards, body of knowledge, and career paths. Part of this uniqueness is derived from the technology itself, which requires special technical skills and the ability to innovate and deal with complexities, risks, uncertainties, and integration. However, managing in technology is much broader and more complex than can be summarized in a few words. Its functions stretch across the whole spectrum of management and all of its subsystems and social interfaces, which will be explored in the rest of this book.

2.3 HISTORY OF MANAGING TECHNOLOGY

How did we get into such a complex world of business? What role did technology play in the past? As shown in Figure 2.2, technology has been around for a long time. Man's quest for survival led to the development of improved tools and techniques for gathering food, building shelters and fighting wars.[2] The earlier civilizations were classified by the technologies they used, such as the *Stone Age, Bronze Age*, and *Iron Age*, while more recent periods have been labeled the *Steam Age, Electricity Age, Nuclear Age, Electronic Age, Space Age, Information Age*, and *Biotechnology Age*, all focusing on the most rapidly advancing technology of their time. Over the past 6,000 years technology transfer has occurred by passing the lessons learned from one generation to the next, with limited transfer of knowledge among wider geographic regions and cultures. Evidence of sophisticated technology management dates back more than 4,000 years, with pyramids, aqueducts, waterworks, chinaware, and war machinery still giving testimony today of the impressive technology-based accomplishments of these times. With the increased application of technology, some formal technological vision and leadership also evolved. Well-recorded projects, from ancient urban planning with water, sewer, and road systems[3] to astronomical discoveries and geographical navigation, provide evidence of early technological leadership. The work of the Greek philosopher Aristotle is a good example of early scientific reasoning and leadership. In his *Metaphysics*, written during the fourth century B.C., he not only summarized the technological knowledge known to mankind at that time but also suggested ways to integrate this knowledge and to establish an attitude for scientific reasoning. Starting in the eighteenth century, the Industrial Revolution spawned an enormous growth in technology with a focus on mass production and economies of scale, with considerable benefits to many sectors of society, such as agriculture, construction, and military operations.

---

[2] For an interesting discussion of the historical development of technology with examples of specific accomplishments see Ray Gehani (1998), pp. 25–49.

[3] Specific examples of these well-planned urban centers are Harappa and Mohan-jo-daro on the Indus River and community centers of the Longshan People along the banks of the Huang He (Yellow River), dating back to about 2500 B.C. These and other examples are discussed by Ray Gehani (1998), pp. 28–36.

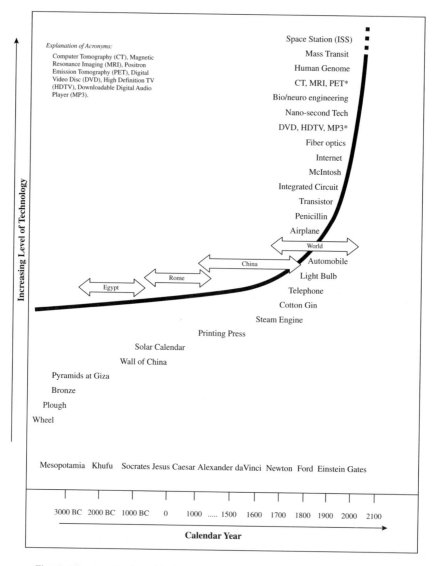

Figure 2.2 Acceleration of technology development showing selected milestones.

However, formal recognition of technology and its importance to business for leveraging resources and competitive positioning did not occur until the 1950s. The next 30 years brought an increasing awareness of the special nature of technology to the business community, recognizing both the benefits for product development, production, and service, *and* the higher risks, uncertainties, and special skill requirements for dealing with technology in the workplace. Until the 1980s, *management focus* was on innovation and technology transfer toward the generation of proprietary knowledge applied to new products, processes, and services. Scholars defined technology primarily in terms of R&D efforts (Parsi 1989), a position adopted by many science and engineering organizations, such as the National Science Foundation that established a metric, still quoted today: "an industry or organization is classified as technology-intensive if more than 3% of its sales revenue is devoted to research and development."

As the world moved toward higher levels of competitiveness and globalization during the 1980s, many industrialized countries increased their R&D efforts. In the United States, R&D expenditures averaged an annual increase of 6 percent, reaching $136 billion in 1987. But regardless of the impressive level of R&D activity, U.S. managers were concerned about decreasing productivity and overall competitiveness against other industrial nations, especially Japan and European countries (U.S. National Research Council 1997 & 2003, Steele 1989). Typical of this concern is the comment by Dr. John Armstrong, head of IBM's R&D center, who was quoted in the *New York Times*[4] discussing these challenges: "The traditional academic view is that you do basic research first, then applied research, then product development, then give it to manufacturing. But, the Japanese have been very effective in rapidly transferring research into successful commercial products by performing several of the conventional development phases in parallel. We at IBM had to change our development processes to generate new products more quickly."

These managerial concerns were further supported by macroeconomic studies that showed a weak correlation between R&D expenditures and competitive performance (McLean and Round 1978). In addition, researchers found that many economically thriving Japanese firms allocated relatively few resources to R&D, but instead spent larger amounts on "productivity enhancements," such as new product developments, automation of manufacturing and testing, and information processing, earning on average significantly higher returns on their investments than U.S. firms. Together, all of these facts point toward the strong possibility that management of technology is not confined to R&D or engineering, but includes many other facets of the enterprise and its environment, a position that is well supported today. It also points to the unique nature of managing technology in today's more complex business environment, as discussed in the previous section. Finally, these historic developments provide the background that helps to explain the specific forces that continue to shape the way we manage our organizations today.

2.4   FORCES THAT DRIVE TECHNOLOGY COMPANIES TODAY

Six major shifts in our business environment influenced technology management and created the ambience that prompted many of the recent changes in the way we conduct business. These paradigm shifts, as summarized in Table 2.1, must be understood for managing effectively in today's technology-based organizations:

Shift from Linear Processes to Dynamic Systems. While in the past, management concepts were based on predominately linear models, typically exemplified by production lines, sequential product developments, scheduled services, and

---

[4] For detailed comments and discussions see John Markoff "A corporate lag in research funds is causing worry," *New York Times*, January 23, 1989, A1.

discovery-oriented R&D, today's management has to operate in a much more dynamic and interactive way, involving complex sets of interrelated, nonlinear, and often difficult-to-define processes. These dynamic organizational environments demand a more sophisticated management style, which relies strongly on group interaction, resource and power sharing, individual accountability, commitment, self-direction, and control. Consequently, management today relies to a considerable extent on member-generated performance norms and evaluations, rather than on hierarchical guidelines, policies, and procedures. While this paradigm shift is driven by changing organizational complexities, capabilities, demands, and cultures, it also leads to a radical departures from traditional management philosophy on organizational structure, motivation, leadership, and project control. As a result, traditional "hard-wired" organizations and processes are replaced by more flexible and nimble networks that are usually derivatives of the conventional matrix organization. However, these networks have more permeable boundaries, more power and resource sharing, and more concurrent operational processes.

Shift from Efficiency toward Effectiveness. Many companies have broadened their focus from *efficient* execution of their operations and projects—emphasizing job skills, teamwork, communications, and resource optimization at the operational level—to *include organizational effectiveness*. This shift responds to the need for better integration of ongoing activities and projects into the overall enterprise, making sure that "we are doing the right thing." As an example, companies are leveraging project management as a core competency, integrating project-oriented activities closely with other functions, such as marketing, R&D, field services, and strategic business planning. While this shift is enhancing the status and value of certain business functions within the enterprise, it raises the overall level of responsibility and accountability, and puts higher demands on previously more autonomous functions, such as R&D and product development, to perform as a full partner within the integrated enterprise system.

Table 2.1   Changes in the Technology-Oriented Business Environment

Shift from . . .

. . . mostly linear work processes to highly dynamic, organic, and integrated management systems

. . . efficiency toward effectiveness

. . . executing projects to enterprise-wide project management

. . . managing information to fully utilizing information technology

. . . managerial control to self-direction and accountability

. . . managing technology as part of a functional specialty to MoT as a distinct skill set and professional status

Shift from Executing Projects to Enterprise-Wide Project Management. Many companies use project management extensively today for far more than just implementing specific projects. These companies leverage the full capabilities of project management, enterprise-wide, as a core competency, achieving accelerated product developments, higher levels of innovation, better quality, and better overall resource utilization. To achieve this level of competency, project operations must be integrated with the strategic planning system and business processes across the total enterprise. Managerial focus has shifted from the mechanics of controlling projects according to established schedules and budgets to optimizing desired results across a wide spectrum of performance measures that span the total enterprise. With its own body of knowledge, norms, and standards accepted worldwide, professional certification, and formal education programs at the master's and Ph.D. levels, project management has established its professional position over the last two decades. Today, the principles of project management apply across industries and around the globe, and are critical for effective execution of virtually all high-tech missions.

Shift from Managing Information to Fully Utilizing Information Technology. Today's technology provides managers in any part of the enterprise with push-button access to critical information on operational status and performance. The availability and promise of technology have led to the development of an enormous variety of powerful IT-based tools and techniques, and the acquisition of these tools by managers at all levels. With the powerful promise for increasing operational effectiveness managers are eager to use these tools in support of their activities, ranging from resource estimating, to scheduling, risk analysis, and decision support. The future challenge is at the systems level. Senior IT managers must look beyond the immediate application, such as project management or manufacturing resource planning, and integrate and *apply* IT to the firm's business process, solving operating problems *and* increasing business efficiency, rather than just replacing traditional forms of communications, interactions, and problem solving.

Shift from Managerial Control to Self-Direction and Accountability. With increasing business complexities, advances in information technology, changing organizational cultures, and new market structures, companies look *beyond* traditional managerial control for effective execution of their projects, operations, and missions. Especially top-down controls, based on centralized command and communications, while critically important, are no longer sufficient for generating satisfactory results. Organizational activities are increasingly project-oriented, relying on technology, innovation, cross-functional teamwork, and decision making, intricate multicompany alliances, and highly complex forms of work integration. The dynamics of these environments foster to a considerable extent member-generated performance norms and work processes, and a shift toward more team ownership, empowerment, and self-control. All of this has a profound impact on the way managers must manage and lead, and analyze the work environment for effective intervention. The methods of communication, decision making, soliciting commitment, and risk sharing are constantly shifting away from a centralized, autocratic management style to a team-centered, more self-directed form of control.

Shift from Managing Technology as Part of a Functional Specialty to Management of Technology and Engineering Management with Distinct Skill Sets and Professional Status. With their own body of knowledge, norms, and standards; professional certification; and formal education programs at the master's and Ph.D. levels, both management of technology and engineering management have established their professional status. Companies can choose from a pool of formally educated MoT and EM professionals, and have access to professional training that follows well-established operational, managerial, quality, and ethical standards.

A complex environment? Yes! But it is just the beginning of understanding the great challenges that managers face in our technology-based businesses environment. It is a starting point for understanding the interaction of organizational, behavioral, technical, and social variables that create the dynamics of this continuously changing landscape.

Balancing and optimizing the set of multiple issues involved will most likely produce the best organizational performance in the long run.

In response to the challenges created by the transformation of our business environment, many companies had to broaden their resource base via alliances, mergers, acquisitions, and joint ventures to achieve operational results and market reach. They also have explored alternate organizational designs, business processes, and leadership styles. Traditional forms of organizational structure and leadership have often been found ineffective in today's environment, but are gradually being replaced by self-directed, self-managed team concepts (Barner 1997, Thamhain and Wilemon 1999). Managing in today's technology-based enterprise requires a great understanding of the organizational system and its environment. It also requires extraordinary skills, talent, and the determination to succeed.

2.5 MEASURING TECHNOLOGY CONTENT AND INTENSITY

Virtually every company today claims to be technology intensive. Managers stress the fact that their organization uses technology in the development, production, and marketing of their products and services. Computers, information, equipment, and communication technology are everywhere. Organizations from R&D to manufacturing, banks, hospitals, legal services, government agencies, and retail stores all rely intensively on technology for their successful operation. Yet, there are differences in the degree of technology that an organization relies on. Obviously, a small grocery store is less technology intensive than a pharmaceutical research company. The challenge is to classify the technological intensity of a business operation. While scholars have classified technologies and technological intensities for their research (Babcock 1991, Khalil 2000, Narayanan 2001), there are no universally accepted metrics for defining the *level of technology*.

2.5.1 Dimensions of Technology Intensity

For comparative purposes, a number of specific dimensions can be suggested and developed into a measurement system for classifying technology intensity. These dimensions can also be used in combination with analytical models, such as *force-field* analysis,[5] for determining *changes* in the *intensity of technological orientation* of a company or industry, thus determining whether an organization, an industry, or society as a whole moves toward a more or less technology-intensive position:

- Capital equipment per employee
- Value added per employee
- Sales of goods and services per employee
- Technology-based skill requirements per employee
- Training requirements per employee per year
- Purchase or acquisition of technical materials and services
- Life cycle (shrinking) of materials, components, and services used in the business process
- Life cycle (shrinking) of products and services generated

Classification of Technology.   In addition to identifying technological intensity, some scholars have suggested classification schemes of technology according to the following categories (see Gaynor 1996)[6]:

- ➤ *State-of-the-Art Technologies.* Technologies equal or superior to competitive offerings
- ➤ *Proprietary Technologies.* Technologies protected by patents, and so forth.
- ➤ *Known Technologies.* Technologies common to many companies but used uniquely
- ➤ *Core Technologies.* Technologies essential for maintaining competitive positions

---

[5] The *forcefield analysis* is an analytical tool based on the pioneering research by Kurt Lewin. It is useful for examining countervailing forces to predict primary trends in complex systems, such as markets, technology, or society. As an example, the forcefield process could be used to determine technology trends within the textile manufacturing industry by identifying and analyzing the drivers and barriers to technological advancement. The dimensions can be used as guidelines for defining more industry-specific technology drivers and barriers. This very common and popular concept of *forcefield analysis* is similar to the Japanese *Ishikawa diagramming technique*.

[6] For further discussion please see Gaynor, Handbook of Technology (1996), pp. 1.26–1.27.

➤*Leveraging Technologies.* Technologies that support several products or classes of products

➤*Supporting Technologies.* Technologies that support core technologies.

➤*Pacing Technologies.* Technologies that control the product or service development

➤*Emerging Technologies.* Technologies under consideration for future application

➤*Scouting Technologies.* Technologies tracked for potential applications

➤*Unknown Technologies.* Technologies currently unknown, but believed of considerable benefit

One of the benefits of this classification is that it provides a framework for discussion, comparison, and strategizing. While few companies find it necessary, or possible, to attach to their businesses specific technology labels, they use these classifications for establishing some common ground for debating strategic business implications. Hence, the true value of these classifications is the gaining of strategic perspective for leveraging the business. As pointed out by Gaynor (1996), not every technology is *core*, *leveraging*, or *supporting*. The classification system allows managers to put technology into perspective. While core technology might be most important in a particular situation, other technologies should not be ignored. A breakthrough development could involve state-of-the-art technology at one end of the business, while another breakthrough could come from known technologies. Real-world examples range from 3M Post-it notes to pharmaceutical products, and can be found daily.

2.6 CASE STUDY

Reprogramming Amazon

At its nine massive distribution centers from Fernley, Nevada, to Bad Hersfeld, Germany, workers scurry around the clock to fill up to 1.7 million orders a day—picking and packing merchandise, routing it onto conveyors, and shipping the boxes to every corner of the world. It's an impressive display, but utterly misleading. The kind of work that will truly determine Amazon's fate is happening in places like the tiny, darkened meeting room at its Seattle headquarters.

Jeffrey A. Wilke, a compact, intense senior vice-president who runs Amazon's worldwide operations and customer service, and an engineering team are trying out a "beta" test version of new software they wrote. When the buying automation program is ready for prime time in mid-2004, Amazon's merchandise buyers will be able to chuck reams of spreadsheets for graphics-rich applications that crunch data for them, so they can more quickly and accurately forecast product demand, find the best suppliers, and more. The effort is one of scores of technology projects under way at Amazon that ultimately may change the entire experience of shopping online—and Amazon itself.

Just as most folks have come to view Amazon as a retailer that happens to sells online, guess what? It's morphing into something new. In ways few people realize, Amazon is becoming more of a technology company—as much as Microsoft and Wal-Mart. "What gets us up in the morning and keeps us here late at night is technology," says founder and chief executive officer Jeffrey P. Bezos. "From where we sit, advanced technology is everything."

No, Amazon isn't selling its own shrink-wrapped software or leaving the retail business behind. But developing technology is becoming at least as important as selling Harry Potter books or The Strokes CDs. Indeed, some analysts say it's possible that in a few years so many other retailers will be using Amazon's tech expertise to sell on its site that they could account for more than half the products sold on Amazon.com. Says Bezos: "Amazon Services could be our most important business."

Already, Amazon's technological efforts have helped it reduce costs and boost sales so much that revenues are expected to surge 32 percent this year, to $5.2 billion. As a result, by the time the glittering ball descends in Times Square on New Year's Eve, Amazon may well reach a milestone some never thought it could: its first full-year profit. No wonder its stock has rocketed 152 percent this year, to $49.34 a share.

But all that is just the start. Building on a raft of tech initiatives, from an ever-richer Web site to new search technology, Bezos aims to reprogram the company into something even more potent. The notion is to create a technology-driven nexus for e-commerce that's as pervasive and powerful as Microsoft's Windows operating software is in computing. That's right: Bezos hopes to create a Windows for e-commerce.

Self-Reinforcing Cycle

Far-fetched? *Not* necessarily. After all, the Amazon.com Web site is already a giant *Web* application. And bit by bit, Amazon is building what techies from Silicon Valley to Redmond call a platform: a stack of software on which thousands or millions of others can build businesses that in turn will bolster the platform in a self-reinforcing cycle.

Since last year, Amazon has been steadily tuning innovations it developed for its own retail site into so-called Amazon Web services. Even independent programmers are getting interested: In just 18 months, up to 35,000 programmers have downloaded software that enables them to pick and choose Amazon services and, much as they do with Windows, write new applications based on them.

One program makes it easy to list products for sale on Amazon. Another lets merchants instantly check prices via a wireless Web device. "I see Amazon not just as a place to sell things but as a provider of technology," says Paul Bausch, programmer and author of *Amazon Hacks: 100 Industrial-Strength Tips & Tools,* a new book on how to use the technology behind Amazon's site.

Amazon doesn't make money directly from the Web services. Merchants and developers can get free access to the services and can use them to sell from any outpost on the Web. However, many of the merchants who use these applications end up selling their wares to the 37 million customers assembled at Amazon.com. When that happens, Amazon takes a commission of about 15 percent and these revenues have much higher margins than Amazon's own retail business.

Busting Out

While Amazon isn't exiting retailing, its tech initiatives could help Amazon to bust out of the conceptual prison of stores and virtual confines of a single Web site. Says e-commerce consultant and author John Hagel III: "It's really breaking apart the whole store metaphor." Into what? Already, Amazon has applied its own technology to forge an identity as an online mall—a piece of business that generates gross margins about double its 25 percent retail margins. Plugging into its massive e-commerce system, thousands of retailers from mom-and-pop shops to Lands' End Inc. and Circuit City Stores Inc. sell through its site. Amazon even runs the Web sites and distribution for the likes of Target Corp. and Toys "R" Us Inc., which are featured on Amazon.com.

If Bezo's new plans work, Amazon could become not just a Web site but a service that would allow anyone, anywhere to find whatever they want to buy—and to sell whatever they want almost as easily. For Amazon, that means its finances could look considerably better than traditional retailers'. Already it turns over inventory 19 times a year, nearly double that of Costco Wholesale Corp. and almost triple that of Wal-Mart.

How much Amazon can expand the narrow margins of retail remains to be seen. But if even the conservative forecasts of analysts are correct, Amazon has a lot of upside in coming years.

Impossible Dream?

Bezos' vision won't be easy to fulfill. Trying to be a world-class retailer, a leading software developer, and a service provider simultaneously strikes some observers as a nearly impossible endeavor. Even some Amazon partners report shortcomings in merchandising and technical support. Others worry about

Amazon's inherent conflict in playing both retailer and mall owner. "They're biting off a lot," says Forrester Research Inc. analyst Carrie A. Johnson. "That's their biggest risk."

Amazon is hardly alone in its ambitions, either. eBay Inc., the Web's largest marketplace, is building its own e-commerce platform. Already, it boasts several million sellers, at least 37 million active buyers, and more than $20 billion in gross sales—quadruple Amazon's—and is far more profitable because eBay doesn't handle goods. Increasingly, merchants of all stripes view the two companies as key channels to online customers. And eBay isn't the only contender. Search upstart Google Inc. and even Microsoft, each with its own Web service initiatives, also aim to be hubs for connecting both shoppers and merchants.

Still, Bezos' bet on technology has paid off so far. Consider what has happened in its much-criticized distribution centers, which Amazon spent $300 million to build. Back in 2000, they were eating up at least 15 percent of sales, partly because processes to pick and pack different items such as books, toys, and CD players weren't very efficient. Chutes holding pending orders got backed up when products didn't arrive on time.

Now, by most accounts, the warehouses hum more like Dell Inc.'s build-to-order factories. With a menu-driven software console, workers can anticipate where bottlenecks are likely to occur and move people around to avoid them. Another program rolled out this year sets priorities, based on current customer demand.

The result: Amazon's distribution centers can handle triple the volume of four years ago and cost half as much to operate relative to revenue, just 7 percent of sales. Wilke believes further software improvements can boost productivity by up to 10 percent a year. "They couldn't do this without very, very good software," says Stephen C. Graves, a professor of management science and engineering systems at Massachusetts Institute of Technology's Sloan School of Management who has helped analyze Amazon's distribution operations.

Even so, Bezos isn't resting easy. While Amazon's spending on technology likely will remain fairly steady for the next year at about $216 million, thanks to declining tech prices, chief technology officer Al Vermeulen says he will hire hundreds more software engineers and computer scientists in the next year to slake Bezos' thirst for tech. "Jeff is a very big driver of the technology," says Amazon director Tom Alberg, managing director of Seattle-based Madrona Venture Group.

For one, Bezos believes that there's still plenty of room for improvement of the Web site itself. To that end, last year he hired what likely is a first for corporate

America: a chief algorithms officer, Udi Manber. His mission: to develop improved algorithms for Amazon's newest tech push: *search*. On Oct. 23, Amazon launched *Search Inside the Book*, a feature that allows visitors to find any word or phrase on 35 million pages in 120,000 books—and let them read entire pages around those keywords. In the week following, average sales growth for those books was nine percentage points higher than for books not in that database.

Manber has bigger plans yet. He now heads Amazon's first Silicon Valley outpost, a subsidiary called A9 that's charged with coming up with cutting-edge search technology. It's not just a defensive shot at search phenom Google, which is testing a shopping search engine it calls Froogle. "We need to help people get everything they need, not just a Web page," says Manber. And whatever he and his team come up with—he's mum now—it won't be confined to Amazon.com. A9 plans to offer search services it creates to other e-commerce sites as well.

Technologies like that in which Amazon is reaching out beyond its own site offer the most intriguing new opportunities—and challenges. Consider Amazon's *Merchant.com* business, which takes over the entire e-commerce operations of other retailers. By all appearances, it's a success. In Toys "R" Us's most recent quarter, in which results were dismal, the one bright spot was a 15 percent jump in sales at Toyrus.com, run by Amazon. Other retailers seem happy, too. "Amazon is the most sophisticated technology provider and service partner on the Internet," says Target vice-chairman Gerald L. Storch.

But how much that business will grow is debatable. As both retailer and mall operator, Amazon has divided loyalties. "Some companies worry about creating the next Wal-Mart that's going to take their business away down the road," says Dave Fry, CEO of Ann Arbor (Mich.)–based e-commerce consultant Fry Inc., which has helped several retailers sell on Amazon. As a result, many are going with rivals such as GSI Commerce Inc., which runs online operations for 61 retailers, from Linens 'n Things to The Sports Authority. Says Forrester's Johnson: "Ultimately, Amazon will be most successful selling their own and other retailers' products on their platform."

Yet, Web services may offer the most expansive potential of all Amazon's tech initiatives. It's still early, but it's possible that Amazon has latched onto one of tech's juiciest dynamics—a self-reinforcing community of supporters. Indeed, it seems to be harnessing the same "viral" nature of the open-source movement that made Linux a contender to Windows. Says Whit Andrews, an analyst at Gartner Inc.: "It creates an enormous community of people interested in making Amazon a success."

Even Microsoft. In Office 2003, people can click on a word or name in any document and be whisked off to Amazon.com so they can buy a related book or other product. Says Gytis Barzdukas, director of Microsoft's Office product management group: "It gives Amazon the ability to market to a whole new set of customers and to become a part of people's work processes."

Amazon, the next Microsoft? Not so fast. For all the promise, building a broad platform is about as tough as a goal gets in the tech business—as nearly every competitor to Microsoft can attest. And unlike most tech companies, Amazon has to contend not just with bits and bytes, but also with the bricks and mortar of warehouses and the fickle fingers of Web shoppers. But for now, at least, Amazon's pursuit of cutting-edge technology has given it time to figure out what comes next.

Source: Excerpted with permission from *Business Week,* December 22, 2003, pp. 82–86, "Reprogramming Amazon," by Robert D. Hof.

2.7 SUMMARY OF KEY POINTS AND CONCLUSIONS

The key points that have been made in this chapter include:

- Virtually every company today sees itself as technology intensive.
- Six business subsystems make management of technology unique and different from other types of management: (1) work, (2) people, (3) work process, (4) managerial tools and techniques, (5) organizational culture, and (6) business environment.
- Technology management dates back for more than 4,000 years, but has accelerated exponentially over the past 100 years. Especially during the past 50 years, technological advances have exceeded all of the advances made during the rest of the history of the human race.
- Six major shifts in our business environment have influenced technology management: shifts from (1) linear processes to dynamic systems, (2) efficiency toward effectiveness, (3) executing projects to enterprise-wide project management, (4) managing information to fully utilizing information technology, (5) managerial control to self-direction and accountability, (6) managing technology as part of a functional specialty to management of technology and engineering management.

2.8 CRITICAL THINKING: QUESTIONS FOR DISCUSSION

1. How can technology companies best prepare their people for effective management and leadership?
2. How can educational institutions help in preparing students for technology leadership positions in industry?
3. Develop an assessment tool (i.e., checklist) for measuring the technological intensity of an enterprise.

4. How can senior management prepare themselves to anticipate changes in the business environment, short range and long range?

5. What strategic decisions and directions at Amazon.com were key to the company's success?

2.9   REFERENCES AND ADDITIONAL READINGS

Andrew J. and Sirkin, H. (2003) "Innovation for cash," *Harvard Business Review*, Vol. 81, No. 4 (September), pp. 50–61.

Babcock, D. L. (1991) *Managing Engineering and Technology*, Englewood Cliffs, NJ: Prentice-Hall.

Barkema, Harry G., Baum, Joel A., and Manix, Elizabeth A. (2002) "Managing challenges in a new time," *Academy of Management Journal*, Vol. 45, No. 5 (October), pp. 916–930.

Barner, R. (1997) "The new millennium workplace," *Engineering Management Review* (IEEE), Vol. 25, No. 3 (Fall), pp. 114–119.

Cairncross, F. (2002) *The Company of the Future*, Cambridge, MA: Harvard Business School Press.

Dertouzos M., Lester R., Solow R., and the MIT Commission on Industrial Productivity (1990) *Made in America: Regaining the Productive Edge*, New York: HarperPerennial.

Dorf, R. C., ed. (2004) *The Technology Management Handbook*, Boca Raton, FL: CRC Press.

Fellenstein, C. and Wood, R. (2000) *Exploring E-Commerce, Global E-Business and E-Societies*, Upper Saddle River, NJ: Prentice-Hall PTR.

Gaynor, G. H. (1986) "Management of technology: description, scope and implications," Chapter 1, *Handbook of Technology Managing* (G. Gaynor, ed.), New York: McGraw-Hill

Gaynor, G. H., ed. (1996) *Handbook of Technology Managing*, New York: McGraw-Hill.

Gehani, R. R. (1998) *Management of Technology and Operations*, New York: Wiley.

Haddad, C. J. (2002) *Managing Technological Change*, Thousand Oaks, CA: Sage Publications.

Hof, R. D. (2003) "Reprogramming Amazon," *Business Week*, December 22, 2003, pp. 82–86.

Karlgaard, R. (2003) "Disruption, high and low," *Forbes*, October 11.

Khalil, T. (2000) *Management of Technology*, New York: McGraw-Hill.

Kuhn, T. S. (1970) *The Structure of Scientific Revolutions*, Chicago, IL: University of Chicago Press.

McLean, I. W. and Round, D. K. (1978) "Research and product innovation in Australian manufacturing industries," *Journal of Industrial Economics*, Vol. 27, pp. 1–12.

Naisbitt, J. (1982) *Megatrends*, New York: Warner Books.

Narayanan, V. K. (2001) *Managing Technology and Innovation for Competitive Advantage*, Englewood Cliffs, NJ: Prentice-Hall.

Parsi, A. J. (1989) "How R&D spending pays off," *Business Week* (August, Bonus Issue), pp. 177–179.

Sakaiya, T. and March W. (1991) *Knowledge Value Revolution*, New York: Kodansha American Publication.

Senge, Peter (1994) *The Fifth Discipline: The Art and Practice of the Learning Organization*, New York: Doubleday/Currency.

Steele, L. W. (1989) *Managing Technology: The Strategic View*, New York: McGraw-Hill.

Thamhain, H. J. (1990) "Managing Technology: The People Factor," *Technical and Skill Training*, August/September, pp. 24–31.

Thamhain, H. J. (2003) "Managing Innovative R&D Teams," *R&D Management*, Vol. 33, No. 3 (June), pp. 297–312.

Thamhain, H. J. and Wilemon, D. L. (1999) "Building effective teams in complex project environments," *Technology Management*, Vol. 5, No. 2 (May), pp. 203–212.

Tofler, A. (1980) *The Third Wave,* New York: William Morrow & Company.

U.S. National Research Council (1997) Committee on Japan, *Maximizing U.S. Interests in Science and Technology Relations with Japan*, Washington, DC: National Academy Press.

U.S. National Research Council, NRC (2003) Committee on Information Technology and Creativity (W. Mitchell, A. Inouye, and M. Blumenthal, Editors), *Beyond Productivity: Information, Technology, Innovation, and Creativity,* Washington, DC: National Academy Press.

U.S. National Science and Technology Council (1996) *Technology in the Nation's Interest,* Washington, DC: Office of Technology Policy, U.S. Department of Commerce.

3

ORGANIZING THE HIGH-TECHNOLOGY ENTERPRISE

GM'S ADVANCED VEHICLE DEVELOPMENT—REDUCING TIME TO MARKET WITH STREAMLINED ORGANIZATIONAL PROCESSES

After getting beaten to the punch year after year by Toyota and Honda, General Motors has stepped up its efforts dramatically to reduce the time to launch new products and to react to market changes. While Toyota and Honda are still faster, GM and the other U.S. Big Two are getting more competitive. "Most of GM's new product programs are on a 24-month schedule," says Rick Spina, executive director, program management, GM North America. He adds, "We are continuing to cut it down, but 24 months is pretty much our norm now. We know, we can do it, but the world doesn't stand still." Most critics agree. U.S. auto makers have come a long way from the 1970s, when government safety and emissions standards turned the company bureaucracy into an organizational albatross, and it took 18 months or more to bring a new car model to market. Today, thanks to management's determination and leadership, and to the enormous advances in design technology and processes, from CAD/CAM to simultaneous engineering, product development cycles continue to shrink. U.S. car makers have new products in the market after 18 months, with some specialty cars, such as the Hummer H2 and Ford GT, being done as quickly as 16 or 14 months. However, time-to-market performance takes more than just technology and project management process templates. It is achieved through improved productivity, elimination of bureaucracy, and focused decision making, as explained by Mark Hogan, GM group vice president, advanced vehicle development: "The new approach is possible because the overall product development organization has continued to increase productivity and become more focused under the *Vehicle Line Executive (VLE) System.* The up-front work that has to be done to determine if the program is viable and profitable is very complex and includes many different variables. It starts with portfolio planning. Our teams orchestrate the work and resources associated with the development of a new product idea, and then see it through to the point where we know the product can be executed as a sound business case." "We are making sure, we have all the stakeholders involved," adds Ron Pniewski, GM North America vice president of planning. "The previous sequential approach, using a lot of hand-offs, is time-consuming and invites communication breakdowns." The current Advanced Vehicle Development process starts off with a core team that includes all the line functions that have been established at GM for decades, such as engineering, design, planning, purchasing, manufacturing, quality, and marketing. Once the new product team and its support groups have been organized, the work integration through the various product development stages is virtually seamless. Yet, managing this process is not a trivial matter, as echoed by Mark Hogan: "In essence, we're managing a virtual organization. Most of the critical players associated with the up-front work are also part of the bigger product development organization. It's our role to ensure the directions and desires of GM's senior product leadership are being effectively applied in creating new products. This requires aligning some of industry's best talents and resources, and to make the right decisions faster through streamlined decision making and better organizational focus."

Source: GMC Press Releases, 2002–2004, *www.prdomain.com/companies/g/gm/news.*

3.1 TODAY'S BUSINESS PROCESSES REQUIRE FLEXIBILITY, SPEED, AND EFFICIENCY

Effective organizational structures are fundamental to business success. This is not only true for General Motors, but also applies to NASA, dot.coms, and community hospitals. As technology changed the competitive landscape, organizations too had to change, to keep up with the demand for greater flexibility, speed, and efficiency. New administrative tools, product development techniques, and project management tools have evolved, especially over the past 20 years. These tools offer better capabilities for executing operations more integrated with the business process, and with greater emphasis on supply chain integration, horizontal decision making, and work/technology transfer. This in turn leads to flatter, leaner, and more change-responsive organizations that can deliver enterprise objectives, such as new product development, by integrating resources effectively across multifunctional organizational segments.

Very noticeably over the past 20 years, activities within the enterprise have become increasingly business-oriented. That is, each functional component of the enterprise is being measured increasingly by its contribution to specific enterprise objectives, rather than by its ability to provide superior functional services in its specialty, such as R&D, marketing, engineering, or manufacturing. This drive toward broader business accountability combined with the pressures for faster, more effective market response have led to many new and innovative organizational designs, such as *simultaneous engineering, concurrent project management, design-build,* and *Stage-Gate processes.* However, none of these new organizational forms can function as a business by itself. They become *overlays to the traditional functional organization*, the baseline of any enterprise, from our ancient beginnings to modern times.

This presents major challenges. The drive toward greater cross-functional efficiency and agility also requires large degrees of *resource and power sharing*, hence diluting central decision making and control toward unified enterprise objectives. It also diminishes the autonomy of functional resource groups to develop and maintain the best functional capabilities needed by the enterprise. When GM's group vice president for advanced vehicle development, Mark Hogan, talked about the new approach to product development under the *Vehicle Line Executive (VLE) System,* he also pointed at the importance of resource alignment, supply chain integration, central organizational focus, and senior leadership. GM's emphasis on central direction and control is very clear when you look at GM's Advanced Vehicle Development Center in Warren, Michigan. The reality is that even GM, with its enormous pressure for agility and speed, is not too eager to give up central control and to empower management at the operational level. However, organizations are continuously evolving to adapt to the changing business environment. In spite of the challenges that flatter, leaner, and faster organizations are presenting, the trends of restructuring toward better cross-functional, horizontal integration are continuing. Let's take a more formal look at the organizational options that exist for today's enterprises.

3.2 ORGANIZATION DESIGNS FOR TECHNOLOGY-BASED ENTERPRISES

How can a company be organized to conduct its business most effectively and yield the greatest value to its stakeholders? Different times in history produced different answers. In 1600 the British East India Company was formed by a group of independent people, joining together for a single business mission: a trading mission to the East Indies. Although the company became one of the most powerful commercial enterprises of its time, and took part in the creation of British India, Hong

Kong, and Singapore, *its stakeholders disbanded each time a mission was completed,* at least for the first fifty years of the company's existence. Three hundred years later, Henry Ford created quite a different organizational model. He demonstrated that a successful company must be both vertically and horizontally integrated, owning virtually all stages in the supply chain and having strong central control. How does this compare to an Internet startup company, or an Intel or General Morors, today? Is there a "norm" of an organizational structure today? Fitting an organizational approach to a company is like fitting cloth to a person, says Alan Glasser.[1] It's a matter of style, personal taste, and circumstances.

How can the activities and functional support systems of an enterprise be organized most effectively to optimize desired results? In today's environment, where companies struggle with issues of complexity, agility, resource efficiency, and interdependence, the need exists for both centralized control *and* decentralized decision making, functional autonomy *and* cross-functional integration. This is a tricky balancing act and a great challenge. Consider the internal work environment of a typical high-tech company. Management has to deal with a broad spectrum of contemporary challenges. Such challenges include time-to-market pressures, accelerating technologies, innovation, resource limitations, technical complexities, social and ethical issues, operational dynamics, risk, and uncertainty, as summarized in Table 3.1. Facing such a dynamic environment often makes it difficult to manage activities through traditional, linear work processes or top down controls. In response to these challenges, many companies and their management have moved from reliance on hierarchy and central control to flatter, more dynamic, and more cross-functionally transparent organizations. Their managements are trying to attend to both dimensions of vertical and horizontal integration. This creates a classic dilemma: how to organize to ensure unified mission control of the enterprise while providing autonomy and flexibility for horizontal integration and delivery. The answer to this dilemma is a delicate power balance between functional resource units and cross-functional business processes that results in some form of a matrix structure.

Table 3.1 High-Tech Business Environment: Today's Characteristics and Challenges

- Changing business models and structures
- Complex business performance measurements
- Complex joint ventures, alliances, and partnerships
- Complex projects
- Complex success criteria
- Different organizational cultures and values
- Global markets
- High risks and uncertainties
- Integrating across functions

[1]See *Research and Development Management*, p. 244, by Alan Glassser (1982).

Table 3.1   (Continued)

- Integrating broad spectrum of functions and support services
- Integrating many business processes
- Many stakeholders
- Multifunctional buy-in and commitment
- Need for continuous improvement
- Need for sophisticated people skills
- Organizational conflict, power and politics
- Resource constraints
- Self-directed teams
- Tight, end-date-driven schedules
- Tough performance requirements
- Virtual organizations, markets, and support systems

3.3   ORGANIZATIONAL LAYERS AND SUBSYSTEMS

The business areas of an enterprise do not operate in a vacuum, but are integrated within the functional support system of the company, which is part of the institutional framework of the enterprise. The three fundamental organizational layers are shown in Figure 3.1:

1. *Institutional Framework.* This is the area of "immortality," providing strategic directions, long-range survival and growth plans, policies, and procedures. This layer is staffed with senior management, corporate officers, and directors, who provide broad guidelines and resource allocations for the enterprise.

2. *Functional System.* This is the traditional organizational framework of the firm. It is an area of slow change and the provider of stability. It is the functional system that positions the enterprise for competitive advantage, growth, and profitability by advancing methods of operation, markets, and supply lines, and by integrating new technologies into the operating areas of the organization. Typical resource groups of the functional system include R&D, engineering, development, manufacturing, marketing, human resources, legal, quality control, and purchasing, just to name a few of the more common functional subsystems.

3. *Operational Areas.* This is the contemporary part of the enterprise that expands and shrinks as needed by the business. The operational areas of the firm are often organized as programs and projects, such as new product developments, contracts, off-the-shelf deliverables, internal maintenance, and field support operations. It is the functional system that provides the needed resources to the operational areas, which leads in most cases to a

matrix organization. In the extreme, the operational areas expand to absorb the entire functional system, creating a pure product organization or projectized or aggregated organization. In this approach the combined functional-operational system exists for the sole purpose of supporting one operation, such as the development of a new car model at GM, a new commercial airliner at Boeing, or a Mars Rover at the Jet Propulsion Laboratory (JPL). It is these operational areas that are most directly responsible for business results to the enterprise. It is also these operational areas that are most visible to the market and customer, and are held most accountable for business results by top management.

Each of the three organizational layers exists in every firm. These layers often overlap significantly and occupy different amounts of space relative to each other. Consider, for example, a high-tech, but relatively undifferentiated, computer assembly plant or a newspaper publisher. These firms can be expected to have a relatively large part of their resources organized along functional lines. Conversely, a consulting firm or aerospace company would be organized with two strong axes of functional and project/operational responsibilities. Yet, another situation exists for companies, such as Boeing, that most likely organize the whole company around product lines, such as 747, 767, 777, etc., hence integrating both functional and operational areas with focus on a particular product or project.

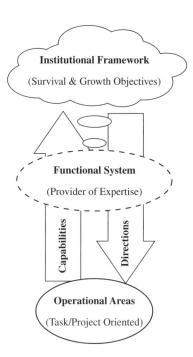

Figure 3.1   The three principal organizational layers.

With the great variety of products, services, markets, and supply chain scenarios, it is not surprising to find a large number of organizational structures and processes in today's world of business. While the resulting organizational structures are often bewildering and confusing, they can be broken down into a few basic components that can be explained in a simple model, such as that shown in Figure 3.2. In virtually every enterprise, responsibilities can be divided into two overlapping categories or axes: (1) responsibilities related to the *management of resources* that provide the traditional *functional organization*, such as engineering, manufacturing, and marketing, and (2) responsibilities related to the *management of the business, its projects, and its operations*, hence providing mission-oriented results of product rollout or project deliverables. As shown in Figure 3.2, the operations axis is an overlay to the functional axis, well positioned to contract with the functional organization for services needed for the integration of specific projects or programs. This is the essence of matrix management. Every enterprise has these two organizational axes to some degree. However, the organizational construct and managerial process vary a great deal, depending on the nature of the business, which will be discussed in the next section of this chapter.

3.4   ORGANIZATIONAL CHOICES

Fundamentally, companies have two choices for structuring their business operations:

- *Functionally Organized.* Resources are grouped by "functional" capabilities and managed via a hierarchical chain of command and control processes.
- *Project Organized.* Resources are allocated to specific projects that are managed autonomous and independently.

However, virtually no company works as a *pure functional* or *pure project* structure. Every company has some functional components that provide support services

Figure 3.2   Two organizational axes: project operations and resource functions.

and infrastructure, and every company has some project activity. As soon as a functionally organized business engages in some project activities, or a project organized business creates some overhead functions, it operates as a matrix organization.

3.4.1   Matrix Organized

This is a hybrid between the functional and the project organization that relies on resource sharing among functional units for producing specific project deliverables. Real-world businesses operate as hybrids. However, the degree of projectized versus functional structure varies a great deal, not only among companies, but also within each company. The matrix provides an effective and convenient framework for structuring any business, because its design depends less on the physical restructuring of organizational components than the management style, policies, procedures, and budgeting processes that determine the sharing of power and responsibilities. Figure 3.3 shows graphically the matrix as part of an organizational continuum, somewhere between the two extremes of pure functional and pure projectized. Further, the "strength" of the matrix, and therefore its location between the two extremes, depends to a large degree on the management style and interaction of people within the organization, as will be discussed next.

3.4.1.1   The Functional Organization

This is the traditional and most fundamental form of organization and management. It has been successfully used since ancient times by governments, military organizations, churches, and commercial enterprises. The trademark characteristics of the functional organization are the separation of "functional responsibilities," such as R&D, engineering, product development, marketing, finance, human resources, and so forth, and its hierarchical structure, which leads to clearly defined chains of command, controls, and communication channels. As summarized in Table 3.2, the *strength* of this organizational form is in effectively utilizing its resources, taking advantage of economies of scale, developing areas of specialization and expertise, and providing the institutional framework for long-range enterprise planning and

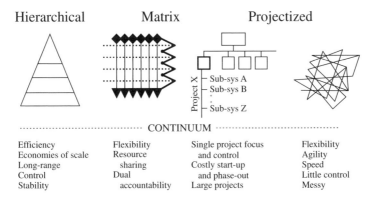

Figure 3.3   The organizational continuum of the matrix.

Table 3.2 Characteristics of the Functional Organization

Strength

- Institutional framework for planning, control, and stability
- Efficient use of collective experiences and facilities
- Concentration of expertise, benefiting most advanced developments
- Long-range preparation for survival and growth
- Effective utilization of operational facilities and systems
- Career continuity and advancement potential
- Well suited for mass production and standardized operations

Weaknesses

- Difficult cross-functional communications and controls
- Limited ability of technology transfer and multidisciplinary integration
- Slow and ineffective decision making at enterprise level
- Limited flexibility in responding to changes in business environment
- Risk adverse, limited entrepreneurial capacity
- Functional structure interferes with business process

control. Still, today, the functional organization provides the basic platform of operation for any enterprise, regardless of its business or environment. This concept of organizational structure has been the backbone for companies that are in a single business, especially those whose products and services come out of a single operational facility and are marketed through a single distribution and sales network. If decentralization of the enterprise is sought, regional or product divisions can be formed, as is typical for the automobile and aircraft industries (i.e., General Motors, Ford, and Boeing). Or, the company can operate in a number of unrelated conglomerated businesses, such as General Electric, Verizon, and ITT. Each of these decentralized businesses has its own organizational systems for development, manufacturing, sales, marketing, and so forth, hence operating its own "functional" business according to its product or regional charters.

Yet, in spite of its robustness and economic benefits, the functional organization has its *limitations and weaknesses*, especially in the dynamic, fast-changing, and market-focused business environment that we have been experiencing for several decades. In fact, the traditional functional organization is most challenged in *complex and technology-intensive environments*, characterized by high speed, great change, and great uncertainty (Shim and Lee 2001, Thamhain 2004, Zhang, Keil, Rai, and Mann 2003). Critical success factors (CSF) span a wide spectrum of cross-functional areas, involving technological, organizational, and interpersonal issues, including gaining and maintaining cohesiveness, commitment, technology transfer, self-direction, rapidly changing technology and requirements, resource limitations, innovation and demands for flexibility, and speedy implementation. In these contemporary business environments, traditional models of organization and management are often not effective. As a result, these traditional hierarchical organizations are augmented with contemporary systems, tools, and techniques that lead to the matrix and projectized organization. However, regardless of these limitations and extensions into other organizational forms, the functional structure provides one of the most fundamental and stabilizing layers in any organization, and is the backbone of any company or institution.

3.4.1.2   *The Projectized Organization*

The *projectized* (also called *aggregated* or *monolithic*) structure essentially organizes the business activities of an enterprise into project groups. In its purest form, the *projectized* organization is the most extreme departure from the functional organization. The enterprise is partitioned into project units (or programs), with resources allocated to specific projects, managed autonomously and independently. Often each project is run as a profit center, like a business unit in a conglomerate or a division of a large company. In fact, the projectized form of organization and management is *similar to the divisionalized structure*. It offers a contemporary approach to building an organization for the purpose of executing a single project. Hence, the organization has a limited life. It builds up with the project and terminates with it! As summarized in Table 3.3, the projectized organization represents the strongest form of project authority. Each project manager has complete control over all support functions needed, people, facilities, and functions necessary to execute the project, start to finish. That is, the projectized business unit contains not only the technical functions, but also the operational and administrative support functions, such as marketing, quality, finance, human resources, and legal.

Table 3.3   Characteristics of the Projectized Organization

Strengths

- Strong control over all project activities by single authority
- Rapid reaction time, time to market
- Schedule, resource, and performance trade-offs
- Large project integration capability
- Personnel loyalty to project, one boss
- Well-defined interfaces to contractors and customers
- Well suited to large, long-term projects

Weaknesses

- Inefficient use of production facilities and capital equipment
- Difficulty of balancing work loads during technology transfers
- Costly project start-up and phase-out
- Limited technology development
- Little opportunity for sharing experiences among projects
- Limited career ladders

The principal advantage of the projectized organization is the high degree of control over each project, its resources, and its interfaces by a single authority, the project manager. For larger projects this translates into (1) more rapid reaction time to market/customer changes, (2) faster time to market, (3) schedule, resource, and performance trade-offs within each project, and (4) the capability to integrate larger projects in comparison to any other organization. Because of the full line of authority, similar to the functional organization, trade-off decisions between performance, cost, and schedule can be made rapidly and effectively with a focus on the end objectives. All people assigned to the project are loyal to the project and its mission objectives. By definition, there are no competing projects or dual accountabilities, and there is only one boss. This type of project organization is often preferred by the customer, because it can mirror the customer's organization, therefore providing one-to-one people interfaces between customer and contractor.

The principal disadvantages of the projectized organization, especially visible for smaller project and projects with shorter life cycles, are (1) considerable start-up cost and time requirement, (2) limited opportunity to share experiences and production elements, and (3) limited opportunity to use economies of scale and to balance workloads.

Because of these disadvantages and limitations, even companies in project-oriented businesses seldom projectize completely, unless their projects are large enough to utlize the dedicated resources, and long enough in duration to justify the organizational setup and phase-out efforts. What is more common, especially in technology-based companies, is a partially projectized organization. That is, project managers may fully control *some resources* that are particularly critical to the project, and can be fully utilized over the project life cycle (resource leveling), while other resources remain under the control of functional managers who share them among many projects as needed. Such resource sharing especially prevails in administrative support functions such as human resources, accounting, and legal. Thus, many enterprises that at first glance look projectized are really hybrids of several organizational types and layers. The matrix organization is in fact such a hybrid organization that operates across a whole spectrum of organizational entities. Stretched in one direction, it behaves very much like a functional organization, and stretched in the other direction, it becomes projectized.

3.4.1.3   The Matrix Organization

The matrix approach offers a compromise solution for project-oriented businesses that cannot dedicate resources over the life cycle of their projects, but must share them among many or all projects. Matrix organizations are overlaid on the functional structure of the firm which, when designed right, combines the strength of both the functional and the projectized organization. The matrix is a relatively straightforward and simple concept. Yet, it has been surrounded by much mystique and confusion.

Here is how it works: When a functionally organized enterprise has to perform several tasks, missions, or programs simultaneously, it automatically operates as a matrix. The matrix is essentially an overlay of contemporary project organizations on the functional resource departments of a company, as symbolically shown in Figures 3.2 and 3.4.

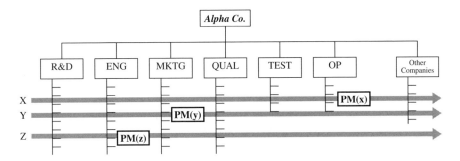

Figure 3.4 The matrix: schematic structure.

All departments and support units retain their functional, hierarchical characteristics. A clear chain of command and control exists for all units within the enterprise. At the operational level of the enterprise, task leaders or project engineers direct and coordinate the work. They are responsible for the implementation of plans that are directed either from the top down or horizontally via project managers, such as PM(x), PM(y), or PM(z) in Figure 3.4. These task leaders essentially have two bosses, upward to their department managers for effective resource utilization and review of work quality, and horizontally to project leaders for the timely and resource-effective implementation of specific projects according to established plans. Similar *dual accountabilities* often exist at the next level. Department managers share with their task leaders responsibilities along the horizontal axis of the organization for resource-effective implementation of projects plans, while vertically they are responsible to senior management for the maintenance and advancement of personnel, facilities, and support technologies. Project managers, such as PM(x), can either report to a functional manager, as shown in Figure 3.4, or into a *Program Management Office (PMO)*, led by a *director of programs*, responsible for all projects conducted within the enterprise. Further up the organizational hierarchy, senior management is responsible for overall capability coordination and resource allocation, strategic planning, overall corporate direction, and leadership, very similar to top-level responsibilities in traditional hierarchical organizations.

The advantages of the matrix organization, summarized in Table 3.4, are similar to both the functional and the projectized organization combined: reasonable reaction time to emerging business opportunities, customer requirements, and changes; effective integration of multidisciplinary activities; and efficient resource utilization. The matrix also enjoys a high concentration of specialized resources, positioning the company for creating the best support technologies and most advanced products. The task leaders and project managers within the organization provide effective interface points to other organizational units and externally to the customer, contractor, and supplier communities. Perhaps one of the most important features of the matrix is its ability to start up and phase out projects quickly and economically, which results in greater agility and flexibility of the enterprise in pursuing emerging business opportunities and in dealing effectively with a mixture of projects that vary in size, duration, and scope.

Table 3.4   Characteristics of the Matrix Organization

Strengths

- Effective resource utilization
- Quick formation of project teams
- High quality of resources and skill sets
- Ability to handle complex projects
- Ability to work on many projects concurrently
- Quick response to business/market needs
- Organizational agility
- Standardized business processes
- Unified business strategy across many operations
- Career development and growth opportunities

Weaknesses

- Power and resource sharing
- Dual accountability
- Leading without authority
- Organizational ambiguity
- Complex organizational interfaces and work transfers
- Priority conflict and interruptions
- Complex cost accounting
- Difficulty of project control
- Resource multiplexing leads to organizational conflict and inefficiency
- Career uncertainty, risk, and anxieties

The weaknesses and limitations of the matrix organization relate to its unconventional structure that relies on resource and power sharing, often a source of organizational conflict, and always a challenge to managerial accountability and control. Furthermore, the additional organizational overlay (two axes) is likely to increase cost overhead, and together with the power and resource sharing, requires a more sophisticated management style than either the traditional functional or the projectized organization.

Yet, in spite of its limitations, the *matrix, when designed properly, combines some of the best features of both the functional and the projectized organization.* Therefore, it is not surprising to find the matrix structure in virtually every enterprise that executes projects or has to deal with multidisciplinary task integration. Technology management is to a large degree synonymous with matrix management!

Four Matrix Categories. Matrix organizations come in many forms and shapes depending on the specific business processes and operational needs within each company. Formally, matrix organizations are classified into *four specific categories*[2] according to their primary purpose and mission objectives:

1. *Project-Function Matrix.* This represents an overlay of the project structure to the functional resource organization. It is the most common type of matrix organization, used for executing a wide variety of projects, ranging from development to service and training. The project or program manager is responsible for the business results, including project acquisition, planning, organizing, integration, and customer interfacing. Resource managers are responsible for the functionality and quality of the project components, and the management of their departmental resources, including the development and effective utilization of personnel, facilities, and technologies.

2. *Product-Function Matrix.* This matrix emphasizes product-oriented businesses. It is an overlay of the product organization on the functional or resource organization. Project or product-line managers are responsible for business results. Typically, they have to work across functional lines to achieve the integration of product design, prototyping, fabrication, sales, and marketing. Resource managers are responsible for the development and effective utilization of company resources and the technical implementation of project components, including their functionality and quality.

3. *Product-Regional Matrix.* This matrix is an overlay of the product organization on the regional network of a company's operation. This is often a sales-oriented structure where product managers are responsible for business results cutting across geographic-regional lines. This type of business operation also leads to multidimensional matrices with multiple layers of business operations.

4. *Multidimensional Matrix.* This organizational structure combines several matrix types, resulting in many layers, such as an overlay of the product organization on the functional organization, within a business division or regional network. Therefore project managers must cross functional lines in various divisions and geographic regions to achieve the desired business results. Similar to other matrix organizations, resource managers are responsible for the development and effective utilization of company resources and the technical implementation of project components.

Additional Matrix Axes. The relatively simple structure of the project-function matrix, with its two basic axes, as shown in Figures 3.2 and 3.4, can be expanded to reflect the realities of more complex business environments, as shown for the multidimensional matrix. The concept can be expanded even

---

[2]Classifications of matrix organizations date back more than 30 years. One of the earlier summary descriptions of different matrix types was published by the Conference Board, New York, in 1979, *Matrix Organizations of Complex Businesses.*

further. Companies that work across geographic, cultural, technological, and industrial boundaries, typical for multinational ventures or large programs, might establish additional matrix axes, which overlap the functional organization. Project charters, management directives, policy directives, and process flow diagrams provide the tools for defining these multidimensional matrices operationally. In addition, personal discussions and interactions, such as team sessions, workshops, focus group meetings, and managerial involvement with the project teams, can help in effectively communicating the matrix process and in clarifying the way the organization actually functions.

3.5   REAL-WORLD HYBRIDS

For real-world technology companies, the choice among functional, divisionalized, projectized, or matrix organization is not simple. A multibusiness company such as General Electric operates many unrelated businesses, largely on a high-technology platform, bound together by a common operating system. Top down the firm looks like a conglomerate, divided into operating companies. However, each of GE's 13 businesses consists of dozens of related, yet highly diversified enterprises with very different organizational needs and structures. Take, for example, GE's Transportation Business, comprising Aircraft Engines and Rail, two industry-leading business units with products and services ranging from jet engines to locomotives, mining trucks, and gas turbines, serving many different markets worldwide. Because these operations involve very large, long-term programs, most of these businesses have a projectized structure to concentrate resources and control on single projects or programs with dedicated customer and vendor interfaces, and specialized technology developments and field services. Yet, internally many resource departments, while dedicated to a single-project profit center, operate along matrix lines, executing subprojects that are eventually integrated with the larger program. In addition, there is a layer of traditionally structured administrative and strategic units that provide services and share resources across several businesses. Furthermore, to deal effectively with customers, suppliers, and regulators, each business must cross geographic, cultural, technological, and industrial boundaries, establishing additional organizational layers or axes that overlap the core business.

Not every company is as big as General Electric. Yet, many companies a lot smaller than GE also deal with a mixture of large and small programs, short range and long term, in different markets and multinational environments. These companies not only have the challenge of choosing the right organizational structure, but also must institutionalize this structure, that is, make it congruent with established business processes, cultures, and value systems.

3.5.1   Managerial Perspective

Choosing an appropriate organizational structure is difficult for any company. It is especially tough for firms that are involved with a mixture of functional and project-related activities varying in size, duration, and markets. Since many technology-based companies fall into this category, it is not surprising that most of these enterprises operate, by design or default, as organizational hybrids. The core of these companies is often organized along matrix lines with hierarchically structured core functions, such as R&D, engineering, testing, manufacturing, marketing, field services, legal, and human resources. These resource functions are shared among the various project operations, providing a relatively high degree of organizational versatility and flexibility. In addition to the basic matrix, we find other organizational layers of *projectized clusters, mini-matrices, individual project organizations,*[3] *staff-project,*[4] and *intermix organizations.*[5] These hybrid structures help management to accommodate the widest possible range of business activities with great flexibility, while retaining much of the traditional functional stability and resource effectiveness.

3.6 UNDERSTANDING THE WORKING ENVIRONMENT

Real-world technology organizations are complex, both on paper and in practice. They must be carefully designed to accommodate the needed infrastructure and support systems for the business process while maintaining effective resource utilization. In many cases, even the managers who "designed" these organizational structures are unable to classify or describe their creations in simple terms, but speak of organizational system overlays, resource sharing, and joint responsibilities.

In fact, if we look at the organizational description of a typical high-technology company, such as given in an annual report or on a company Web site, we find that most high-tech businesses are very complex in terms of their internal operations and outside interfaces. Many of these companies conduct project-related businesses, and their core structure is some form of a matrix with an unconventional array of interfaces and reporting relationships.

To work effectively, people must understand where they fit into the enterprise and what their responsibilities are. Especially with the complex workings and intricacies of modern technology companies, it is important to define the management process together with the command and control structure, which includes responsibilities,

---

[3]An *individual project organization* is a one-person program office, chartered with the coordination of a single project across functional lines, according to established plans and objectives. It is simple, efficient, and quick to establish and to dissolve. It often exists as an overlay on other major organizational structures, such as the matrix or projectized organization.

[4]A *staff-project organization* is a mini-matrix, similar in characteristics to the *individual project organization,* but broader in terms of resources and responsibilities. As the name implies, the staff-project organization consists of a project leader with a small number of staff people directly reporting to him or her.

[5]Intermix organizations are created by splitting off resources from functional departments, such as individuals, teams, or complete operational groups, including facilities. These resources are temporarily transferred to the new project organization (intermix), to perform specific, usually short-range assignments, such as working on bid proposals, mergers, or feasibility assessments. The unique feature of the intermix organization is that it can be "created instantaneously"—but at the expense of other ongoing activities within the enterprise.

reporting relations, and interfaces for all organizational components. The tools come from conventional management practices. They provide the basis for managerial direction, communication, and control, as well as the informational infrastructure for teamwork, technology transfer and decision making at the operating level of the company. The principal tools are:

Policy Directive. This top-level document describes the overall philosophy and principles of managing the business within a particular organizational unit. A sample policy directive is shown in Appendix 1.1 on page 357 for managing the *engineering activities* within a company.

Procedure. This operational guideline describes the various components of the business process, including how specific work and projects are to be executed, and how supporting activities, such as cost estimating, scheduling, testing, and training, are to be conducted.

Charter of Key Positions. This policy document defines the operational framework for a particular department, business unit, or program office. It clearly describes its mission, scope, broad responsibilities, authority, organizational structure, interfaces, and reporting relations. Charters should be revised if the scope or mission of the position changes, as is the case for positions such as program offices or new product development organizations. Sample charters are provided for a *product design manager* and an *engineering team leader* in Appendix 1.2 and 1.3 on page 360.

Organization Chart. Regardless of the intricacies of the organization, its structure, and its terminology, a simple organization chart of the core organization defines the major reporting and authority relations, interfaces, and communication channels. In spite of its static nature, the organization chart provides a useful bird's eye view of principal organizational structure.

Responsibility Matrix. This chart is especially useful for defining the interfunctional responsibilities and multiple accountabilities for project-related activities. The chart shows who is responsible for what. Its application can be expanded from covering a particular department to the whole company, and even go beyond, to reach into supplier and customer environments.

Job Description. Similar to the charter, but more detailed and specifically for an individual position (or class of positions), the job description defines (1) principal duties, (2) qualifications, (3) responsibilities, (4) reporting relations, and (5) basic authorities. Job descriptions should be developed for all key personnel, such as managers, lead engineers, scientists, and project managers. Job descriptions are modular and portable among similar positions. Once established, job descriptions become building blocks for staffing, professional development, and performance evaluation and rewards. A sample job description for an R&D project manager is provided in Appendix 1.4 on page 361.

While the Exhibits in Appendix 1 provide examples of policy and procedural tools for defining complex organizations, it should be emphasized that these tools must be modified to fit the specific organizational needs, specs, cultures, and business processes. The objective of showing these samples in Appendix 1 is to provide a framework for managers for developing their own tools.

3.6.1 Make Interdisciplinary Relations Work for You

Technology-intensive organizations are, by and large, interdisciplinary, that is matrix-based with shared power and resources. When functional resource personnel are assigned to specific projects, team members are likely to maintain strong ties to their home functions. These ties are very normal and predictable as part of the existing cultural network. They are also desirable and necessary for broad conceptual thinking and ultimately cross-functional integration of the work. Yet, these functional ties are often seen by the project team leaders as "disloyalty" and lack of full commitment to the project effort they are assigned to. However, research shows that fighting these ties of team members to the home office is counterproductive. It leads to personal anxieties and mistrust. Project team members realize that their job security and career advancement come to a large degree via their home office and its management. At the same time, these home office connections are valuable linkages from the project office to the functional resource organization that should be carefully cultivated and maintained.

Commitment of assigned personnel to the project is a separate issue. Research shows that project ownership has little to do with team member alliance with home offices, but depends instead on the personal involvement and pride team members have in the project, and the professional excitement they experience in the team environment. These are dimensions that can be influenced by the project team leader via recognition, project visibility, and management involvement. Further, making accomplishments visible and providing feedback to functional managers will influence the reward process administered though the functional organization. This will build a strong team member commitment to the project, with the realization that many of the elements that contribute to their professional excitement and career advancement come through the project organization. These are the influences that help in building and sustaining commitment and ownership to a project and its mission objectives.

3.7 SUMMARY OF KEY POINTS AND CONCLUSIONS

The key points that have been made in this chapter include:

- The need for broader business accountability and pressures for faster, more effective market response led to many new and innovative organizational designs, such as *simultaneous engineering, concurrent project management, design-build*, and *Stage-Gate processes.*

- Three fundamental organizational layers exist in every enterprise: (1) institutional framework, (2) functional system, and (3) operational areas.

- Companies have two choices for structuring their business operations: *functionally and project organized.* However, virtually no company works in a *pure* mode. Real-world businesses operate as hybrids between functional and project organization, relying on resource sharing among functional units for producing specific project deliverables,. This is known as the matrix organizational structure.

- The matrix structure exists within an organizational continuum, somewhere between the two extremes of pure functional and pure projectized structures. Its exact operational location between the two extremes depends on the management style and interaction of people within the organization.

- The projectized form of organization is *similar to the divisionalized structure.* It represents the strongest form of project authority and an effective approach to executing large project. It requires considerable start-up cost and time, and offers limited opportunity for sharing experiences and resources.

- Matrix organizations offer a compromise solution for project-oriented businesses. The matrix is an overlay on the functional structure of the firm, combining the strength of both the functional and the projectized organization.

- Matrix organizations rely on resource and power sharing, often a source of organizational conflict, and always a challenge to managerial accountability and control, requiring a more sophisticated management style.

- It is important to define the management process, together with its command-and-control structure, including responsibilities, reporting relations, and organization. The principal tools are policy directives, procedures, the charter of key positions, the organization chart, the responsibility matrix, and job descriptions.

- Commitment of assigned personnel to a project is crucial to team performance. Commitment and project ownership can be enhanced through personal involvement, generating professional excitement and encouraging pride in the project activities.

3.8 CRITICAL THINKING: QUESTIONS FOR DISCUSSION

1. Describe a business environment that would benefit from a predominately projectized internal organization.

2. What advice would you give to your project managers in a matrix organization to minimize "matrix conflict"?

3. How do you build and sustain project ownership among the team members assigned from other resource departments?

4. Many managers see the matrix as a messy, sloppy organizational structure that is unworkable. Assuming that you cannot projectize, what alternatives do you see to matrix management?

5. Develop a list of dos and don'ts for effective matrix management.
6. Develop a policy or operational guideline for conducting project activities in your company.
7. Project leaders must often step across functional lines and deal with personnel over whom they have little or no formal authority. How can these leaders "earn" the authority they need to function effectively in a matrix environment?
8. How could senior management help project leaders to manage effectively in a matrix environment?

3.9 REFERENCES AND ADDITIONAL READINGS

Anderson, C. and Fleming, M. (1990). Management control in an engineering matrix organization," *Industrial Management*, Vol. 32, No. 2 (March/April), pp. 8–13.

Anderson, Erling (2003) "Understanding your project organization's charter," *Project Management Journal*, Vol. 34, No. 4 (December), pp. 4–11.

Bahrami, Homa (1992) "The emerging flexible organization: perspectives from Silicon Valley," *California Management Review*, Vol. 34, No. 4 (Summer), pp. 33–52.

Bailetti, Antonio, J., Callahan, John R., and DiPietro, Pat (1994) "A coordination structure approach to the management of projects," *IEEE Transactions on Engineering Management*, Vol. 41, No. 4, (November), pp. 394–403.

Barner, R. (1997) "The new millennium workplace," *Engineering Management Review* (IEEE), Vol. 25, No. 3 (Fall), pp. 114–119.

Bishop, Suzanne K. (1999) "Cross-functional project teams in functionally aligned organizations," *Project Management Journal*, Vol. 30, No. 3 (September), pp. 6–12

Clark, Kim B. and Wheelwright, Steven C. (1992) "Creating product plans to focus product development," *Harvard Business Review*, Vol. 70, No. 2 (March/April 1992), pp. 70–82.

Clark, Kim B. and Wheelwright, Steven C. (1992) "Organizing and leading heavyweight development teams," *California Management Review*, Vol. 34, No. 3 (Spring), pp. 9–28.

Cleland, David I. (1991) "The age of project management," *Project Management Journal*, Vol. 22, No. 1 (March), pp. 21–32.

Cooper, Robert G. and Kleinschmidt, Elko J. (1993) "Stage-Gate systems for new product success," *Marketing Management*, Vol. 1, No. 4, pp. 20–29.

Deschamps, Jean-Philippe and Nayak, P. Ranganath (1995) "Implementing world-class process," Chapter 5 in *Product Juggernauts*, Cambridge, MA: Harvard University Press.

Glasser, A. (1982) *Research and Development Management*, Englewood Cliffs, NJ: Prentice-Hall.

Grover, V. (1999) "From business reengineering to business process change management," *IEEE Transactions on Engineering Management*, Vol. 46, No. 1 (February), pp. 36–46.

Iansiti, Marco, and MacCormack, Alan (1997) "Developing product on internet time," *Harvard Business Review*, (September/October), pp. 108–117.

Nayak, P. Ranganath (1991) "Managing rapid technological development," *Prism* (Second Quarter), pp. 19–39.

Shim, D. and Lee, M. (2001) "Upward influence styles of R&D project leaders," *IEEE Transactions on Engineering Management*, Vol. 48, No. 4 (November), pp. 394–413.

Sobek, Durwald K., Liker, Jeffrey K., and Ward, Allen C. (1998) "Another look at how Toyota integrates product development," *Harvard Business Review* (July/August), pp. 36–49.

Thamhain, Hans J., (1994) "Designing project management systems for a radically changing world," *Project Management Journal*, Vol. 25, No. 4 (December), pp 6–7.

Thamhain, Hans (2000) "Accelerating new product developments via stage-gate processes," Proceedings, *31st Annual Symposium of the Project Management Institute—PMI 2000*, Philadelphia, October 7–16.

Thamhain, Hans J. and Wilemon, David L. (1998) "Building effective teams for complex project environments," *Technology Management*, Vol. 4, pp. 203–212.

Thamhain, H. (2004) "Team leadership effectiveness in technology-based project environments," Project Management Journal, Vol. 35, No. 4 (December), pp. 35–46.

Zhang, P., Keil, M., Rai, A. and Mann, J. (2003) "Predicting information technology project escalation," *Journal of Operations Research*, Vol. 146, No. 1, pp. 115–129.

Zirger, B. J. and Hartley, Janet L. (1996) "The effect of acceleration techniques on product development times," *IEEE Transactions on Engineering Management*, Vol. 43, No. 2 (May), pp. 143–152.

4

CONCURRENT ENGINEERING AND INTEGRATED PRODUCT DEVELOPMENT

THE PENTAGON RECONSTRUCTION PROJECT[ab]

When hijacked American Airlines flight 77 slammed into the west face of the Pentagon on September 11, 2001, over 400,000 square feet of office space was destroyed, together with the communications, command, and control technology infrastructure, and an additional 1.6 million square feet of working facilities was damaged. The very same day, project "Phoenix" was formed with the mission of reconstructing the west face of the Pentagon. "Without any preplanning, budgets, or contract approvals; it was an unconventional start of an extraordinary project," says former Phoenix Deputy Program Manager Mike Sullivan. The primary constraint for the completion of this project was a one-year deadline requested by the U.S. government. Fortunately for the Phoenix Project, the nearly completed *Wedge 1 Project (*part of the 20-year *Pentagon Renovation Program, PenRen)* was still active. On the very day of the attack, resources from *PenRen* were quickly reallocated to form the Phoenix integrated project team (IPT). "With all of the stakeholders around the table," says Clark Sheakley, a client liaison with General Dynamics, "it was possible to create and update plans on the fly and to keep everyone informed." The schedule and other parts of the project plan were very informal during the first month after the attack. Virtually overnight the budget to reconstruct the Pentagon was estimated, based on historic cost data from the ongoing *PenRen Project,* and submitted to the U.S. Congress, which authorized $700 million in emergency funds. Once the Federal Bureau of Investigation released the site, the project plan went through a more formal development.

The project's tight schedule led to concurrent scope and plan development in parallel to project execution. "Before plans were finalized or even in draft form, the project was already under way," says Sheakley. To accelerate the schedule, an "ultra-fast-track schedule" was developed. For example, reconstruction was divided into three horizontal stages. The structure of the three outer rings of the building complex were sequenced independently, allowing early installation of the limestone façade, main electric and communication vaults, and mechanical and roofing systems, concurrently. In addition, smaller sections of concrete were poured so that crews could work simultaneously, rotating through each of the construction sequences, rather than waiting for an entire floor to be completed before sending in the subsequent crew.

Due to the fast pace of the project, keeping the schedule current was difficult. Many times it served more as a benchmark than a primary guide. "At times, the schedule could not keep up with the work because it was so far ahead," says Mike Sullivan, deputy program manager.

Yet another dimension of the fast-track process included the procurement plan. The Government Program Office was exempt from systems integration responsibilities. Contractors were fully responsible for the integration of systems, subsystems, equipment, and support equipment, and they had to validate full system performance after integration. This approach decreased implementation time, increased product quality, and reduced engineering changes, program office staff, and overall project cost. "It was a challenge to manage all the resources at any specific point in the project, given the number of resources," Sullivan says. "But, the bottom line is that you hire good people, and you let them do their jobs."

The *design-build approach* adopted here allowed design and construction to operate as a single entity under one contract, contrasting with the standard government approach of design-bid-build, which can create startup delays, and conflict between contractors and owners.

Taken together, the 3,000-member project team completed the demolition and reconstruction of the damaged west face section of the Pentagon 28 days ahead of schedule and $194 million under budget. The Phoenix management team says that success comes down to two central elements: people and procedure. Project management molded the team into a tangible process conducive to optimum utilization of all available resources. According to Deputy Program

Manager Mike Sullivan, "the lesson is that if you get a bunch of people who are dedicated and committed to achieve a clearly defined goal, within a well-defined work process, you can accomplish almost anything!"

Source: Background and additional information on this project can be found at the Pentagon reconstruction Web site, *www.pentagonreconstruction.gov.*

[ab] Excerpts from Natalie Bauer, "Rising from the Ashes," *pmNetwork*, Vol. 18, No. 5 (May 2004), pp. 24-32. Excerpts reprinted by permission.

4.1.   THE NEED FOR EFFECTIVE MANAGEMENT PROCESSES

Not every project is as urgent and time-critical as the reconstruction of the nation's center of defense. Yet, many enterprise missions are under similar pressure to perform against end-date-driven schedules. They bear the same necessity for accelerating development projects, effectively utilizing resources, and minimizing implementation risks. The message is clear: virtually every organization in our fiercely competitive business environment is under pressure to do more things faster, better, and cheaper. Speed has become one of the great equalizers in competitiveness and is a key performance measure. New technologies, especially in computers and communications, have removed many of the protective barriers to business, created enormous opportunities, and transformed our global economy into a hypercompetitive enterprise system. To survive and prosper, the new breed of business leaders must deal effectively with time-to-market pressures, innovation, cost, and risks in an increasingly fast-changing global business environment. Concurrent engineering has gradually become the norm for developing and introducing new products, systems, and services (Haque et al. 2003, Yam et al. 2003).

4.1.1   From Idea to Market

Whether we look at the implementation of a new product, process, or service or we want to rebuild the Pentagon, create a new movie, or win a campaign, reducing project cycle time translates most likely into cost savings, risk reduction, market advantages, and strategic benefits. Bottom line, *"time is money,"* to paraphrase Benjamin Franklin.

Project management has traditionally provided the tools and techniques for executing specific missions, on time and resource efficiently. These tools and techniques have been around since the dawn of civilization, leading to impressive results ranging from Noah's Ark and the ancient pyramids and military campaigns to the Brooklyn Bridge and Ford's Model-T automobile. While the first formal project management processes emerged during the Industrial Revolution of the eighteenth century, with focus on mass production, agriculture, construction, and military operations, the recognition of project management as a business discipline and profession did not occur until the 1950s with the emergence of formal organizational concepts such as the matrix, projectized organizations, life cycles, and phased approaches (Morris 1997).

These concepts established the organizational framework for many of the project-oriented management systems in use today, providing a platform for delivering mission-specific results. Yet, the dramatic changes in today's business environment often required the process of project management to be reengineered to deal effectively with the challenges (Denker et al. 2001, Nee and Ong 2001, Rigby 1995, Thamhain 2001) and to balance efficiency, speed, and quality (Atnahene-Gimo, 2003). As a result, many new project management tools and delivery systems evolved in recent years under the umbrella of *integrated product development (IPD)*. These systems, however, are not just limited to product development, but can be found in a wide spectrum of modern projects, ranging from construction to research, foreign assistance programs, election campaigns, and IT systems installation (Koufteros et al. 2002, Nellore and Balachandra 2001). The focus that all of these IPD applications have in common is the effective, integrated, and often concurrent multidisciplinary project team effort toward specific deliverables, the very essence of *concurrent engineering processes*.

4.2   A SPECTRUM OF CONTEMPORARY MANAGEMENT SYSTEMS

Driven by the need for effective multidisciplinary integration and the associated economic benefits, many contemporary project execution methods have evolved akin to both IPD and CE. These methods often focus on specific project environments such as manufacturing, marketing, software development, or field services (Gerwin and Barrowman 2002). As a result, many mission-specific project management platforms emerged under what is today characterized as the IPD umbrella, long before IPD had been recognized as a formal concept. These well-established platforms include systems such as *Design for Manufacture (DMF)*, *Just-in-Time (JIT)*, *Continuous Process Improvement (CPI)*, *Integrated Product and Process Development (IPPD)*, *Structured Systems Design (SSD)*, *Rolling Wave (RW) Concept* (see Githens 1998), *Phased-Developments (PD)*, *Stage-Gate Processes* (see Cooper and Kleinschmidt 1993), *Integrated Phase-Reviews (IPR)*, and *Voice-of-the-Customer (VOC)*, just to name a few of the more popular concepts. What all of these systems have in common is the emphasis on effective cross-functional integration, and incremental, iterative implementation of project plans. However, one of the oldest of these contemporary project management concepts, and perhaps one of the most widely used IPD concepts, today, is *concurrent engineering (CE)*.

4.3   CONCURRENT ENGINEERING — A UNIQUE PROJECT MANAGEMENT CONCEPT

Concurrent engineering, is an extension of the multiphased approach to project management. At the heart of its concept is the concurrent execution of task segments, which creates overlap and interaction among the various project teams. It also increases the need for strong cross-functional integration and team involvement, which creates both managerial benefits and challenges (Wu, Fuh, and Nee 2002). While *concurrent engineering* was originally seen as a method for primarily reducing project cycle time and accelerating product development (Prasad et al. 1998, Prasad 1998), today the concept refers quite generally to the most resource- and time-efficient execution of multidisciplinary undertakings. Moreover, the CE concept has been expanded from its original engineering focus to a wide range of projects, ranging from construction and field installations to medical procedures, theater productions, and financial services (Pham, Dimov, and Setchi 1999; Pilkinton and Dyerson 2002; Skelton and Thamhain 1993). The operational and strategic values of concurrent engineering are much broader than just a gain in lead time and resource effectiveness; they include a wide range of benefits to the enterprise, as summarized in Table 4.1. These benefits are primarily derived from effective cross-functional collaboration and full integration of the project management process with the total enterprise and its supply chain (Prasad et al. 1998, Prasad 2003). In this context, concurrent engineering provides a process template for effectively managing projects. Virtually any project can benefit from this approach as pointed out by the *Society for Concurrent Product Development (www.soce.org)*. As a working definition, the following statement brings the management philosophy of concurrent engineering into perspective:

> *Concurrent engineering* provides the managerial framework for *effective, systematic, and concurrent integration of all functional disciplines necessary for producing the desirable project deliverables, in the least amount of time and resource requirements, considering all elements of the product life cycle.*

In essence, concurrent engineering is a systematic approach to integrated project execution that emphasizes parallel, integrated execution of project phases, replacing the traditional linear process of serial engineering and expensive design-build-roll-out rework. The process also requires strong attention to the human side, focusing on multidisciplinary teamwork, power sharing, and team values of cooperation, trust, respect, and consensus building, engaging all stakeholders in the sharing of information and decision making, starting during the early project formation stages and continuing over the project life cycle.

Table 4.1   Potential Benefits of Concurrent Engineering

- Better cross-functional *communication* and *integration*
- Decreased *time to market*
- Early detection of *design problems*, fewer *design errors*
- Emphasizes human side of *multidisciplinary teamwork*
- Encourages *power sharing, cooperation, trust, respect,* and *consensus building*
- Engages all stakeholders in *information sharing* and *decision making*
- Enhances ability to support *multisite manufacturing*
- Enhances ability to cope with *changing requirements, technology, and markets*

Table 4.1   (Continued)

- Enhances ability to execute *complex projects* and *long-range* undertakings
- Enhances *supplier communication*
- Fewer *engineering changes*
- High-level of *organizational transparency*, R&D-to-marketing
- Higher *resource efficiency* and *personnel productivity;* more resource-effective project implementation
- Higher *project quality*, measured by customer satisfaction
- Minimizes "downstream" *uncertainty, risk and complications*; makes the project *outcome more predictable*
- Minimizes design-build-rollout *reworks*
- Ongoing *recognition and visibility of team accomplishments*
- Promotes *total project life cycle thinking*
- Provides a *template or roadmap* for guiding multiphased projects from concept to final delivery
- Provides *systematic approach* to multiphased project execution
- Shorter *project life cycle* and execution time
- *Validation of work in progress* and deliverables

The concurrent engineering process is graphically shown in Figure 4.1, depicting a typical product development. In its basic form, the process provides a template or roadmap for guiding multiphased projects from concept to final delivery. One of the prime objectives for using concurrent engineering is to minimize "downstream" uncertainty, risk, and complications, and hence make the project outcome more predictable (Iansiti and MacCormack 1997; Moffat 1998; Noori and Deszca 1997; O'Connor 1994; Sobek et al. 1998). However, concurrent execution and integration of activities does not just happen by drawing timelines in parallel, but is the result of carefully defined cross-functional linkages and skillfully orchestrated teamwork. Moreover, concurrent phase execution makes several assumptions regarding the organizational system and its people, as summarized in Table 4.2 and discussed in the next section.

4.4   CRITERIA FOR SUCCESS

For simplicity, concurrent engineering is often shown as a linear process, with overlapping activity phases, scheduled for concurrent execution, as shown in Figure 4.1. However, to make such concurrent project phasing possible, the organizational process must be designed to meet specific criteria that establish the conditions conducive to concurrent, incremental implementation of phased activities, as summarized in Table 4.2. These conditions are very similar to conventional modern project management, which provides the operating platform for concurrent engineering. By its very definition, concurrent engineering is synonymous with cross-disciplinary cooperation, involving all project teams and support groups of the enterprise, internally and externally, throughout the project life cycle. The CE process relies on organizational linkages and integrators that help in identifying problems early, networking information, transferring technology, satisfying the needs of all stakeholders, and unifying the project team. It is important to include all project stakeholders in the project team and its management—not only enterprise-internal components, such as R&D, engineering, manufacturing, marketing, and administrative support functions, but also external stakeholders, such as customers, suppliers, regulators and other business partners.

Table 4.2   Criteria for Successful Management of Concurrent Engineering Projects

Concurrent engineering teams and their leaders must be able to:

- Allocate sufficient time and resources for up-front planning
- Identify major task teams, their mission, and interfaces at the beginning of the project cycle
- Work out the logistics and protocol for concurrent phase implementation
- Lay out the master project plan (top level) covering the project life cycle
- Establish a consensus on a project plan among project team members
- Be willing to work with partial incremental inputs, evolving requirements, and it's life cycle
- Identify all project-internal and -external "customers" of its work and establish effective communication linkages and ongoing working relationships with these customers
- Work flexibly with team members and customers, adjusting to evolving needs and requirements
- Share information and partial results regularly during the project implementation
- Identify the specific deliverables needed by other teams (and individuals) as inputs for their part of the project, including the timing for such deliverables
- Establish effective cross-functional communication channels and specific methods for work transfer
- Establish techniques and protocols for validating the work and its appropriateness to its "customers" on an ongoing basis
- Work with partial results (deliverables) and incremental updates from upstream developments
- Reiterate or modify tasks and deliverables to accommodate emerging needs of downstream task teams and to optimize the evolving project outcome
- Prepare for its mission prior to receiving mission details (e.g., manufacturing is expected to work on pilot production setup prior to receiving full product specs or prototypes)
- Work as an integrated part of a unified and agreed-on project plan
- Have tolerance for ambiguity and uncertainty
- Establish reward systems that promote cross-functional cooperation, collaboration, and joint ownership of results
- Have top management buy in and support the concurrent engineering process
- Have established a uniform project management system throughout the concurrent engineering team/organization

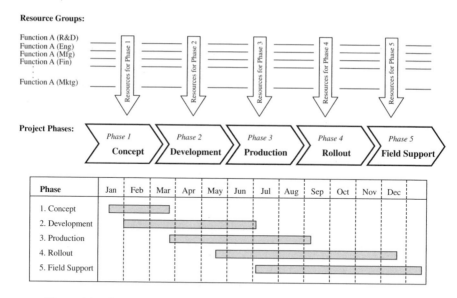

Figure 4.1   Graphical representation of concurrent execution of project phases.

Taken together, the core ingredient of successful concurrent engineering is the development and effective management of organizational interfaces. For most organizations, these challenges include strong human components that are more difficult to harness and to control than the operational processes of project implementation (Prased 1998). They involve many complex and constantly changing variables that are hard to measure and even more difficult to manage, especially within self-directed team environments that are often required for realizing the concurrent engineering process (Bauly and Nee 2000, Hull et al. 1996). Procedures are important. They provide (1) the baseline and infrastructure necessary to connect and integrate the various pieces of the multidisciplinary work process, and (2) an important starting point for defining the communication channels necessary for effectively linking the core team with all of its support functions. Yet, it should be realized that the resulting process is only as good as the people who implement it.

4.5   DEFINING THE PROCESS TO THE TEAM

After reaching a principle agreement with major stakeholders, the concurrent engineering process should be defined, showing the major activity phases or stages of the project to be executed. Even more advantageous for future projects would be the ability to define phases that may be common to a *class of projects* that are being executed by the enterprise over time. To illustrate, let us use the example of a new product development, shown in Figure 4.1. The concurrent engineering process proceeds through five project phases: (1) concept development, (2) detailed development, (3) pilot production, (4) product rollout, launch, and marketing, and (5) field support.

Each phase or stage is defined in terms of principle scope, objectives, activities, and deliverables, as well as functional responsibilities. Each project phase must also include cross-functional interface protocols, defining the specific collaborations and organizational linkages needed for the concurrent development. While the principal cross-functional interfaces can be summarized graphically, as shown in the upper part of Figure 4.1, more sophisticated group technology tools, such as the Quality Function Deployment (QFD) Matrix shown in Figure 4.2, are usually needed for defining (1) the specific cross-functional requirements, (2) the methods of work transfer (often referred to as technology transfer), and (3) the stakeholder interactions necessary for capturing and effectively dealing with the changes that ripple through the product design process.

The best time for setting up these interface protocols is during the definition phase of a specific project when the team organization is most flexible regarding lines of responsibility and authority. To illustrate, Figure 4.2 shows the specific inputs and outputs required during the various phases of a product development process. Each arrow indicates that a specific input/output requirement exists for that particular interface. Most likely, some interface requirements exist from each project phase to each of the others. In our example, the total number of potential interfaces is defined by the 5×5 matrix which equals 25 interfaces (this explains why the QFD Matrix is also referred to as N-Squared Chart). The QFD Matrix is a useful tool for identifying specific interface personnel and input/output requirements. That is, for each interface, key personnel from both teams have to establish personal contacts and negotiate the specific type and timing of deliverables needed. In many cases, multiple interfaces exist simultaneously, necessitating complex multiteam agreements over project integration issues. An additional challenge is the incremental nature of deliverables resulting from the concurrent project execution. For downstream phases, such as production, to start their work concurrently with earlier project phases, such as product development, it is necessary for all interfaces to define

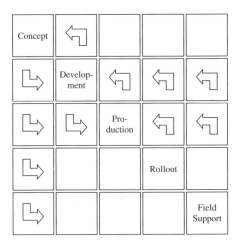

Figure 4.2   Quality Function Deployment (QFD) Matrix, a tool for defining interfaces.

and negotiate (1) what part of the phase deliverables can be transferred "early," (2) the exact schedule for the partial deliverable, and (3) the validation, iteration, and integration process for these partial deliverables.

Yet, another important condition for concurrent engineering to work is the ability of "downstream task leaders" to guide the "upstream" design process toward desired results, and to define the "upstream" gate criteria on which they depend as "customers." This interdisciplinary integration is often accomplished by participating in project and design reviews, by soliciting and providing feedback on work in progress, and by cross-functional involvement with interface definitions and technology transfer processes. Interface diagrams, such as the QFD Matrix shown in Figure 4.2, can help to define the cross-functional roadmap for establishing and sustaining the required linkages. for each task group.

4.6   UNDERSTANDING THE CHALLENGES

Concurrent engineering is about organizational development and business process improvement. Potential benefits, such as higher quality, speed, resource effectiveness, and cycle time, are very attractive to management, but they are also associated with high implementation risk, a natural by-product of the organizational development and change process. Often, the new process does not work as well as the old one, or worse, is a complete disaster. That's why few companies go into a major reorganization lightly. At best, introducing a new system is painful, disruptive, and costly. At worst, it can destroy existing operational effectiveness and the ability to compete successfully in the marketplace. Seasoned managers understand the many challenges of working with these contemporary business systems, such as:

- Strong autonomy of task teams makes system integration a shared responsibility of all team leaders.
- Power and resource sharing among task leaders and resource managers requires a more sophisticated management style and the ability to deal with inevitable organizational conflict and politics.
- The focus on process templates (i.e., C.E.) tends to make project execution ridgid.
- CE process templates tend to isolates task teams within their program activities.
- Objectives focus on phase outputs, such as deliverables, rather than on overall project and mission objectives.
- The need for self-directed teamwork limits top-down and central control.
- The need to work with incremental, partial inputs and outputs at the task level, makes project planning, measurements, and progress reporting more difficult.
- Intricate work processes such as concurrent engineering require additional administrative support, management, and resources.
- Implementing and sustaining concurrent engineering requires senior management involvement and support, and long-term organizational commitment.

Experienced managers evaluate these challenges against current business scenarios. They also root-cause any known problems and failures of *others,* before they dive into reorganizing their own business process.

Some of the toughest challenges of concurrent engineering (CE), and its broader umbrella of integrated product development (IPD), relate to compatibility issues with the organizational culture. Concurrent engineering requires a collaborative culture with a great deal of organizational resource and power sharing. This is often not present to the degree needed for concurrent engineering to succeed. Designing, customizing, and implementing a new project management system usually affects many organizational subsystems and processes, from innovation to decision making, and from cross-functional communications to the ability to deal effectively with risk and organizational conflict. Hence, integrating concurrent engineering with the business process and its physical, informational, managerial, and psychological subsystems, without compromising business performance, is an important challenge that must be dealt with during the implementation phase. Strong involvement of people accross the organization is required for concurrent engineering to become acceptable by the people, and hence to become institutionalized and to succeed as a viable business process in an organization.

4.7 UNDERSTANDING THE ORGANIZATIONAL COMPONENTS

While the above challenges are more visible to managers, the benefits (Table 4.1) are not always obvious just by looking at the basic concept of concurrent engineering. In fact, these benefits are often derivatives of more subtle organizational processes that unfold within a well-executed concurrent project management system. These processes need to be understood and skillfully exploited by project leaders and managers to gain the full benefits of concurrent engineering. Breaking the benefits down and relating them to organizational process, might help in gaining additional insight into the workings of concurrent engineering:

1. *Uniform Process Model.* The concurrent engineering concept provides a uniform process model or template for organizing and executing a predefined class of projects, such as specific new product developments.

   Primary Benefit: Time and resource savings during the project/product planning and startup phase.

   Secondary Benefits: A standardized process model breaks the project cycle into smaller, predefined modules or phases, resolving some of the project complexities, predefining potential risks and areas requiring managerial interactions and support. Standardized platform for project execution provides basis for continuous process improvement and organizational learning.

2. *Integrated Product Development, IPD.* Because of its focus on cross-functional cooperation, concurrent engineering promotes an integrated approach to product development and other project work.

Primary Benefit. Promotes unified, collective understanding of project challenges and search for innovative solutions. Helps in team integration: identifying organizational interfaces, lowering risks, and reducing cycle time.

Secondary Benefits. Responsibilities for team and functional support personnel are more visible.

3. *Gate Functions.* The concurrent engineering platform is similar to other multiphased project management concepts, such as Stage-Gate®, Structured Systems Design, or Rolling Wave Concepts, hence encouraging the integration of predefined gates, providing for performance reviews, sign-off criteria, checkpoints, and early warning systems.

Primary Benefit. Ensures incremental guidance of the product/project execution and early problem detection; provides cross-functional accountability; helps in identifying risk and problem areas; minimizes rework; highlights organizational interfaces and responsibilities.

Secondary Benefits. Stimulates cross-functional involvement and visibility; identifies internal customers, promotes full life-cycle planning, focuses on win strategy.

4. *Standard Project Management Process.* The concurrent engineering concept is compatible with the standard project management process, its tools, techniques, and standards. Predefined gates provide performance and sign-off criteria, checkpoints, and early warning systems, ensuring incremental guidance of the product development process and early problem detection.

Primary Benefit. Provides cross-functional accountability; helps in identifying risk and problem areas; minimizes rework; highlights organizational interfaces and responsibilities.

Secondary Benefits. Stimulates cross-functional involvement and visibility; identifies internal customers, promotes full life-cycle planning, focuses on win strategy.

5. *QFD Approach.* Using the Quality Function Deployment (QFD) concept, built into the concurrent engineering process, helps you to define cross-functional interfaces and pushes both the performing and the receiving organization toward closer cooperation and "upstream" guidance of the product development.

Primary Benefit. Provides an input/output model for identifying workflow throughout the project/product development process, and identifies organizational interfaces and responsibilities.

Secondary Benefits. Stimulates cross-functional involvement and visibility; identifies internal customers, promotes full life-cycle planning.

6. *Early Testing.* Concurrent engineering encourages early testing of overall project or product functionality, features, and performance. These tests are driven by team members of both "downstream" and "upstream" project phases. "Downstream" members seek assurances for problem-free transfer of the work into their units, and "upstream" members seek smooth transfer and sign-off for their work completed.

Primary Benefit. Early problem detection and risk identification, opportunity to "fail early and cheap," less rework.

Secondary Benefits. Stimulates cross-functional involvement and cooperation, assists system integration.

7. *Total Organizational Involvement and Transparency.* Because of its emphasis on mutual dependencies among the various phase teams, strong cross-functional involvement and teamwork is encouraged, enhancing the level of visibility and organizational transparency.

Primary Benefit. Total development cycle/system thinking; enhanced cross-functional innovation; effective teamwork, enhanced cross-functional communications and product integration; early warning system, improved problem detection and risk identification; enhanced flexibility toward changing requirements.

Secondary Benefits. Total team recognition; enhanced team spirit and motivation; conducive to self-direction and self-control.

Taken together, the top benefits of concurrent engineering refer to time, resource, and risk issues that ultimately translate into increased project performance: (1) reducing project startup time, (2) reducing project cycle time, (3) detecting and resolving problems early, (4) promoting system integration, (5) promoting early concept testing, (6) minimizing rework, (7) handling more complex projects with higher levels of implementation uncertainty, (8) working more resource effectively, and (8) gaining higher levels of customer satisfaction.

4.8 RECOMMENDATIONS FOR EFFECTIVE MANAGEMENT

A number of specific suggestions may help managers to understand the complex interaction of organizational and behavioral variables involved in establishing a concurrent engineering process and help in managing projects effectively with such a system. The sequence of recommendations follows to some degree the chronology of an organizational system design and its implementation and managerial application. Although each organization is unique with regard to its business, operation, culture, and management style, field studies show a general agreement on the types of factors that are critical to effectively organizing and managing projects in concurrent multi-phase environments (Denker et al. 2001, Harkins 1998, Nellore and Balachandra 2001, Pillai et al. 2002, Prasad 1977, Thamhain and Wilemon 1998).

Phase I: Organizational System Design

Take a Systems Approach. The concurrent engineering system must eventually function as a fully interconnected subsystem of the organization, and should be designed as an integrated part of the total enterprise (Haque, Pawar, and Barson

2003). Field studies emphasize consistently that management systems function suboptimally, at best, or fail due to a poor understanding of the interfaces that connect the new system with the total business process (Kerzner 2001, Moffat 1998). System thinking, as described by Senge and Carstedt (2001), Checkland (1999), and Emery and Trist (1965), provides a useful approach for front-end analysis and organization design.

Build on Existing Management Systems. Radically new methods are usually greeted with anxiety and suspicion. If possible, the introduction of a new organizational system, such as concurrent engineering, should be consistent with already established project management processes and practices within the organization. The more congruent the new operation is with the already existing practices, procedures, and distributed knowledge within the organization, the higher will be the cooperation with management in implementing the new system. The highest level of acceptance and success is found in areas where new procedures and tools are added incrementally to already existing management systems. These situations should be identified and addressed first. Building upon an existing project management system also facilitates the incremental enhancement, testing, and fine-tuning of the new concurrent engineering process. Particular attention should be paid to the cross-functional workability of the new process.

Custom-Design. Even for apparently simple situations, a new concurrent engineering process should be customized to fit the host organization and its culture, needs, norms, and processes (Hull, Collins, and Liker 1996). For reasons discussed in the previous paragraph, the new system has a better chance for smooth implementation and for gaining organizational acceptance if the new process appears consistent with already established values, principles, and practices, rather than a new order to be imposed without reference to the existing organizational history, values, or culture (see Swink, Sandvig and Mabert 1996; Kerzner 2001).

Phase II: System Implementation

Define Implementation Plan. Implementation of the new concurrent engineering system is by itself a complex, multidisciplinary project that requires a clear plan with specific milestones, resource allocations, responsibilities, and performance metrics. Further, implementation plans should be designed for measurability, early problem detection and resolution, and visibility of accomplishments, providing the basis for recognition and rewards.

Pretest the New Technique. Preferably, any new management system should be pilot tested on small projects with an experienced project team. Asking a team to test, evaluate, and fine-tune a new concurrent engineering process is often seen as an honor and a professional challenge. It also starts the implementation with a positive attitude, creating an environment of open communication and candor, and a focus on actions leading toward success.

Ensure Good Management Direction and Leadership. Organizational change, such as the implementation of a concurrent engineering system, requires top-down leadership and support to succeed. Team members will be more likely to help implement the concurrent engineering system, and cooperate with the necessary organizational requirements, if management clearly articulates its criticality to business performance and the benefits to the organization and its members. People in the organization must perceive the objectives of the intervention to be attainable and have a clear sense of direction and purpose for reaching these goals. Senior management involvement and encouragement are often seen as an endorsement of the team's competence and recognition of their efforts and accomplishments (Thamhain and Wilemon 1998). Throughout the implementation phase, senior management can influence the attitude and commitment of their people toward the new concept of concurrent engineering. Concern for project team members, assistance with the use of the tool, enthusiasm for the project and its administrative support systems, proper funding, help in attracting and retaining the right personnel, and support from other resource groups, all will foster a climate of high involvement, motivation, open communications, and desire to make the new concurrent engineering system successful.

Involve People Affected by the New System. The implementation of a new management system involves considerable organizational change with all the expected anxieties and challenges. Proper involvement of relevant organizational members is often critical to success (Barlett 2002, Nellore and Balachandra 2001). Key project personnel and managers from all functions and levels of the organization should be involved in assessing the situation, evaluating the new tool, and customizing its application. While direct participation in decision making is the most effective way to obtain buy-in toward a new system (Pham, Dimov, and Setchi 1999), it is not always possible, especially in large organizations. Critical factor analysis, focus groups, and process action teams are good vehicles for team involvement and collective decision making, leading to ownership, greater acceptance, and willingness to work toward successful implementation of the new management process (Thamhain 2001).

Anticipate Anxieties and Conflicts. A new management system, such as concurrent engineering, is often perceived as imposing new management controls, being disruptive to the work process, and creating new rules and administrative requirements. People's responses to such new systems range from personal discomfort with skill requirements to dysfunctional anxieties over the impact of tools on work processes and performance evaluations (Sundaramurthy and Lewis 2003). Effective managers seem to understand these challenges intuitively, anticipating the problems and attacking them aggressively as early as possible. Managers can help in developing guidelines for dealing with problems, and establishing conflict resolution processes, such as informational meetings, management briefings, and workshops, featuring the experiences of early adopters. They can also work with the system implementers to foster an environment of mutual trust and cooperation. Buy-in to the new process and its tools can be expected only if its use is relatively risk-free (Stum 2001). Unnecessary references to performance appraisals, tight supervision, reduced personal freedom and autonomy, and overhead requirements should be avoided, and specific concerns dealt with promptly on a personal level.

Detect Problems Early and Resolve. Cross-functional processes, such as concurrent engineering, are often highly disruptive to the core functions and business process of a company (Denker, Steward, and Browning 2001; Haque et al. 2003). Problems, conflict, and anxieties over technical, personal, or organizational issues are very natural and can be even healthy in fine-tuning and validating the new system. In their early stages, these problems are easy to solve, but usually hard to detect. Management must keep an eye on the organizational process and their people to detect and facilitate resolution of dysfunctional problems. Roundtable discussions, open-door policies, focus groups, process action teams, and management by wandering around are good vehicles for team involvement, leading to organizational transparency and a favorable ambience for collective problem identification, analysis, and resolution.

Encourage Project Teams to Fine-Tune the Process. Successful implementation of a concurrent engineering system often requires modifications of organizational processes, policies, and practices. In many of the most effective organizations, project teams have the power and are encouraged to make changes to existing organizational procedures, reporting relations, and decision and work processes. It is crucial, however, that these team initiatives be integrated with the overall business process and supported by management. True integration, acceptance by the people, and sustaining of the new organizational process will only occur through the collective understanding of all the issues and a positive feeling that the process is helpful to the work to be performed. To optimize the benefits of concurrent engineering, it must be perceived by all the parties as a win-win proposition. Providing people with an active role in the implementation and utilization process helps to build such a favorable image for participant buy-in and commitment. Focus teams, review panels, open discussion meetings, suggestion systems, pilot test groups and management reviews are examples for providing such stakeholder involvement.

Invest Time and Resources. Management must invest time and resources for developing a new organizational system. An intricate system, such as concurrent engineering, cannot be effectively implemented just via management directives or procedures, but instead requires the broad involvement of all user groups, helping to define metrics and project controls. System designers and project leaders must work together with upper management toward implementation. This demonstrates management confidence in, ownership of, and commitment to the new management process. This will also help to integrate the new system with the overall business process. As part of the implementation plan, management must allow time for the people to familiarize themselves with the new vision and process. Training programs, pilot runs, internal consulting support, fully leveraged communication tools such as groupware, and best-practice reviews are examples of action tools that can help in both institutionalizing and fine-tuning the new management system. These tools also help in building the necessary user competencies, management skills, organizational culture, and personal attitudes required for concurrent engineering to succeed.

Phase III: Managing in Concurrent Engineering

Plan the Project Effectively. As with any other project management system, effective project planning and team involvement are crucial to success. This is especially important in the concurrent engineering environment, where parallel task execution depends on continuous cross-functional cooperation for dealing with the incremental work flow and partial result transfers. Team involvement, early in the project life cycle, will also have a favorable impact on the team environment, building enthusiasm toward the assignment, team morale, and ultimately team effectiveness. Because project leaders have to integrate various tasks across many functional lines, proper planning requires the participation of all stakeholders, including support departments, subcontractors, and management. Modern project management techniques, such as phased project planning and Stage-Gate concepts, plus established standards, such as PMBOK, provide the conceptional framework and tools for effective cross-functional planning and organization of the work for effective execution.

Define Work Process and Team Structure. Successful project management in concurrent engineering requires an infrastructure conducive to cross-functional teamwork and technology transfer. This includes properly defined interfaces, task responsibilities, reporting relations, communication channels, and work transfer protocols. The tools for systematically describing the work process and team structure come from the conventional project management system; they include (1) *project charter*, defining the mission and overall responsibilities of the project organization, including performance measures and key interfaces, (2) *project organization chart,* defining the major reporting and authority relationships, (3) *responsibility matrix* or *task roster,* (4) *project interface chart*, such as the N-*squared chart* discussed earlier, and (5) *job descriptions.*

Develop Organizational Interfaces. Overall success of concurrent engineering depends on effective cross-functional integration. Each task team should clearly understand its task inputs and outputs, interface personnel, and work transfer mechanism. Team-based reward systems can help to facilitate cooperation with cross-functional partners. Team members should be encouraged to check out early feasibility and system integration. QFD concepts, *N*-square charting, and well-defined phase-gate criteria can be useful tools for developing cross-functional linkages and promoting interdisciplinary cooperation and alliances. It is critically important to include in these interfaces all of the support organizations, such as purchasing, product assurance, and legal services, as well as outside contractors and suppliers.

Staff and Organize the Project Team. Project staffing is a major activity, usually conducted during the project formation phase. Because of time pressures, staffing is often done hastily and prior to defining the basic work to be performed. The result is often team personnel that are suboptimally matched to the job requirements, resulting in conflict, low morale, suboptimum decision making, and ultimately poor project performance. While this deficiency will cause problems for any project organization, it is especially unfavorable in a concurrent engineering project environment that relies on strong cross-functional teamwork and shared decision making, built on mutual trust, respect, and credibility. Team personnel with poorly matched skill sets for the job requirements are seen as incompetent, affecting their trust, respect, and credibility, and ultimately their "concurrent team performance." For best results, project leaders should *negotiate the work assignment* with their team members one to one, at the outset of the project. These negotiations should include the overall task, its scope, its objectives, and performance measures. A thorough understanding of the task requirements develops often as the result of personal involvement in the front-end activities, such as requirements analysis, bid proposals, project planning, interface definition, or the concurrent engineering system development. This early involvement also has positive effects on the buy-in toward project objectives, plan acceptance, and the unification of the task team.

Communicate Organizational Goals and Objectives. Management must communicate and update the organizational goals and project objectives. The relationship and contribution of individual work to overall business plans and their goals, as well as of individual project objectives and their importance to the organizational mission, must be clear to all team personnel. Senior management can help in unifying the team behind the project objectives by developing a "priority image," through their personal involvement, visible support, and emphasis on project goals and mission objectives.

Define Work Interfaces and Effective Communication Channels. Interdisciplinary transparency is a critical prerequisite for concurrent engineering. Effective cross-functional communication is crucial to the intensive information sharing and joint decision making needed during a concurrent project execution. In addition to modern technology, such as voice mail, e-mail, electronic bulletin boards, and conferencing, and process tools, such as the *N*-Square Chart, management can facilitate the free flow of information, both horizontally and vertically, by workspace design, regular meetings, reviews, and information sessions (Hauptman and Hirji 1999).

Ensure Senior Management Support and Leadership. You must win the hearts and minds of your work team before it supports and uses a new management system. Senior management, by their enthusiasm, commitment, sense of purpose, and urgency, provides a strong emotional driver toward acceptance of a new organizational process. This visible management involvement and support also lowers the anxieties and doubts over the organizational value and stability of the new tool and helps to unify the team behind its implementation and effective utilization.

Manage Conflict and Problems. Conflict is inevitable in the concurrent engineering environment with its complex dynamics of power and resource sharing and incremental decision making. Project managers should focus their efforts on problem avoidance. They should recognize potential problems and conflicts early in their development, and deal with them before they become big and their resolutions consume a large amount of time and effort. Focus team sessions, brainstorming, experience exchanges, and process action teams can be powerful tools for developing the concurrent workgroup into an effective, fully integrated, and unified project.

Encourage Continuous Fine-Tuning and Improvement. In addition to the implementation phase, the actual working process of concurrent engineering provides excellent opportunities for continuous improvement of the system and its fine-tuning to the specific project work at hand and its organizational host environment. Managers can encourage project leaders and their teams to examine the new process for weaknesses and potential improvements. It is important to establish support systems, such as discussion groups, action teams, suggestion systems, and post project review sessions, to capture and leverage the lessons learned, and to fine-tune the system as part of a continuous improvement process.

4.9 CONCLUSION

In today's dynamic and hypercompetitive environment, proper implementation and use of concurrent engineering is critical for expedient and resource-effective project execution. The full range of benefits of concurrent engineering is in fact much broader than just a gain in lead time and resource effectiveness; it includes a wide spectrum of competitive advantages to the enterprise, ranging from increased quality of project deliverables to the ability to execute more complex projects and higher levels of customer satisfaction. These benefits are primarily derived from effective cross-functional collaboration and full integration of the project management process with the total enterprise and its supply chain. However, these benefits do not occur automatically!

Designing, implementing, and managing in concurrent engineering requires more than just writing a new procedure, delivering a best-practice workshop, or installing new information technology. It requires the ability to engage the organization in a systematic evaluation of specific competencies, such as concurrent engineering, assessing opportunities for improvement, and designing a project management system that is fully integrated with the overall enterprise system and its strategy. Too many managers end up disappointed that the latest management technique did not produce the desired result. Regardless of its conceptual sophistication, concurrent engineering is just a framework for processing project data and aligning organizational strategy, structure, and people. To produce benefits for the firm, these tools must be fully customized to fit the specific business process and be congruent with the organization's system and its culture.

One of the most striking findings, from both the practice and the research of concurrent engineering, is the strong influence of human factors on project performance. The organizational system and its underlying process of concurrent engineering are equally critical, but must be effectively integrated with the human side of the enterprise. Effective managers understand the complex interaction of organizational and behavioral variables. During the *design and implementation* of the concurrent engineering system, they can work with the various resource organizations and senior management, creating a win-win situation between the people affected by the intervention and senior management. They can shake up conventional thinking and create a vision without upsetting established cultures and values. To be successful both *implementing* concurrent engineering and *managing* projects through the system requires proactive participation and commitment of all stakeholders. It also requires congruency of the system with the overall business process and its management system.

Taken together, leaders must pay attention to the human side. To enhance cooperation among the stakeholders, managers must foster a work environment where people see the significance of the intervention for the enterprise, and personal threats and work interferences are minimized.

One of the strongest catalysts, to both the *implementation of concurrent engineering* and the *management of projects* through its system, is professional pride and excitement of the people, fueled by visibility and recognition of work accomplishments. Such a professionally stimulating environment seems to lower anxieties over organizational change, reduce communications barriers and conflict, and enhance the desire of personnel to cooperate and to succeed, a condition critically important for developing the necessary linkages for effective cross-functional project integration.

While no single set of broad guidelines exists that guarantees success for managing in concurrent engineering, project management is not a random process! A solid understanding of modern project management concepts and their tools, support systems, and organizational dynamics is one of the threshold competencies for leveraging the concurrent engineering process. It can help managers both in developing better project management systems and in leading projects most effectively through these systems.

4.10   SUMMARY OF KEY POINTS AND CONCLUSIONS

The key points that have been made in this chapter include:

- Concurrent engineering (CE) is an extension of the multiphased approach to project management.

- Concurrent engineering provides the managerial framework for concurrent integration of all activities necessary for producing the desirable results in the shortest time and best resource utilization.

- Many mission-specific project management platforms, similar to concurrent engineering, emerged under the *Integrated Product Development (IPD)* umbrella, such as Design for Manufacture (DMF), Just-in-Time (JIT), Continuous Process Improvement (CPI), Integrated Product and Process Development (IPPD), Structured Systems Design (SSD), Rolling Wave (RW) Concept, Phased-Developments (PD), Stage-Gate™ Processes, Integrated Phase-Reviews (IPR), and Voice-of-the-Customer (VOC).

- All of these systems emphasize effective cross-functional integration and incremental, iterative implementation of project plans.

- The top benefits of concurrent engineering refer to time, resource, and risk issues that ultimately translate into increased project performance.

- The core ingredient of successful concurrent engineering is the development and effective management of organizational interfaces.

- The best time for setting up these interface protocols is during the definition phase of a specific project.

- The QFD Matrix is a useful tool for identifying specific interface personnel and input/output requirements.

- To work as a management system, concurrent engineering requires a collaborative culture and a great deal of organizational power sharing.

- Strong involvement of people from all organizational levels is required to institutionalize concurrent engineering and ensure its use by the project team.

- The introduction of a new concurrent engineering system should be consistent with already established project management processes and practices within the organization.

- Even for apparently simple situations, a new concurrent engineering process should be customized to fit the host organization and its culture, needs, norms, and processes.

- Any new concurrent engineering management system should be pilot tested on small projects with an experienced project team.

- Cross-functional processes, such as concurrent engineering, are highly disruptive to the core functions and business process of a company. Problems, conflict, and anxieties over technical, personal, or organizational issues are very natural and can be even helpful in fine-tuning and validating the new system.

- The tools for systematically describing the work process and team structure come from the conventional project management system; they include project charter, project organization chart, responsibility matrix or task roster, project interface chart, and job descriptions.

- Management must make provisions for fine-tuning and continuous improvement of the concurrent engineering process.

- Successful implementation and management of concurrent engineering requires proactive participation and commitment of all stakeholders. It also requires congruency of the system with the overall business process and its management system.
- A solid understanding of modern project management concepts and their tools, support systems, and organizational dynamics is one of the threshold competencies for leveraging concurrent engineering and managing projects effectively through its process.

4.11   CRITICAL THINKING: QUESTIONS FOR DISCUSSION

1. What are the reasons behind concurrent engineering's ability to accelerate project execution and increase resource effectiveness?
2. Discuss methods for enhancing cross-functional cooperation and integration.
3. Develop a procedure for defining work interfaces based on the *N*-squared chart.
4. What are the major challenges managing projects within a concurrent engineering process?
5. Discuss managerial actions conducive to concurrent project execution.
6. Discuss managerial leadership styles most conducive to concurrent project execution.
7. How can top management facilitate concurrent engineering and concurrent project management?

4.12   REFERENCES AND ADDITIONAL READINGS

Atnahene-Gimo, K. (2003) "The effects of centrifugal and centripetal forces on product development speed and quality," *Academy of Management Journal*, Vol. 43, No. 3 (June), pp. 359–373.

Barlett, J. (2002) "Risk concept mapping," Proceedings, *Fifth European Project Management Conference,* Cannes, France (June 19–20).

Bauly, J. and Nee, A. (2000) "New product development: implementing best practices, dissemination and human factors," *International Journal of Manufacturing Technology and Management*, Vol. 2, No. 1/7, pp. 961–982.

Bishop, S. (1999) "Cross-functional project teams in functionally alligned organizations," *Project Management Journal*, Vol. 30, No. 3 (September), pp. 6–12.

Chambers, C. (1996) "Transforming new product development," *Research Technology Management*, Vol. 39, No. 6 (November/December), pp. 32–38.

Checkland, P. (1999) *Systems Thinking, Systems Practice*, New York: Wiley & Sons.

Cooper, R. and Kleinschmidt, E. (1993) "Stage-Gate systems for new product success," *Marketing Management*, Vol. 1, No. 4, pp. 20–29.

Denker, S., Steward, D., and Browning, T. (2001) "Planning concurrency and managing iteration in projects," *Project Management Journal*, Vol. 32, No. 3 (September), pp. 31–38.

Emery, F. (1969) *Systems Thinking*, New York: Penguin.

Emery, F. and Trist, E. (1965) "The causal texture of organizational environments," *Human Relations*, Vol. 18, No. 1, pp. 21–32.

Gerwin, D. and Barrowman, N. (2002) "An evaluation of research on integrated product development," *Management Science*, Vol. 48, No.. 7 (July), pp. 938-954.

Githens, G. (1998) "Rolling wave project planning," Proceedings, 29th Annual Symposium of the Project Management Institute, Long Beach, CA (October 9–15).

Goldenberg, J., Horowitz, R., and Levav, A. (2003) "Finding your innovation sweet spot," *Harvard Business Review*, Vol. 81, No. 3 (March), pp. 120–128.

Haddad, Carol (1996) "Operationalizing the concept of concurrent engineering," *IEEE Transactions on Engineering Management*, Vol. 43, No. 2 (May), pp. 124–132.

Haque, B., Pawar, K., and Barson, R. (2003) "The application of business process modeling to organizational analysis of concurrent engineering environments," *Technovation*, Vol. 23, No. 2 (February), pp. 147–162.

Harkins, J. (1998) "Making management tools work," *Machine Design*, Vol. 70, No. 12 (July 9), pp. 210–211.

Hauptman, O. and Hirji, K. (1999) "Managing integration and coordination in cross-functional teams," *R&D Management*, Vol. 29, No. 2 (April), pp. 179–191.

Hull, F., Collins, P., and Liker, J. (1996) "Composite form of organization as a strategy for concurrent engineering effectiveness," *IEEE Transactions on Engineering Management*, Vol. 43, No. 2 (May), pp. 133–143.

Iansiti, M. and MacCormack, A. (1997) "Developing products on Internet time," *Harvard Business Review* (September–October), pp. 108-117.

Kerzner, H. (2001) *The Project Management Maturity Model*, New York: Wiley & Sons.

Koufteros, X., Vonderembse, M., and Doll, M. (2002) "Integrated product development practices and competitive capabilities: The effects of uncertainty, equivocality, and platform strategy," *Journal of Operations Management*, Vol. 20, No. 4 (August), pp. 331–355.

LaPlante, A. and Alter, A. (1994) "Corning Inc.: the stage-gate innovation process," *Computerworld*, Vol. 28, No. 44 (October 31), pp. 81–84.

Litsikas, M. (1997) "Break old boundaries with concurrent engineering," *Quality*, Vol. 36, No. 4 (April), pp. 54–56.

Moffat, L. (1998), "Tools and teams: competing models of integrating product development projects," *Journal of Engineering and Technology Management*, Vol. 1, No. 1 (March), pp. 55–85.

Morris, P. (1997) *The Management of Projects*, London: Thomas Telford.

Nee, A. and Ong, S. (2001) "Philosophies for integrated product development," *International Journal of Technology Management*, Vol. 21, No. 3, pp. 221–239.

Nellore, R and Balachandra, R. (2001) "Factors influencing success in integrated product development (IPD) projects," *IEEE Transactions on Engineering Management*, Vol. 48, No. 2 (May). pp. 164–174.

Neves, T., Summe, G. L., and Uttal, B. (1990) "Commercializing technology: what the best companies do," *Harvard Business Review* (May/June), pp. 154–163.

Noori, H. and Deszca, G. (1997) "Managing the P/SDI process: best-in-class principles and leading practices," *Journal of Technology Management*, Vol. 13, No. 3, pp. 245–268.

O'Connor, P. (1994) "Implementing a stage-gate process: a multi-company perspective," *Journal of Product Innovation Management*, Vol. 11, No. 3 (June), pp. 183–200.

Paashuis, V. and Pham, D. (1998) *The Organisation of Integrated Product Development*, Berlin: Springer Verlag.

Pham, D., Dimov, S., and Setchi, R (1999) "Concurrent engineering: a tool for collaborative working," *Human Systems Management*, Vol. 18, No. 3/4, pp. 213–224.

Pilkinton, A. and Dyerson, R. (2002) "Extending simultaneous engineering: electric vehicle supply chain and new product development," *International Journal of Technology Management*, Vol. 23, No. 1, pp. 74–88.

Pillai, A., Joshi, A., and Raoi, K. (2002) "Performance measurement of R&D projects in a multi-project, concurrent engineering environment," *International Journal of Project Management*, Vol. 20, No. 2 (February), pp. 165–172.

Prasad, B. (1976) *Concurrent Engineering Fundamentals, Volume 1: Integrated Product and Process Organization*, Engelwood Cliffs, New Jersey: Prentice Hall.

Prasad, B. (1977) *Concurrent Engineering Fundamentals, Volume 2: Integrated Product Development*, Engelwood Cliffs, New Jersey: Prentice Hall.

Prasad, B. (1998) "Decentralized cooperation: a distributed approach to team design in a concurrent engineering organization," *Team Performance Management*, Vol. 4, No. 4, pp. 138–146.

Prasad, B. (2002) "Toward life-cycle measures and metrics for concurrent product development," *International Journal of Computer Applications in Technology*, Vol. 15, No. 1/3, pp. 1–8.

Prasad, B. (2003) "Development of innovative products in a small and medium size enterprise," *International Journal of Computer Applications in Technology*, Vol. 17, No. 4, pp. 187–201.

Prasad, B., Wang, F, and Deng, J. (1998), "A concurrent workflow management process for integrated product development," *Journal of Engineering Design*, Vol. 9, No. 2 (June), pp. 121–136.

Rasiel, E. (1999). *The McKinsey Way*, New York: McGraw-Hill.

Rigby, Darrel K. (1995) "Managing the management tools," *Engineering Management Review (IEEE)*, Vol. 23, No. 1 (Spring), pp. 88–92.

Senge, P. and Carstedt, G. (2001) "Innovating our way to the next industrial revolution," *Sloan Management Review*, Vol. 42, No. 2, pp. 24–38.

Senge, Peter M. P. (1990) *The Fifth Discipline: The Art and Practice of the Learning Organization*, New York: Doubleday/Currency.

Shabayek, A. (1999) "New trends in technology management for the 21st century," *International Journal of Management*, Vol. 16, No. 1 (March), pp. 71–76.

Skelton, Terrance and Thamhain, Hans J. (1993) "Concurrent project management: a tool for technology transfer, R&D-to-market," *Project Management Journal*, Vol. 24, No. 4 (December), pp.41–48

Sobek, Durwald K., Jeffrey, K., Liker, K., Allen, C., and Ward, A. (1998) "Another look at how Toyota integrates product development," *Harvard Business Review* (July/August), pp. 36–49.

Stum, D. (2001) "Maslow revisited: building the employee commitment pyramid," *Strategy and Leadership*, Vol. 29, No. 4 (July/August), pp. 4–9.

Sundaramurthy, C. and Lewis, M. (2003) "Control and collaboration: paradoxes of governance," *Academy of Management Review*, Vol. 28, No. 3 (July), pp. 397–415.

Swink, M., Sandvig, J., and Marbert, V. (1996) "Customizing concurrent engineering processes: five case studies," *Journal of Product Innovation Management*, Vol. 13, No. 3 (May), pp. 229–245.

Thamhain, H. (1994) "A manager's guide to effective concurrent project management," *Project Management Network*, Vol. 8, No. 11 (November), pp. 6–10.

Thamhain, H. (1996) "Best practices for controlling technology-based projects," *Project Management Journal*, Vol. 27, No. 4 (December), pp. 37–48.

Thamhain, H. (2001) "The changing role of project management," Chapter 5 in *Research in Management Consulting* (Anthony Buono, ed.), Greenwich, CT: Information Age Publishing.

Thamhain, H. (2001) "Leading R&D Projects without Formal Authority," *Management of Technology: Selected Topics* (T. Khalil, ed.), Oxford, UK: Elsevier Science.

Thamhain, H. (2002) "Criteria for effective leadership in technology-oriented project teams," Chapter 16 in *The Frontiers of Project Management Research* (Slevin, Cleland, and Pinto, eds.), Newton Square, PA: Project Management Institute, pp. 259–270.

Thamhain, H. (2003) "Managing innovative R&D teamsteams," *R&D Management*, Vol. 33, No. 3 (June), pp. 297–311.

Thamhain, H. and Wilemon, D. (1998) "Building effective teams for complex project environments," *Technology Management*, Vol. 4, pp. 203–212.

Wu, S., Fuh, J., and Nee, A. (2002) "Concurrent process planning and scheduling in distributed virtual manufacturing," *IIE Transactions*, Vol. 34, No. 1 (January), pp. 77–89.

Yam, R., Lo, W., Sun, H., and Tang, P. (2003) "Enhancement of global competitiveness for Hong Kong/China manufacturing industries," *International Journal of Technology Management*, Vol. 26, No. 1, pp. 88–102.

5

MANAGING PEOPLE AND ORGANIZATIONS

STRONG PEOPLE FOCUS, EMPHASIZING BROAD AND DEEP SKILLS AT GE

"I am very proud of our team," says Jeffrey Immelt, chairman of the board and chief operating officer, in GE's 2003 Annual Report. "They are passionate and committed. I spend about one-third of my time with my partner Bill Conaty, GE's human resources leader. We recruit, we train, we develop, we improve, we think about people constantly. I am happy to report that retention remains high, with voluntary turnover among our leaders of less than 3%.

Historically, we have been known as a company that developed *professional managers.* These are broad problem-solvers with experience in multiple businesses or functions. However, I want to raise a generation of growth leaders— people with market depth, customer touch and technical understanding. This change emphasizes depth.

We are expecting people to spend more time in business or a job. We think this will help our team develop 'market instincts,' so important to growth and for growth, and the confidence to grow global businesses. Ultimately, careers should be broad and deep, giving our leaders the confidence to solve problems and the experience to drive growth. But today, to get the right balance, we are emphasizing depth.

About a year ago, one of our executive development classes suggested that we reformulate our values to capture the spirit of GE as a growth company. Values can't just be words on a page. To be effective, they must shape actions. We looked to make them simpler, more inclusive and aspirational.

With this in mind, we reshaped GE values around four core actions: *Imagine, Solve, Build* and *Lead. Imagine* at GE is the freedom to dream and the power to make it real. This requires the values of passion and curiosity. *Solve* reflects GE's unique ability to tackle the world's toughest problems and expresses our values of resourcefulness and accountability. *Build* requires a performance culture that creates customer and shareowner value, and the word captures our values of teamwork and commitment. *Lead* reflects our spirit of optimism that embraces change, and our values of openness and energy; it's what it will take to win.

Jim Campbell, who led our Consumer Products business in 2003, is a great example of GE values. In 2003, we asked Jim to grow earnings by double digits, generate $1 billion in cash, restore GE as a clear leader in innovation, rebuild relationships with our customers, reduce structural cost by integrating Lighting with Appliances and persuade his production associates to ratify a new contract. He hit them all.

Jim had to *imagine* new approaches for innovative products in an old industry. He had to *solve* real problems in our Lighting business, where we have underperformed. He had to *build* new relationships with customers like Home Depot, Lowe's and Wal-Mart. He had to *lead* in a difficult market . . . one in which tough actions like restructuring were required to fund growth. He did all this with a remarkable feel for the market and our team. He is both a learner and a winner. He lives the GE values.

I love our GE team. I am thankful for their loyalty, commitment and hard work. It is an honor to lead them."

Source: "Growth Platforms," General Electric 2003 Annual Report, pp. 9–11. For additional information visit *www.ge.com.*

5.1   CHANGING ROLES OF MANAGERIAL LEADERSHIP

In their quest to stay competitive in our challenging business environment, managers have to work effectively with people. People are our most valuable asset. They are the heart and soul of the company's core competency, critical to the successful implementation of any strategic plan, operational initiative, or specific project undertaking. The real value of this human asset is measured in terms of integrated skill sets, attitudes, ambitions, and compassions for the business. With the dramatic rise of our globalized economy, businesses face the enormous task of transforming themselves to the next level of competitiveness. To develop a business team that has all of the right ingredients, matched with the needs of the enterprise, is not a simple task, but requires sophisticated leadership and skillful orchestration from the top as reflected by Jeff Immelt's statement. The mandate for managers is clear: They must weave together the best practices and programs for continuously developing their people to achieve the highest possible performance. However, even the best practices and most sophisticated teaching methods do not guarantee success. They must be carefully integrated with the business process, and its culture and value system. The challenges are especially felt in today's technology organizations, which have become highly complex internally and externally, with a bewildering array of multi-faceted activities, requiring sophisticated cross-functional cooperation, integration, and joint decision making, as discussed already in Chapters 1 and 2 of this book. Because of these dynamics, technology organizations are seldom structured along traditional functional lines; rather, they operate as matrices or hybrid organizations with a great deal of power and resource sharing. In addition, lines of authority and responsibility blur among formal management functions, project personnel, and other subject experts. This more empowered and self-directed work force needs a much higher bandwidth of skills to solve operational problems consistent with current and future market needs. Collectively, the people of the enterprise must have a "market instinct," as Immelt emphasized. Similarly to the way that GE reshaped its values around four core actions, *imagine, solve, build*, and *lead*, many other technology companies define their own dimensions of effective leadership.

5.1.1 Core Management Issues

In spite of the many differences among companies, technology organizations must be able to deal effectively with the following set of management issues:

1. *Manage Technical Work Content.* Every job has technical content, such as electrical engineering, biochemistry, market research, or financial analysis. Especially in technology organizations, a lot of jobs have a broad and complex technological context. The ability to manage the technical work on an individual job level and collectively throughout the organization is a *threshold competency*, critically important to the success of any business. Its managerial components relate to staffing, skill sets, professional development, support technologies, experiential learning, and in-depth management.

2. *Manage Talent.* Businesses do not produce great results because of their equipment, buildings, and infrastructure, but because their people, ideas, and actions bring the system alive. For many technology companies, talent is everything! The type of talent, and its fit with the business needs and organizational culture, determines everything from idea generation to problem resolution and business results. Talent does not occur at random. Nor should it be taken as granted. It needs to be searched out, attracted, developed, and maintained. An

organization's personnel policies and award systems must be consistent with these talent objectives. Loosing a top talent is a sin! Companies like GE conduct postmortems on every top talent loss, and hold their management accountable for those losses.

3. *Manage Knowledge.* Technology companies are knowledge factories. In essence, they buy, trade, transfer, and sell knowledge. Their value lies increasingly in the collective knowledge that becomes the basis for creating new ideas, concepts, products, and services. The emphasis must be on orchestrated management of this collective knowledge. New products and services usually do not come from a single brilliant idea, but are the result of broad-based collaborative efforts throughout the organization. They are the result of an intricately connected, vast knowledge network with high inter-connectivity and low cross-organizational impedance. Setting up effective support systems and managing the development, processing, filtering, sharing, and transferring of knowledge to achieve added value is a very important and challenging task, which, by and large, involves people and sophisticated people skills.

4. *Manage Information.* Similarly to knowledge management, information management has a strong human side, which often does not receive sufficient attention. Regardless of the available technology, people are involved in gathering, transferring, interpreting, sharing, and acting upon information. To be effective, management of information must include the important human side.

5. *Manage Communications* Communication is the backbone of a firm's command-and-control structure. It is the catalyst toward crucial integration of organizational efforts toward achieving unified results. This is especially critical in technology firms with their unconventional organizational structures and strong need for cross-functional coordination. Communication systems are much broader than information systems. They also include a wider range of human issues, such as interaction dynamics, power, politics, trust, respect, and credibility. While greatly supported by IT, some of the most powerful, effective, and expedient forms of communication are conducted in group meetings that include various forms of verbal and nonverbal information flows.

6. *Manage Collaboration and Commitment* With the complexities of technological undertakings, broad-scale collaboration is often necessary to solve a problem or achieve a mission objective. This includes company internal as well as external forms of joint ventures, ranging from informal cross-functional agreements to multicompany consortia, codevelopment projects, and joint ventures.

7. *Build a Supportive Organizational Environment.* Successful companies have cultures and environments that support their people. These companies provide visibility and recognition to their people. They also show the impact of these accomplishments on the company's mission and project-related objectives.

This creates an ambiance where people are interested in and excited about their work, which produces higher levels of ownership, cross-functional communication, cooperation, and some tolerance for risk and conflict.

8. *Ensure Direction and Leadership.* Managers themselves are change agents. Their concern for people, assistance with problem solving, and enthusiasm for the enterprise's mission and objectives can foster a climate of high motivation, work involvement, commitment, open communication, and willingness to cooperate across the organization.

Technology managers often describe their organizational environments as "unorthodox," with ambiguous authority and responsibility relations. They argue that such environments require broader skill sets and more sophisticated leadership than traditional business situations. In fact, in today's more open, dynamic, and utterly competitive business environment, management performance is based to a large degree on teamwork. Yet, attention to individuals, and their competence, accountability, commitment, and sense for self-direction is crucial for organizations to function properly at higher levels, where the various individual efforts are integrated into deliverable systems. All of this points toward a more open, adaptable management style that is often referred to as a "team-centered" style. Such a management style is based on the thorough understanding of the motivational forces and their interaction with the enterprise environment. It is a style that Immelt referred to as typical for *growth leaders*, requiring breadth and depth of both technical and business skills. As a precondition and part of the *growth leader's* task, this assumes that the leader can actually build a team of fully competent, motivated, and committed people.

Research shows[1] that the work environment provides very important conditioning for effective leadership. In particular professionally stimulating work and collegial involvement have a favorable impact on team performance and are conducive to team-centered leadership, which will be explored further in this chapter.

5.2   MOTIVATION AND TECHNOLOGY PERFORMANCE

Understanding people is important in any management situation. It is especially critical in today's technology-based organizations. Leaders who succeed within these often unstructured work environments must confront untried problems to manage their complex tasks. They have to learn how to move across various organizational lines to gain services from personnel not reporting directly to them. They must build multidisciplinary teams into cohesive groups and deal with a variety of networks, such as line departments, staff groups, team members, clients, and senior management, each with different cultures, interests, expectations, and charters. To get results, these engineering and technology managers must relate socially as well as

---

[1]For a more detailed discussion of field investigation and its results see Thamhain (2004) "Leading technology-based project teams," *Engineering Management Journal,* Vol. 16, No. 1, pp. 22–31.

technically and must understand the culture and value system of the organization in which they work. The days of the manager who gets by with only technical expertise or pure administrative skills are gone.

What works best? Observations of best-in-class practices show consistently and measurably two important characteristics of high performers in technology organizations: (1) they enjoy work and are excited about the contributions they make to their company and society, and (2) they have strong needs, both professionally and personally, and use their professional work as vehicle to fulfill these needs. Specifically, field research studies have identified 16 professional needs that are strongly associated with high-tech job performance.

5.2.1   Sixteen Professional Needs That Affect Technology-Based Performance

Research studies show that the fulfillment of certain professional needs can drive professional people to higher performance; conversely, the inability to fulfill these needs may become a barrier to individual performance and teamwork.[2] The rationale for this important correlation is found in the complex interaction of organizational and behavioral elements. Effective team management involves three primary components: (1) people skills, (2) organizational structure, and (3) management style. All three components are influenced by the specific task to be performed and the surrounding environment. That is, the degree of satisfaction of any of the needs is a function of (1) having the right mix of people with appropriate skills and traits, (2) organizing the people and resources according to the tasks to be performed, and (3) adopting the right leadership style. The 16 professional needs critical to technology-based performance are:

1. *Interesting and Challenging Work.* This is an intrinsic motivator that satisfies professional esteem needs and helps to integrate personal goals with the objective of the organization.

2. *Professionally Stimulating Work Environment.* This is conducive to professional involvement, creativity, and interdisciplinary support. It also fosters team building and is conducive to effective communication, conflict resolution, and commitment toward the organization's structure, facilities, and management style.

3. *Professional Growth.* This is measured by promotional opportunities, salary advances, the learning of new skills and techniques, and professional recognition. A particular challenge exists for management in limited-growth or zero-growth businesses to compensate for lack of promotional opportunities

---

[2]Several field studies have been conducted over the past 15 years, investigating the needs and conditions most favorably associated with team performance in R&D and high-technology product environments. The findings are being used here to draw conclusions on professional needs and leadership style effectiveness. For details of these field studies see Thamhain (1983, 1990, 2004) and Thamhain and Wilemon (1987, 1999).

by offering more intrinsic professional growth in terms of job satisfaction and skill building.

4. *Overall Leadership.* This involves dealing effectively with individual contributors, managers, and support personnel within a specific functional discipline as well as across organizational lines. It includes technical expertise, information-processing skills, effective communication, and decision-making skills. Taken together, leadership means satisfying the need for clear direction and unified guidance toward established objectives.

5. *Tangible Rewards.* This include salary increases, bonuses, and incentives, as well as promotions, recognition, better offices, and educational opportunities. Although extrinsic, these financial rewards are necessary to sustain strong long-term efforts and motivation. Furthermore, they validate the "softer" intrinsic rewards, such as recognition and praise, and reassure people that higher goals are attainable.

6. *Technical Expertise.* People must have all the necessary interdisciplinary skills and expertise available within their team to perform the required task. Technical expertise includes understanding the technicalities of the work; the technology of underlying concepts, theories, and principles; design methods and techniques; and the functioning and interrelationship of the various components that make up the total system.

7. *Assisting in Problem Solving.* Examples include facilitating solutions to technical, administrative, and personal problems. It is a very important need, which, if not satisfied, often leads to frustration, conflict, and poor-quality work.

8. *Clearly Defined Objectives.* The goals, objectives, and expected outcomes of an effort must be clearly communicated to all affected personnel. Conflict can develop over ambiguities or missing information.

9. *Management Control.* People like to see a certain degree of structure, direction, and control of their work, without being micromanaged. This is also important for effective team performance. Managers must understand the interaction of organizational and behavior variables in order to exert the direction, leadership, and control required to steer the work effort toward established organizational goals without stifling innovation and creativity.

10. *Job Security.* This is one of the very fundamental needs that must be satisfied before people consider higher-order growth needs.

11. *Senior Management Support.* Support is needed in four major areas: (1) financial resources, (2) administrative support, including an effective operating charter, policies, and procedures, (3) work support from functional resource groups, and (4) provision of necessary facilities and equipment. Management support is particularly crucial to larger, more complex undertakings.

12. *Good Interpersonal Relations.* Components include trust, respect, and credibility. These conditions are favorable to effective individual work as well as teamwork; they foster a stimulating work environment with low conflict, high productivity, great involvement, and motivated personnel.

13. *Proper Planning.* This is absolutely essential for the successful management of multidisciplinary activities. It requires communications and information-processing skills to define the actual resource requirements and administrative support necessary. It also requires the ability to negotiate for resources and commitment from key personnel in various support groups across organizational lines.

14. *Clear Role Definition.* This helps to minimize role conflict and power struggles among team members and/or supporting organizations. Clear charters, plans, and good management direction are some of the powerful tools used to facilitate clear role definition.

15. *Open Communication.* This helps to satisfy the need for a free flow of information both horizontally and vertically, keeping personnel informed and functioning as a pervasive integrator of the overall project effort.

16. *Minimum Changes.* Although technology managers have to live with constant change, their team members often see change as an unnecessary condition that impedes their creativity and timely performance. Advanced planning and proper communication can help to minimize changes and lessen their negative impact.

5.2.2 Implications for Organizational Performance

The significance of assessing these motivational forces lies in several areas. First, the previous listing provides insight into the broad needs that technology-oriented professionals seem to have. These needs must be satisfied *continuously* before the people can reach high levels of performance. The emphasis is on *continuously satisfying these needs*, or at least providing a strong potential for such satisfaction. This is consistent with findings from other studies, which show that in technology-based environments a significant correlation exists between professional satisfaction and organizational performance.[3] From the previous listing we know on what areas we should focus our attention. The listing also provides a model for *benchmarking*; that is, it provides managers with a framework for monitoring, defining, and assessing the needs of their people in specific ways, and ultimately building a work environment that is conducive to high performance.

In most cases, managers have a lot more freedom in satisfying employee needs than they recognize. As an example, managers have many options and degrees of freedom in assigning "professionally interesting work." Although top down, the total work structure and organizational goals might be fixed and not negotiable, managers

---

[3]Some of the earlier studies on needs and performance in technology-intensive work environments go back to the 1970s, such as Gary Gemmill and Hans Thamhain, "Influence styles of project managers: some project performance correlates," *Academy of Management Journal*, Vol. 17, No. 2 (June 1974), pp. 216–224. Later, H. Thamhain found strong correlations between professional satisfaction of project team personnel and project performance toward high-tech product developments: "Managing technologically innovative team efforts toward new product success," *Journal of Product Innovation Management*, Vol. 7, No. 1 (March 1990), pp. 5–18.

have a great deal of freedom and control over the way the work is distributed and assigned. Similar operational flexibility exists also in other identified need areas. Second, the previous listing of needs provides a topology for actually measuring organizational effectiveness as a function of these needs and the degree to which they seem to be satisfied.

In fact, I completed several field studies involving over 500 professionals and their managers in technology-based environments.[4] The results of these studies provide some interesting insights into the role of professional needs in achieving technical performance. We measured the degree of satisfaction for each of the 16 needs, as perceived by individual contributors. Then we obtained measures of the group's performance from their immediate supervisors. These performance measures included the following components:

- High levels of energy
- High ability to handle conflict and open communications
- High levels of innovation and creativity
- Commitment and ownership
- Willingness to take risks
- Team-oriented behavior
- High tolerance for stress, conflict, and change
- Cooperation and cross-functional linkages

The degree of association between professional need satisfaction and organizational performance is shown in Table 5.1. Specifically, the table indicates the strength of association between each need and performance factor, using Kendall's Tau rank-ordered correlation measures. Although *all* need satisfactions are positively correlated with group performance, certain factors have a particularly strong association. Here is a summary of conclusions from this analysis, and their management implications:

Professional Needs Perception. The 16 needs listed in Table 5.1 were perceived by technology personnel as particularly important to their ability to work professionally, efficiently, and in a personally satisfied manner.

Performance Correlates. All 16 needs correlated positively to overall work performance and other indirect measures of organizational effectiveness, such as innovation, communications, commitment, personal involvement and energy, organizational interface, willingness to change, and capacity to resolve conflict. The stronger the need satisfaction of individuals in a technology-oriented group, the higher their group's performance.

Most Significant Factors. As indicated by their strong positive association, all needs seem to be important to group performance. However, several factors appear to be especially important, as shown in Table 5.1 by an *xHI-mark*. For

[4] For details see Thamhain (1996, 2003, 2004).

instance, having professionally interesting and challenging work and having overall leadership seemed to influence group performance more strongly than any other factors. Having a professionally stimulating work environment and open communication also can be included with the set of factors that are statistically most strongly associated with positive group performance measures.

Intercorrelations. As one might expect, all need factors cross-correlate to some degree. That is, people who find their work professionally challenging also find the organizational environment stimulating, perceive good potential for professional growth, find their superiors have sufficient leadership qualities, an so forth. In fact, this cross-correlation has been tested via a Kruskal-Wallis analysis of variance by ranks at a confidence level of $p = 96\%$, which indicates a very strong cross-correlation.

It is clear, however, that there are costs associated with satisfying these needs, managing flexibly, and adjusting to changes. What we are lacking is a good method for comparing the relative costs of organizational change with the gain in productivity, such as design or engineering productivity. The challenge is for managers to develop enough understanding of the dynamics of their organizations that they are able to diagnose potential problems and the need for change. Only by understanding

Table 5.1   Effects of Professional Needs Satisfaction on Technology Work Group Performance

Team Member Need Satisfaction	Correlation to Technology Work Group Performance							
	Innovation & Creativity	Communications	Commitment	Involvement and Energy	Cross-org Effectiveness	Willingness to Change	Conflict Handling	*Overall Effectiveness*
1. Work Challenge	*xHI*	*xHI*	*xHI*	*xHI*	HI	HI	HI	*xHI*
2. Stimulating Environment	HI	*xHI*	HI	*xHI*	HI	M	HI	HI
3. Growth	M	HI	HI	HI	M	HI	M	HI
4. Leadership	*xHI*	HI	HI	HI	HI	HI	*xHI*	*xHI*
5. Rewards	M	M	HI	HI	M	HI	M	HI
6. Technical Experience	HI	HI	HI	HI	M	M	M	HI
7. Assistance and Help	M	M	HI	M	HI	HI	HI	HI
8. Clear Objectives	*xHI*	M	HI	HI	HI	M	M	*xHI*
9. Management Control	HI	M	M	HI	HI	HI	HI	HI
10. Job Security	HI	M	M	M	M	M	M	HI
11. Senior management support	HI	HI	*xHI*	HI	HI	HI	HI	HI
12. Good interpersonal relations	M	HI	M	HI	HI	HI	*xHI*	*xHI*
13. Proper planning	HI	HI	HI	*xHI*	HI	M	M	HI
14. Clear role definition	M	M	HI	M	HI	HI	HI	HI
15. Open communications	HI	*xHI*	HI	HI	HI	*xHI*	*xHI*	HI
16. Minimum changes	M	M	HI	HI	M	HI	M	HI

Legend: *xHI:* Extra high Kendall's Tau correlation at $p \leq .01$ (significance level > 99%)
HI: High Kendall's Tau correlation at $p \leq .05$ (significance level > 95%)
M: Medium-high Kendall's Tau correlation at $p \leq .1$ (significance level > 85%)

those variables that influence technology results and performance can one fine-tune the organization to achieve maximum long-range productivity and overall cost-effectiveness. In a technology environment, one of these variables is the degree of need-satisfaction perceived by professional personnel. In fact, the organizational characteristics stimulated by satisfying these needs are precisely the types of traits necessary to work effectively in an environment characterized by technical complexities and rapid changes regarding technology, markets, regulations, and socioeconomic factors. It is also a work environment where traditional methods of authority-based direction, performance measures, and controls have limited effectiveness.

5.2.3   Motivation as a Function of Need Satisfaction

So far we have identified a number of needs that appear important to technology professionals. Their fulfillment leads to satisfaction and can be considered a reward. The motivational process can be explained with a simple concept, the *Inducement-Contribution Model*, shown in Figure 5.1. Let's take the example of a designer named Carol Edwards, who wants to advance her career. Career advancement is her personal goal, which, when reached, would satisfy her needs for more money and prestige, interesting work, job security, and so forth. These satisfiers are, by and large, just the individual's perceptions of the benefits associated with reaching the goal. Such perceptions may, for instance, include more money, which may not be guaranteed. It may even include erroneous ideas, such as getting a reserved parking spot or the key to the executive washroom (which may not exist in Carol's company). However, regardless of the actual reality, it is the *perceived* outcome that motivates the person to work toward the goal (i.e., professional growth). Furthermore, since the employer's organization provides opportunities to move toward the goal, the designer is induced to contribute to the organization. Therefore, this concept is called the Inducement-Contribution Model.

5.2.3.1   *Management's Reaction to Employee's Needs*
Why should management pay attention to employees' needs or goals? We found from the needs analysis discussed earlier in this chapter that professional people perform at higher levels of efficiency if they can satisfy their needs within the organizational environment and in their work itself. If the probability of reaching a given

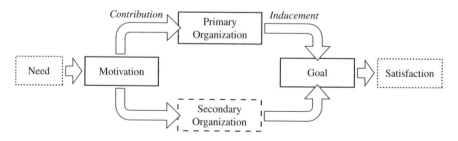

Figure 5.1   The Inducement-Contribution Model of motivation.

goal becomes very low, the person's inducement to contribute to the organization by working toward the goal is low. In such a situation, the person, in our example Carol, is likely to exercise one of several behavioral options or a combination of them. Option one: Carol may give up on her original goal and either manifest the "sour grapes" syndrome or rationalize that her goal is unattainable. In either case, the inducement toward organizational contribution disappears. Option two: Carol terminates her activities with her current organization and seeks employment with another organization, where the opportunity for reaching the desired goal appears better. Option three: Carol may stay with the current organization, but may try to satisfy her needs by working toward her goals through an "outside" organization. She may involve herself in volunteer work or recreational activities. While this may help to satisfy the employee's need for interesting and stimulating activities, her efforts and energy are directed within a secondary organization, which provides inducement and receives her contributions, which are being "leaked away" from her employer.

From this scenario, we can see that the employee's need satisfaction is, in the long term, as important to the healthy functioning of the organization as it is to the employee. Good managers should be closely involved with the staff and their work. The manager should also identify any *unrealistic* goals and correct them, change situations that impede the attainment of realistic goals, and support people in reaching their goals and cheer them on, which is recognition. This will help to refuel the individuals' desire to reach the goal and keep the employees' energies channeled through the primary organization. The tools that help the manager to facilitate professional satisfaction are work sign-on, delegation, career counseling and development, job training and skill development, and effective managerial direction and leadership with proper recognition and visibility of individual and team accomplishments.

5.2.4 Motivation as a Function of Risks and Challenges

Additional insight into motivational drive and its dynamics can be gained by considering motivational strength as a function of the probability and desire to achieve the goal. We can push others, or ourselves, toward success or failure because of our mental predispositions, called *self-fulfilling prophecies*. Our motivational drive and personal efforts increase or decrease relative to the likelihood of the expected outcome. Personal motivation toward reaching a goal changes with the probability of success (perception of doability) and challenge. Figure 5.2 expresses this relationship graphically. A person's motivation is very low if the probability of achieving the goal is very low or zero. As the probability of reaching the goal increases, so does its motivational strength. However, this increase continues only up to a certain level; when success is more or less assured, motivation decreases. This is an area where the work is often perceived as routine, uninteresting, and holding little potential for professional growth.

5.2.4.1 Success as a Function of Motivation

The saying "The harder you work, the luckier you get" expresses the effect of motivation in pragmatic terms. People who have a "can-do" attitude, who are confident

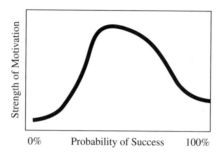

Figure 5.2   The Pygmalion effect, relating strength of motivation to the probability of success.

and motivated, are more likely to succeed in their mission. Winning is in the attitude. This is the essence of the self-fulfilling prophecy. In fact, high-performing technology teams often have a very positive image of their capabilities. As a result, they are more determined to produce the desired results, often against high risks and odds. This has been demonstrated via many studies, such as the classic study at UCLA in the early 1960s, discussed next.

A Field Experiment on Self-Motivation and Fulfillment.   A class of high-school students was randomly divided into two groups. Group A was as bright and academically prepared as Group B. Two teachers were given the assignment to teach a given subject as effectively as possible. Teacher A was told that Group A was an extremely bright group of students, who were selected specifically for this experiment. Teacher B was assigned to Group B and told that this group was a particularly slow group, but that the directors expected that the teacher would do the best she could do with this slow group. At the end of a given time period, both groups were tested on their knowledge of the subject. Maybe it does not surprise the reader that Group A tested considerably higher in the knowledge gained in the subject. The experiment was repeated hundreds of times with different groups, different teachers, different subjects, and different schools. The outcome remained the same: Group A was ahead of B. What you expect from others is likely to happen. Also, what you expect from yourself is likely to happen. This is a self-fulfilling prophecy.[5]

In another study, Dennis Waitley investigated Olympic athletes. He found that most of the athletes who broke a record knew ahead of time that that was the particular event where they could do it. This is an example of the Pygmalion effect. The name goes back in history to the saga of King Pygmalios of Cyprus, who sculpted a beautiful woman out of marble. He then fell in love with his own creation and wished that the statue could be brought to life. He prayed intensely to Aphrodite, and indeed the statue came to life as Galatea, who then married Pygmalios. Hence, if you

---

[5]The impact of superiors' expectations of subordinates was studied by S. Livingston. See "Pygmalion in Management," *Harvard Business Review*, July/August 1969, pp. 81–89. The study concluded that poorer subordinates' performance is associated with superiors having lower personal standards for themselves and lower expectations of their subordinates.

believe in your ability, you can even bring stone to life, at least according to the saga, or, equally challenging, design a new computer system or obtain FDA sign-offs on a new pharmaceutical product or launch a communications satellite into geosynchronous orbit.

5.2.5 Manage in the Range of High Motivation

This simple Pygmalion-Effect Model can provide considerable insight into the dynamics of motivation and can help in guiding the manager's efforts toward building a productive work force. Long before managers used this model, many school systems paid attention to motivation, when they grouped students of each grade into classes of similar aptitude levels. The intent of this homogeneous grouping was to place each pupil in a class that matched his or her academic ability and background. This was done to avoid over- or underchallenging students. If work was too challenging, it became threatening and mentally overwhelming, resulting in low motivation of students. Similarly, if the work was too easy, it appeared boring and dull, also resulting in low motivation to study.

5.2.5.1 *Suggestions for keeping People Motivated*

Applying the above scenario to the high-technology workplace suggests that managers should ensure proper matching of their people to their jobs. Further, managers must foster a work environment and direct their personnel in a manner that promotes "can-do" attitudes and facilitates continuous assistance and guidance toward achieving successful task completion. The process involves four primary issues, as discussed below in terms of *suggestions for stimulation and sustaining motivation in technology-oriented professionals*:

Work Assignments

- Explain the assignment, its importance to the company, and the type of contributions expected.
- Understand the employee's professional interests, desires, and ambitions and try to accommodate to them.
- Understand the employee's limitations, anxieties, and fears. Often these factors are unjustly perceived and can be removed in a face-to-face discussion.
- Develop the employee's interest in an assignment by pointing out its importance to the company and possible benefits to the employee.
- Ensure assistance where needed and share risks.
- Show how to be successful. Develop a "can-do" attitude.
- If possible, involve the employee in the definition phase of the work assignment, for instance, via up-front planning, a feasibility analysis, needs assessment, or a bid proposal.

Team Organization

- Select the team members for each task or project carefully, ensuring the necessary support skills and interpersonal compatibility.
- Select the team members for each task or project carefully, ensuring the necessary support skills and interpersonal compatibility.
- Plan each technical project properly to ensure clear directions, objectives, and task charters.
- Ensure leadership within each task group.
- Sign on key personnel on a one-on-one basis, according to the guidelines discussed in Item 1.

Skill Development

- Plan the capabilities needed in your department for the long range. Direct your staffing and development activities accordingly.
- Encourage people to keep abreast in their professional field.
- Provide for on-the-job experimental training via selected work assignments and managerial guidance.
- Provide the opportunity for some formal training via seminars, courses, conferences, and professional society activities.
- Use career counseling sessions and performance reviews to help in guiding skill development and matching it with personal and organizational objectives.

Management

- Develop interest in the work itself by showing its importance to the company and the potential for professional rewards and growth.
- Promote project visibility, team spirit, and upper-management involvement.
- Assign technically and managerially competent task leaders for each team, and provide top-down leadership for each project and its support functions.
- Manage the quality of the work via regular task reviews and by staying involved with the project team, without infringing on their autonomy and accountability.
- Plan your projects up front. Conduct a feasibility study and requirements analysis first.
- Ensure the involvement of the key players during these early phases.
- Break activities or projects into phases and define measurable milestones with specific results. Involve personnel in the definition phase. Obtain their commitment.
- Try to detect and correct technical problems early in their development.

- Foster a professionally stimulating work environment.
- Unify the task team behind the overall objectives. Stimulate the sense of belonging and mutual interdependence.
- Refuel commitment and interest in the work by recognizing accomplishments frequently.
- Assist in problem solving and group decision making.
- Provide the proper resources.
- Keep the visibility and priority of the project high. No interruptions.
- Avoid threats. Deal with fear, anxieties, mistrust, and conflicts.
- Facilitate skill development and technical competency.
- Manage and lead.

5.3 FORMAL MODELS OF MOTIVATION

Employee motivation has been a major concern of managers throughout history. However, it was only with the beginning of the industrial era at the turn of the century that researchers and practitioners began to formalize concepts and theories of human behavior. These theories can be used to model motivation and leadership behavior and therefore help to understand human behavior.

No one theoretical model, no matter how complex, can describe human behavior accurately for any situation, and certainly not for all situations. However, each model allows us to look at human behavior in a unique way, exploring different facets of the intricate and complex human behavioral system. It is especially the simple model that provides powerful insight and helps us, in concert with other models, to understand people and manage them effectively. This section discusses a selected number of concepts that should provide an integrated perspective on motivation and leadership. It also provides the basis for applying the specific management techniques to high-technology situations, discussed toward the end of this chapter.

5.3.1 The Nature of Motivation

Motivation is an inner drive that transforms activities into desired results. More specifically, people have needs and choose goals that they believe will satisfy these needs. In the effort to reach the goal, the individual is motivated (inducement) to engage in certain activities, which in turn will make contributions to the host organization. After attaining part of all of the goal, the individual consciously or unconsciously judges whether the effort has been worthwhile. Depending on the conclusion reached, the individual will continue, repeat, or stop the particular behavior. As discussed earlier, Figure 5.1 illustrates this basic process graphically. Although somewhat simplistic, the model illustrates the basic concept of motivation and its dynamics. Over the years, more elaborate and detailed models have been developed to reflect the intricacies of human motivation more realistically.

5.3.2 Early Theories of Motivation

Early motivational theories, such as those developed by Frederick Taylor at the beginning of the twentieth century, assumed that work is inherently unpleasant and that incentives such as money must be provided.[6] In these early theories money often is seen as the primary goal in the motivational process. These early theories further assumed that managers knew much more about the job than the workers, who needed to be told exactly what to do and how to do it.

The Hawthorne Studies, conducted by Elton Mayo at Western Electric between 1927 and 1932, provided additional insight into motivation.[7] The studies gave birth to the *human relations approach*, complementing the earlier scientific theories. The human relations approach assumes that people want to participate in management decisions and get involved with the organizational process. They enjoy belonging to their organization and want to feel important and useful. These assumptions led to the recommendations that managers should keep their people informed and involved, allow them self-direction and control, and utilize the informal social process for motivating employees toward organizational goals.

More recently, the so-called human resource approach to motivation evolved. It focuses intensely on human needs. The approach is based on the belief not only that people enjoy the feeling of being part of the organization and its decision-making process, but also that specific contributions and accomplishments are highly desired and valuable to the worker. This work itself can be a principal motivator and inducement to higher productivity.

Each theory makes many assumptions, which underlie its validity. Modern concepts of motivation and leadership recognize the situational nature and the dynamics of each case. Many of the specific concepts deal only with a particular aspect of the overall model, such as the content, process, or reinforcement of motivation. Let us explore some of the most popular concepts.

5.3.3 Maslow's Hierarchy of Needs

Abraham Maslow is known as the father of the need hierarchy,[8] a theory that focuses on content. As shown in Figure 5.3, there are five hierarchical levels of human needs.

1. *Physiological needs* are most important. They include the basic human needs for survival and items necessary for biological functioning, such as food, air, and sex.

---

[6]See Frederick Taylor, *Principles of Scientific Management*, New York: Harper and Brothers, 1911.

[7]See Elton Mayo, *The Human Problems of an Industrial Civilization*. New York: Macmillan, 1933.

[8]The concept of the needs hierarchy was first advanced by W. C. Langer, but Abraham Maslow is best known for refinement of the theory. For details see W. C. Langer, *Psychology of Human Living*, New York: Appleton, Century, Crofts, 1937, and A. H. Maslow, "A Theory of Human Motivation," *Psychology Review*, Vol. 50, 1943, pp. 370–396.

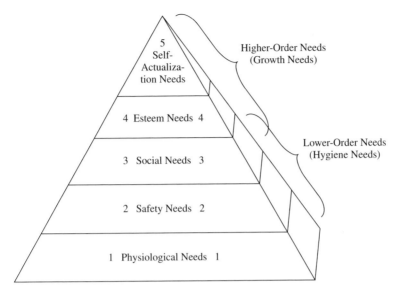

Figure 5.3 Maslow's hierarchy of needs.

Managerial Influences Include:

- Fair salary or wages
- Comfortable work environment (basics)
- Basic job security

2. *Safety needs* come next. They include the need for clothing, shelter, and a secure physical and emotional environment. They also include job security and money.

Managerial Influences Include:

- Base salary or wages
- Job security
- Safe working conditions
- Fringe benefits

3. *Social needs* are the needs of belonging, including love, affection, peer acceptance, information sharing, discussion, participation, and general peer-group support.

Managerial Influences Include:

- Work interaction, teamwork
- Social interactions
- Team spirit
- Sense of belonging

4. *Esteem needs* represent ego needs, which include recognition, respect from others, self-image, and self-respect.

Managerial Influences Include:

- Recognition, visibility
- Accomplishments
- Part of important project or mission
- Earned authority, respect
- Merit pay, bonus

5. *Self-actualization needs* represent the highest-order needs. They include high-level achievements.

Managerial Influences Include:

- Professionally stimulating work
- Work challenges
- Creativity and innovation
- Advancement opportunities
- Autonomy
- High achievements

The lower-order needs are often referred to as "hygiene needs," while the higher-order needs are called "growth needs." Maslow's model suggests that lower-order needs must be fulfilled first before the next higher-order need can occur. Furthermore, people are motivated only if that need is not already fulfilled; only deficit needs are motivators. For example, people must have their physiological needs fulfilled first before they have safety needs. A person who makes so little money that he and his family are starving is not motivated by the opportunity to negotiate a no-layoff agreement. Because of its intuitive logic, Maslow's need hierarchy enjoys a certain popularity and acceptance among managers. However, researchers have found many flaws in the theory. The five levels may not always be present and may occur in different order, including the recurrence of lower-order needs after higher needs have been filled.[9] Yet, because of its simplicity, the model provides a very important framework for guiding the management practitioner and for more detailed analysis and research regarding human behavior.

5.3.4 Herzberg's Two-Factor Theory

Another popular content-oriented theory of motivation was developed by Frederick Herzberg in the 1950s. Based on his interview with accountants and engineers, Herzberg concluded that people can be motivated by two sets of factors[10]:

- *Satisfiers*, which motivate intrinsically; their presence increases satisfaction. Typical satisfying motivators are:

  Achievement

  Recognition, praise

  The work itself

  Job enrichment

  Responsibility

  Advancement

  Personal growth
- *Hygiene factors*, which motivate extrinsically, or superficially; their presence decreases dissatisfaction. Typical hygiene motivators are:

  Technical supervision

  Working conditions

  Interpersonal relationships

  Personal life

  Pay and security

  Company policy and administration

  Status

Notice that the true motivators or satisfiers are derived from the work content, while the hygiene factors are related to the work context and work environment.

There are management implications of Herzberg's theory in several areas. One, employees are motivated in two stages. Managers must provide hygiene factors at an

---

[9]One of the best known modifications of Maslow's theory was developed by Clayton Alderfer at Yale University. Called the ERG theory, the model recognizes three levels of needs: (E) existence. (R) relatedness, and (G) growth. The theory suggests that several levels of needs can cause motivation at the same time. It further suggests that frustration at one level may cause regression to a lower level, which becomes a renewed motivator, while the higher-level need disappears, at least for some time, before resurfacing again.

[10]For a good summary of Herzberg's two-factor theory and its managerial applications, see Frederick Herzberg, "One More Time. How Do You Motivate Workers?" *Harvard Business Review,* January/February 1968, pp. 53–62.

appropriate level. Working conditions must be satisfactory and safe; proper rules, guidelines, and procedures must exist to conduct business in an orderly way; the manager must provide sufficient technical supervision; and the pay and job security must be fair and acceptable. However, none of these hygiene factors really excites and stimulates the employee; they merely ensure that his or her dissatisfaction is being kept to a minimum. After the hygiene factors have been provided for, the manager should concentrate on stage two, fostering a work environment rich in intrinsic motivators, such as professionally interesting and stimulating work, helping people to achieve desired results, recognizing their accomplishments, and facilitating personal and professional growth. Managers also should try job enrichment to provide higher levels of motivations.

Implication number two is that intrinsically motivated people are less concerned about hygiene factors; that is, people who have work challenges, have achievements and recognition, and enjoy their work may need less supervision, procedures, and even pay and security to minimize their level of dissatisfaction.

A third implication is that motivational factors can be enhanced or even created by developing a sense of ownership in the organization and pride and value in the work itself. Japanese-style management is a typical example of fostering such an environment. Many examples in our own country, such as the turnaround in attitudes and values in the automotive industry, show an application of Herzberg's theory. Another facet of this enhancement concept is that motivation can be stimulated at all levels of the work force, not only at the professional or highly skilled level. However, this requires some magic and creative thinking on the part of management. Today an increasing number of management researchers and practitioners in the United States and elsewhere stress the importance of creating this sense of pride and ownership for high-performance management, especially in complex and high-tech work environments.

Yet, for all its conceptual insight and managerial applications, Herzberg's theory is under considerable fire. His critics argue that the theory works only with people who have at least the potential for intrinsic motivation, an assumption that is probably correct for knowledge workers, such as engineering and technology professionals, but doesn't always hold in general. Further, some people argue that the emphasis is too much on how people feel, assuming that productivity will follow automatically when they are motivated, which may not be true. However, Herzberg's theory contributed considerably to managerial understanding of motivation and had a major impact on modern management practices and research. Especially in the management of engineering and technology, Herzberg's concepts are very stimulating and provide some guidance for effective management practices.

5.3.5 Vroom's Expectancy Theory

Victor Vroom developed a framework for evaluating a person's level of motivation.[11] As illustrated in Figure 5.4, the model focuses on how people choose their goals and

[11]See Victor Vroom, *Work and Motivation,* New York: Wiley, 1964.

what factors influence their motivational levels. The Vroom model predicts the level of motivation, M, based on three factors: (1) the probability of reaching a goal, called *expectancy, E,* (2) the probability that the goal will actually produce the desired results, called *instrumentality, I,* and (3) the probability that the results are indeed desirable and satisfying, called *valence, V.* Mathematically this relationship is expresses as: $M = E \times I \times V$. An example may help to clarify this. Our designer Carol Edwards, introduced earlier, reads an advertisement run by your company, looking for a new vice president of engineering. Carol knows that winning the new job would results in benefits, such as more money, status, and so forth. She also values these results highly. Therefore, both instrumentality and valence are high. However, her expectancy of getting the job, that is, of reaching the goal, is very low. As a result, her motivation is very low. She does not apply for the job.

A different situation exists when Carol's boss asks her to rerun a lab test because the original test results got lost, Carol knows that she can repeat the test and recreate the data. Both her expectancy and her instrumentality are high. However, the outcome does not create any satisfaction. On the contrary, the work interferes with her desired activity. Valence is low, and Carol's motivation to do the assignment is low. Yet another situation exists when Carol discusses a new, challenging product assignment with her boss. The boss assures Carol that the project is very important to the company. Its successful completion will be highly visible and lead to company-wide recognition and further growth potential. Carol finds the potential outcome very desirable; she has reasonable assurances that the outcome will actually occur, and feels confident that she can handle the job. All three probabilities are high. Carol is highly motivated.

In summary, Vroom's model suggests that for motivation to occur, three conditions must be met. The individual must expect that his or her effort will actually lead to the goal, that the goal will actually produce results, and that these results are desirable.

The assumptions underlying the Vroom's model are: (1) people have a choice among alternative plans and activities, and they exercise that choice, (2) people consciously or unconsciously evaluate the outcome of their future efforts and adjust the level of effort accordingly, (3) only three factors are the primary determinants of

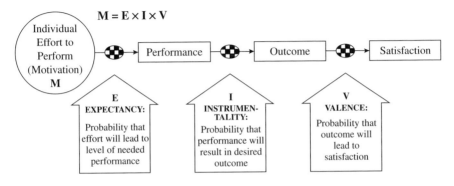

Figure 5.4   Vroom's Expectancy Theory model of motivation.

motivation, and (4) people can actually separate various goals and outcomes in real-world situations and analyze them one at a time, rather than orchestrating all of their goals and needs as an integrated system.

Vroom's theory provides a useful insight into the dynamics of human motivation. The following suggestions for managing effectively in engineering and technology environments are derived from the broader context of Vroom's theory. Engineering and technology manager should:

- Be aware of your employees' desires and ambitions.
- Assign tasks in a way that aligns organizational goals with personal goals.
- Make sure that subordinates have the ability and organizational support to succeed on the job. Develop a "can-do" mentality.
- Provide a clear and workable picture for successfully achieving goals and show the impact on the business, society, and so forth.
- Catch technical problems and personal limitations early in their development and rectify them.
- Recognize accomplishments on an ongoing basis; provide visibility.
- Provide tangible rewards to validate the perception of success.

Simply put, equity theory states that people will be motivated toward certain efforts only if they perceive the outcome to be fair relative to others. Equity theory is applied by Shakespeare, who says through one of his characters, "He is well paid that is well satisfied." Obviously, the benefit from a job well done is more than just pay. It includes recognition, skill development, promotional potential, social relations, freedom of action, and the esteem that follows achievement. Equity theory assumes that people evaluate their job performance and rewards by comparison to others. Accordingly, a person is motivated in proportion to the perceived fairness of the rewards received.[12] The challenge to managers is to administer rewards fairly and equitably. Since most technology work is based on group efforts, people continuously compare their efforts and rewards to those of others. They have their own biases regarding fairness and may value rewards differently from the way that the boss thinks they do. Managers must be close to their people to know what makes them tick and address and administer awards accordingly. Managers must also understand the dynamics of the team and know how the individual compares himself or herself to others. When it comes to bonuses and salary adjustments, managers should realize that their people are likely to be biased. They perceive inequities even though the manager does not. This is one of the reasons that management usually tries to keep salary actions confidential, although they don't always succeed. Distributing salary adjustments over different times for different people makes a direct comparison more difficult. A one-person salary adjustment is a less newsworthy and less "comparable" event than a unilateral salary action across the whole team.

---

[12]For further discussion see R. C. Huseman, J. D. Hatfield, and E. W. Miles, "A New Perspective on Equity Theory," *Academy of Management Review*, November 12, 1987, pp. 232–234.

Equity theory provides another framework for analyzing human behavior and formulating management actions. It provides an especially valuable insight for managing team efforts, which are so important in engineering and technology. Every model and theory has its limitations. Seasoned managers recognize the models' shortcomings but use those parts that are relevant to and useful in their situations.

5.3.6 Reinforcement Theory

People try to maximize pleasure and minimize pain. Reinforcement theory is based on the assumption that behavior that results in a desired outcome (reward) is likely to be repeated. On the contrary, behavior that is punished will be discontinued. This theory was originally tested on animals, but researchers such as B. F. Skinner have shown that it also applies to human beings.[13] Although many critics argue about the relevance of animal experiments to human beings, the inherent logic of the theory and its simplicity have led to a wide acceptance by managers, who use it as a basis for some of their actions. According to reinforcement theory, desired managerial results can be obtained by using one or more of four specific methods: (1) positive reinforcement, (2) avoidance, (3) negative reinforcement as punishment, or (4) extinction. Examples are listed below:

Positive Reinforcement (Reward). Provide a reward when desired results have been achieved. Rewards can include money, promotion, praise, or the work itself.

Avoidance. People may work toward a goal to avoid punishment. For example, an employee may be motivated to come to work on time to avoid a reprimand.

Negative Reinforcement (Punishment). Managers may choose to issue a reprimand, withhold pay, and so forth, to punish an employee who has engaged in an undesirable behavior such as tardiness. The employee might modify his or her behavior to avoid future punishment.

Extinction. The manager can take actions to eliminate an undesired behavior or the cause of it. Examples may be the firing of an employee whose tardiness is a negative influence on others, or closing the cafeteria at certain hours to eliminate undesirable coffee breaks.

5.3.7 Operant Conditioning

In addition to providing a stimulus to people after they have engaged in a desired or undesired behavior, the manager can *condition* his or her people by telling them ahead of time what reward or punishment they can expect. For example, suppose that the boss needs the results from a field test urgently by Wednesday at noon. The boss tells the engineer of the urgent need and provides incentives for the on-time delivery

---

[13]For a comprehensive treatment of this subject, see B. F. Skinner, *Science and Human Behavior*, New York: Macmillan. 1953.

by offering the afternoon off from work if the employee delivers on time. In addition the boss points out, however, that she will expect the engineer to work overtime without pay if he runs into difficulties. Hence, we condition the employee with a combination of stimuli—one positive and one negative reinforcement.

5.3.8   How to Manage

Motivation is of great concern to managers because it directly affects the level of performance, including innovative, creative behavior, risk taking, personal drive and effort, and the willingness to collaborate with others. Behavioral models provide the manager with some insight as to why people behave in a particular way, what energizes them, and how their behavior can be channeled to produce desired results. Yet, none of the models or theories really tells task leaders and managers what to do in specific situations. However, certain methods seem to produce more favorable overall behavior in an engineering and technology environment, where we cannot always define specific objectives, don't have all the answers, arrive at solutions incrementally, must deal with frequent changes, and want to encourage risk taking to achieve innovative results. What seems to work best in such an environment are approaches that fulfill esteem needs of professional people, leading to encouragement, pride, work challenge, recognition, and accomplishment, while minimizing personal risk, threats, fear, conflict, and confusion. Specific recommendations will be made at the end of this chapter.

5.4   LEADERSHIP IN TECHNOLOGY

Leadership is the most talked about, written about, and researched topic in the field of management. Yet, in spite of the great deal of research, management theorists cannot come up with a perfect model that explains the situational nature of leadership and helps to guide managers toward effective leadership behavior that is universally applicable.

So, then, why should managers pay attention to leadership theories? Although they may not be useful in defining a cookbook approach to effective management in all situations, leadership theories may help in analyzing why, in a given situation, particular behavior produces certain results, so we can learn from past experiences and the experiences of others, and hence fine-tune our style at interacting with people.

Early leadership concepts focused on particular traits such as intelligence, vocabulary, confidence, attractiveness, height, and gender. Literally hundreds of studies were conducted during the first two decades of the twentieth century. For the most part, the results were disappointing. These studies could not define a common set of traits for successful leaders and effective management. Neither could these studies explain the enormous trait and style differences that exist among successful leaders. The traits, styles, values, and contributions to society differed considerably from each other.

Based on the lack of success in identifying personal traits of effective leaders, research began to focus on other variables, especially the behavior and actions of leaders.

5.4.1   Rensis Likert and the Michigan Studies

In the late 1940s, studies at the University of Michigan under Rensis Likert identified two basic styles of leader behavior:

Job-Centered Leader Behavior. The leader relies primarily on authority, reward, and punishment power to closely supervise and control work and subordinates. This style follows the characteristics of McGregor's Theory X behavior.[14]

Employee-Centered Leader Behavior. The leader tries to achieve high performance by building an effective workgroup. The leader pays attention to workers' needs and helps to achieve them. Focus is on team development, participation, individual accountability, and self-direction and control McGregor's Theory Y would be characterize this behavior.[15]

In assessing the relevance of these two concepts for technology management situations, it might be tempting to discard the job-centered style as inappropriate because it appears too simplistic. In addition, an authoritarian style seems to be inappropriate to engineering and technology management.

However, a closer look shows that many effective engineering and technology managers, in fact, pay considerable attention to the planning, tracking, and controlling of the work, while staying close to their people, building a coherent task team, and fostering a work environment that emphasizes individual responsibility, recognition, and accomplishment. Hence, the manager's style can combine both dimensions, as concluded by the well-known Ohio State University research on leadership behavior[16] and later on presented in the form of a managerial grid.

5.4.2   The Managerial Grid

Developed by Robert Blake and Jane Mouton, the Managerial Grid has become a widely recognized framework for examining leadership styles.[17] As illustrated in

---

[14]Originally Likert's studies identified four systems of management: (1) autocratic and (2) benevolent critic, both relying on job-centered styles of management, and (3) consultive and (4) participative groups of management, which are employee-centered. For details see Rewsis Liker, *New Patterns of Management*, New York: McGraw-Hill, 1967.

[15]Douglas McGregor classified people for motivational purposes into two categories: (1) those who follow "theory X" dislike work, are lazy, must be controlled, and prefer clear direction and established norms, and those who follow "theory Y" and enjoy their work, need professional challenges, freedom, and gross potential, and are also willing to take responsibility and exercise leadership. For details see Douglas McGregor, *The Human Side of the Enterprise*, New York: McGraw-Hill, 1960.

[16]The Ohio State University studies on leadership behavior go back to 1940 and have many published research papers. For an informative summary of these studies, see Edwin A. Fleishman, "Twenty Years of Consideration and Structure," in *Current Development in the Study of Leadership* (E. A. Fleishman and J. G. Hunt, eds.), Carbondale, Illinois: Southern Illinois University Press, 1973.

[17]See Robert R. Blake and Jane S. Mouton, *The Managerial Grid*, Houston: Gulf Publishing, 1964, and Robert R. Blake and Anne Adams McCanse, *Leadership Dilemmas-Grid Solutions*, Houston: Gulf-Publishing, 1991, which refers to the Managerial Grid as the "Leadership Grid."

Figure 5.5, the grid displays leadership qualities in a two-dimensional matrix. Its horizontal axis represents concern for *tasks and productivity*, namely:

- Desire for achieving better results
- Cost effectiveness, resource utilization
- On-time performance
- Satisfaction of organizational objectives
- Getting the job done

The second dimension, shown on the vertical axis, represents *concern for people*, which is defined as:

- Promoting cooperation and friendship
- Helping people to achieve their personal goals
- Building trust, friendship, and respect
- Minimizing conflicts
- Facilitating to achieve results

By scaling each axis from 1 to 9, Blake and Mouton created a framework for measuring identifying leadership behavior, which they classified into five major styles.

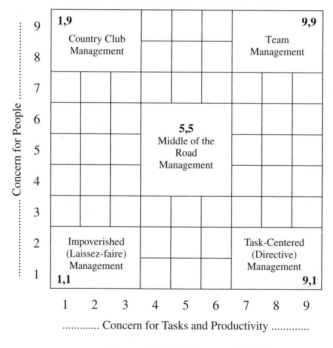

Figure 5.5 The Managerial Grid.

1,1: Impoverished management. Minimal concern for either productivity or people

5,5: Middle of the Road Management. Moderate concern for both productivity and people, to maintain status quo

9,1: Authority-Compliance Management. Primary concern for productivity; people are secondary

1,9: Country Club Management. Primary concern for people; productivity is secondary

9,9: Team Management. High concern for both productivity and people

According to Blake and Mouton, 9,9 is the ideal style of management for most work environments. This finding seems to be especially true in a technology-oriented environments, where managers have to lead people toward innovative results in an uncertain environment. In addition, technology managers often have to step across functional lines and build a work team consisting of people over whom they have limited authority and control. To get results, the manager must build trust and respect and direct the activities via team commitment and self-control. Many training programs have been developed that focus on the Leadership Grid with the objective of assisting managers in achieving a leadership style that is most effective for their specific work environment.

5.4.2.1 GE's Commitment-Competence Grid

An interesting derivative of the Managerial Grid is Commitment-Competence Grid, introduced by Jack Welch, former CEO of General Electric. The grid's two dimensions (1) commitment and (2) competence divide managers, or employees in general, into four categories:

Category IV. People are *highly competent* in producing desired results and fully *committed* to the organization. These are the top 20 percenters, who should be strongly rewarded and promoted.

Category III. People are highly competent but not fully committed. According to GE, they are valuable contributors, vital to the organization, but not necessarily the group that produces the company's future top leadership.

Category II. People are highly committed, with lots of energy and passion, but short on competence, as measured by the ability to energize others, make tough decisions, and deliver desired results. According to GE, these people need specific training and development of some form to stay as "valuable contributors" with the company.

Category I. People who are neither competent nor committed. According to GE, these are the bottom 10 percenters, the weakest players, who should be removed from the organization.

At GE, the Commitment-Competence Grid eventually evolved into the well-publicized "Vital Curve," which uses the four Es: (1) high levels of *energy*, (2) ability to *energize* others, (3) *edge* to make tough decisions, and (4) ability to consistently *execute* and deliver promised results.[18]

5.4.3   Situational Leadership Theories

Fascinated by the results of effective leadership, researchers have attempted to identify the traits and ingredients of successful leadership behavior for a long time. Yet after decades of formal studies, both management scholars and practitioners are convinced that there is no one best behavior profile or style that fits all situations. Leadership effectiveness seems to be a very intricate function of many variables. At the minimum it depends on the complex integration of variables from the leader, the followers, and the work situation. *The contingency theory of leadership*, first described by Fred Fiedler,[19] recognizes these complexities and provides a fresh insight into leadership effectiveness. Some of other popular situational leadership theories include path goal leadership theory, the Vroom-Yetton model, the attributional model, and the operant conditioning approach.

5.4.3.1   Fiedler's Contingency Theory of Leadership

Many situational leadership theories have been proposed to date. However, Fiedler's theory is the one that has been most thoroughly tested so far. It is the result of 30 years of research by Fred E. Fiedler and his associates into the situational effectiveness of task-oriented versus relationship-oriented leadership. Fiedler characterizes the *work situation* using three variables with the following ranges:

- Leader-Member Relations. Range: Good to poor
- Task Structure. Range: Structured to unstructured
- Position Power of the Manager. Range: High to low

Fiedler concluded that a task-oriented leadership style is most effective when all three situational variables are favorable, that is, leader-member relations are good, the task is structured, and the manager's position power is high. Fiedler also found that a task-oriented leadership style was the most effective when leader-member relations were poor, the task was unstructured, and the manager's power was low (i.e., all three variables were unfavorable). He further found that a relationship-oriented leadership style was most effective when the manager faced a mixed situation, in which some variables were favorable but others were unfavorable.

While critics point to the small number of variables and their descriptors in Fiedler's theory, as well as the difficulty of measuring situational components, as severe limitations of the theory, the theory does provide important lessons for managers in technology-based environments:

---

[18]For more discussion on GE's methods of identifying and developing people for leadership positions, Jack Welch, *Jack: Straight from the Gut*, New York: AOL Time Warner Company, pp. 155–168.

[19]See Fred Fiedler, *A Theory of Leadership Effectiveness*, New York: McGraw-Hill, 1967.

- Engineering and technology managers typically work in situations with moderate task structure in which they have low position power but good relationships with their team members. Fiedler's model would suggest a relationship-oriented style of leadership as optimal for such situations. This is consistent with many other research studies of technical and innovation management.

- The startup phase of a new technical task is often either all favorable or unfavorable regarding member relations, task structure, and power. In that case, a directive, more task-oriented style would be most effective, a contention that is supported independently by other research.

- A contingency, change, or problem that suddenly surfaces is likely to weaken the manager's position power and strain member relations. If this contingency, change, or problem is combined with an unstructured task, the manger might have to resort to a more task-oriented or directive style to maintain effectiveness. If prior to the contingency all the situational variables were highly favorable (and therefore the manager did not need much people orientation) more attention must be given to the staff as a result of the contingency, changes, or problem, and the leader must give more attention to facilitating conflict resolution, rebuilding trust, motivating people, and refueling their task commitment.

- The model may also be useful for a situational "what-if analysis," in which it may be used for making managers aware of potential situational changes and assessing whether the particular situation calls for more emphasis on relationships or direction.

While scholars may argue that Fiedler's theory assumes that leadership behavior is a personality trait that cannot be changed to fit a particular situation, the model provides an important framework for assessing leadership effectiveness in situations involving many components. Such a topology has been used in various field studies for determining specific professional training and development needs.[20]

5.4.3.2   *The Project Life Cycle Model*

Another popular situational leadership theory was developed by Paul Hersey and Kenneth Blanchard.[21] As an extension of the managerial grid, the model shown in Figure 5.6 provides insight into leadership effectiveness as a function of workgroup maturity. Maturity is defined as the team members' ability and willingness to assume responsibility for their group assignment. This includes many variables, including skill level, experience, knowledge, confidence, motivation, mutual trust and respect, and commitment to agreed-on results.

Leadership style is defined in terms of managerial grid dimensions: X for task-oriented behavior and Y for relationship-oriented behavior. However, unlike the

[20]For specific methods and results, see H. Thamhain and D. Wilemon, "Leadership, Conflict, and Project Management Effectiveness," *Sloan Management Review*, Vol. 19, No.1 (Fall 1977), and "Managing Technologically Innovative Team Efforts Towards New Product Success," *Journal of Product Innovation Management*, Vol. 7, No. 1, March 1990.

[21]See P. Hersey and K. Blanchard, *Management of Organizational Behavior and Utilizing Human Resources*, 4th ed., Englewood Cliffs, NJ: Prentice-Hall, 1982.

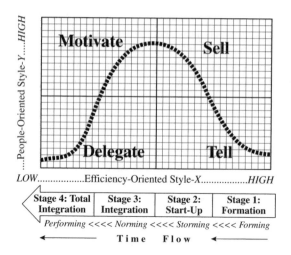

Figure 5.6 The Project Life Cycle approach to team leadership.

two-dimensional approach suggested by Blake and Mouton, the life-cycle situational model provides a more flexible interpretation of the grid components, resulting in four styles of management, which are related in their effectiveness to group maturity:

Telling Style. This high task-oriented, low relationship-oriented style focuses on goal setting, planning, organizing, guiding, directing, and controlling. It is suggested as an effective style with groups of low maturity, such as a newly formed project team, or during the initial phases of project definition. The manager provides specific instructions and close supervision.

Selling Style. This high task-oriented, high relationship-oriented style (equivalent to a 9-9 style in grid jargon) focuses on gaining team acceptance and buy-in to the assignment. It further facilitates interaction among team members, communication with other departments, and feedback. This type of leadership is suggested for workgroups that are well into their formation stage, are beginning to behave as a team, but still have a lot of concerns about their specific roles and missions and are still unsure of their ability to perform the given task as a team.

Participating Style. This low task-oriented, high relationship-oriented style focuses on team building by providing encouragement, support, visibility, and recognition to team members. The managerial objective is to strengthen team integration and self-actualization. The leader is a facilitator who provides guidance with a minimum of direct supervision and control. The participative leadership style is recommended for moderately to highly mature teams. Most technical project teams and task groups are expected to be in this maturity state during the majority of the activity phases.

Delegating Style. This low task-oriented, low relationship-oriented style relies on a fully developed and integrated team for self-management and self-actualization of activities according to agreed-on plans. The manager's role is reduced to delegation, observing, and monitoring, while the team is fully committed to and capable of producing the desired results and achieving self-development.

The Project Life Cycle model of leadership is particularly interesting and useful in project-oriented situations where managers have to deal with the formation and development of multidisciplinary teams throughout the project's life cycle. Also, the nature of technology-based work does not always guarantee a linear progression of team maturity toward the "high" end. Work interruptions, interfunctional transitions, problems, and changing requirements can make it necessary to alter the team's composition, which affects the team's characteristics regarding maturity, readiness, and ability. Situational leadership is a reality in engineering and technology management. The effective manager is a social architect who understands the dynamics of his or her organization, the work, and the people. These managers are able to diagnose potential problems and recognize the need for change.

5.4.4 Perspectives on Engineering and Technology Management

Technology-intensive management is risky business in more than one sense. Activities are highly complex, and their precise outcome often is unpredictable. Organizational structures do not always follow classical concepts of clear lines of authority, communications, and command, the format used in traditional theories of motivation and leadership. Technical people are a special breed who do not always follow the conventional motivation and leadership models. In addition, managers of technology-based organizations must deal with social, ethical, and environmental challenges.

Thus, managers of research, development, and engineering organizations must maintain and integrate two cultures, the corporate-organizational-business culture and the scientific-technical culture. For effective role performance, technology managers must be skilled in dealing with complex administrative tasks, people, and technical challenges. We are entering a far more dynamic era, requiring greater professionalism in the management of research, design, engineering, and other technology areas. Tomorrow's manager, in addition to being technically qualified, must be respected, skilled in dealing with people, and comfortable with the latest business practices and administrative techniques. Hence, the role of the technology manager is to:

1. Translate long-range strategic business objectives into operational objectives and program areas, such as research, development, engineering, manufacturing, and field support.
2. Find, hire, and encourage creative people to pursue new technical knowledge and to make technical inventions.

3. Ensure a free and creative atmosphere for research focused on economic benefits.

4. Implement research and development resulting in product, process, and service creation and improvement, consistent with the business of the corporation.

To perform this role effectively, technology managers must exhibit a style of leadership that balances managerial and technical values. Specific aspects of effective technical leadership are listed below from a major field study[22]:

- Clarity of management direction
- Understanding of the technology leaders are working with
- Understanding of the organization and its interfaces
- Ability to satisfy professional needs and motivate
- Good planning and administrative skills
- Credibility within the organization and its support groups
- Clear written and oral communication channels
- Delineation of goals and objectives
- Providing methods to aid group decision making
- Providing assistance in problem solving
- Helping to resolve conflict

Many of these leadership components become *substitutes* for traditional leadership influences, such as authority and positional power. These leadership substitutes were formally studied by Kerr and Jermier,[23] who characterized them in seven areas: (1) team member ability, (2) experience, (3) training, (4) knowledge, (5) need for independence, (6) professional orientation, and (7) indifference to organizational regard. For example, an engineer who has the skills and abilities to perform her job and also has a high need for independence and professional satisfaction may not need, or in fact may resent, a leader who provides direction and structure. By a similar argument, leadership substitutes come into effect for an engineer who performs some routine, highly structured tasks, but finds the process and the results professionally stimulating and challenging. Here, the task itself provides the subordinate with an adequate level of intrinsic motivation so that close supervision is unnecessary and may even be counterproductive.

In summary, managers of technology must span the boundaries between the technical and the business worlds. In a modern corporation, much of the technology-based

---

[22]Hans J. Thamhain, "Managing Engineers Effectively," *IEEE Transactions on Engineering Management*, Vol. 30, No. 4, November 1983.

[23]For additional discussions that also include a study on managerial style effectiveness with engineers and technicians, see S. Kerr, R. House, and A. Filley, "Relation of Leader Consideration and Initiating Structure to Research and Development Subordinate Satisfaction," *Administrative Science Quarterly*, Vol. 16, No. 1, 1971, pp. 19–30.

work and its supporting technologies cross boundaries both within and outside the organization. Because their managers are trying to maintain and integrate two cultures, the corporate economic culture and the scientific-technical culture, it is important that the technology manager exhibit a style of leadership that balances managerial and technical values. Technology-oriented personnel enjoy technical work and value expertise. They also expect certain things from management—such as providing leadership and support for technical creativity. With the right leadership, the potential productivity gains from technology are great. Management research provides some insight into the dynamics of leadership and its effectiveness in high-technology situations. However, no management model is perfect, and probably none ever will be. We must work with the tools and knowledge that exist today, seeking better ways to manage the resources available to us. Leveraging technology toward productivity improvement requires awareness, skills, commitment, ingenuity, action, and perseverance. Maybe these are the ingredients of the highly praised transformational leadership style.[24]

5.5   THE POWER SPECTRUM IN TECHNOLOGY MANAGEMENT

5.5.1   Motivation, Managerial Power, and Performance

Why do people comply with the requests or demands of others, for example, their superiors? One reason is that they see the other people as being able facilitate the fulfillment of their needs. People comply with the requests of others if they perceive them to be able to influence specific outcomes. These outcomes could be desirable, such as a salary increase, or undesirable, such as a reprimand or demotion. This influence over others is referred to as *managerial power*. Managers, as well as everyone else, use this power to achieve interpersonal influence, which is leadership in its applied form. Therefore, power is the force that, when successfully activated, motivates others toward desired results. In the next section, we will look specifically into the power spectrum that is available to leaders in a technology-oriented environment. Further, we will discuss the situational effectiveness of various leadership styles.

5.5.2   Power Sharing and Dual Accountability

Technology managers must often cross functional lines to get required support. This is especially true for managers who operate within a matrix structure. Regardless of organizational structure, technology managers must build multidisciplinary teams into cohesive workgroups and successfully deal with a variety of interfaces such as functional departments, staff groups, team members, clients, and senior management. This is a work environment in which managerial power is shared by many individuals. In contrast to the traditional organization, which provides position power largely in the form of legitimate authority, the power of engineering and technology

---

[24]For an essay on transformational leadership see Noel M. Tichy and David O. Ulrich, "The Leadership Challenge: A call for the Transformation Leader," *Sloan Management Review*, Fall 1984, pp. 59–68.

managers needs to be supported extensively with knowledge that comes from expertise and credibility that comes from the image of a sound decision maker.

Like many other components of the management system, leadership style has also undergone changes over time. With increasing task complexity, increasing dynamics in the organizational environment, and the evolution of new organizational systems, such as the matrix, a more adaptive and skill-oriented management style has evolved. This style complements the organizationally derived power bases such as authority, reward, and punishment with bases developed by the individual manager. Examples are technical and managerial expertise, friendship, work challenge, promotional opportunities, fund allocations, charisma, personal favor, project goal identification, recognition, and visibility. This so-called System II management style[25] evolved particularly with the matrix. A descriptive summary of both styles is presented in Figure 5.7. Effective technology management combines both styles I and II.

Various research studies by Gemmill, Thamhain, and Wilemon provide an insight into the power spectrum available to project managers.[26] Figure 5.8 indicates the relative importance of nine influence bases in gaining support from subordinates and assigned personnel in technology environments. Technical and managerial expertise, challenging work, and influence over salary are the most important influences that technology managers seem to have, while authority, fund allocations, and penalties appear least important in gaining support from subordinates.

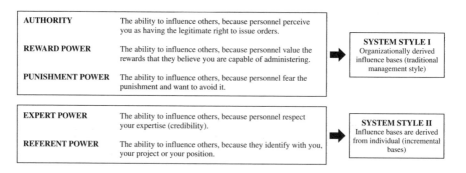

Figure 5.7 Commonly recognized bases of managerial influence (power).

---

[25]According to Likert (1967), this knowledge- and relationship-oriented style of leadership is also classified as style IV.

[26]The original study was published by G. Gemmill and H. Thamhain in "Influence Styles of Project Managers: Some Project Performance Correlates," *Academy of Management Journal* (June 1974). This study used nonparametric Kendall's rank-order correlation techniques to measure the association between the strength of a power base on which project managers seemed to rely and certain performance measures, such as project support, communications effectiveness, personal involvement and commitment, and overall management performance. The findings were later validated for engineering work environments in general.

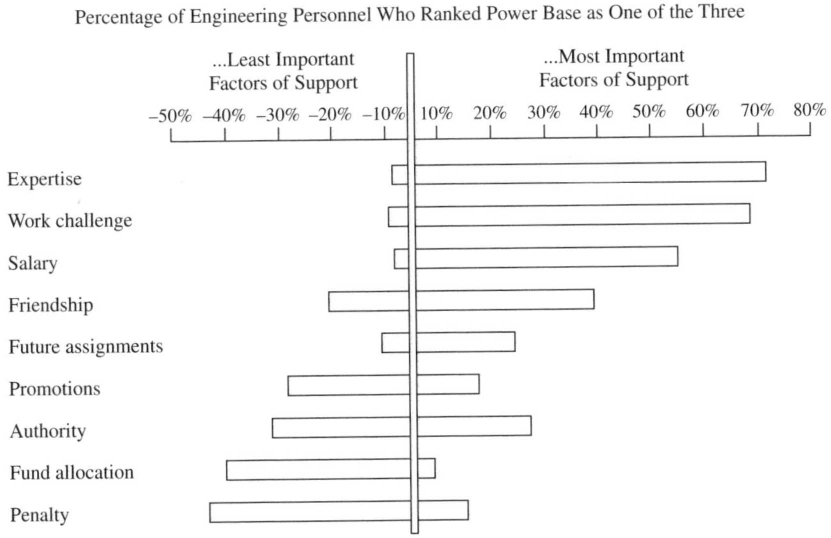

Percentage of Engineering Personnel Who Ranked Power Base as One of the Three

Figure 5.8   Power style profile. Ranking of most and least important factors for gaining support from technology personnel.

5.5.3   Role of Salary

Salary plays a very special role in the managerial power spectrum. From field studies, it is interesting to learn that engineering and technology mangers do not always realize the relatively high importance of salary as an influence base. In fact, on average, mangers rank salary second to last in importance, while their subordinates rank it third highest, as shown in Figure 5.8. In analyzing this difference between managers' and subordinates' perceptions, it was found that professionals indeed make extra efforts for many reasons other than an increase in salary. Frequently technology personnel were asked why they stayed after hours to fix a problem or why they worked casual overtime during weekends. The answers are invariably related to what Maslow would call the fulfillment of growth needs, such as challenging work, recognition, project goal identification, or professional pride. Few people would make these extra efforts because they got paid last Friday or because they are trying to position themselves for an 8 percent raise for the next year.

However, this argument holds only if the personnel perceive fair and adequate compensation. Otherwise salary becomes a barrier to effective teamwork, a handicap for attracting and holding high-quality people, and a source of steady conflict. To illustrate, a person who is motivated to make an extra effort might indeed enjoy the praise and recognition that come with the well-done job. The person may further infer from the manager's action that the job was important to the company and that if the employee continues to receive praise, the recognition and visibility for high performance during the year, he or she may certainly be in line for a raise or promotion. Now, let's suppose that instead, the employee receives a zero increase. The normal response of the employee would be disbelief in the sincerity or value of praise, recognition, or other intrinsic rewards received in the past or anticipated for the future. The employee would most likely be angry and frustrated over being manipulated. He or she might also feel cheated and confused. Overall, this is *not* a situation that leads to long-range motivation, sustained personal drive, and high morale. To the contrary, one should expect resignation, game playing, mistrust, and conflict, the very characteristics of low-performance personnel. Salary is a very important power base, which must be used judiciously, but which also must be in line with the employee's output and efforts and the salary expectations built by management over time.

5.5.4   Correlation to Technology Management Performance

The perceived importance of certain power bases, such as expertise, authority, and work challenge, does not permit by itself any conclusion regarding their effectiveness. This was originally investigated by Gemmill and Thamhain.[27]

Expanded into a more general technology-oriented work environment, the effectiveness of each managerial power base, as seen by subordinates, has been correlated to four performance indicators: (1) degree of support received by managers, (2) communication effectiveness as indicated by subordinates, (3) the degree of personal drive, involvement, and commitment of personnel, and (4) the technology mangers' overall performance as seen by their superiors. Table 5.2 summarizes the performance correlates, which are based on Kendall tau rank-order techniques.[28] Table 5.2 indicates that two influence methods are particularly favorable associated with

Table 5.2   Managerial Style versus Effectiveness in Engineering and Technology Environments, Correlations

Manager's Influence Method as Perceived by Project Personnel	Team Member's			Manager's Effectiveness Rating
	Degree of Support	Willingness to Disagree	Project Involvement	
Expertise	0.15	0.30*	0	0.40*
Work Challenge	0.10	0.25*	0.45*	0.25
Salary	−0.20	−0.10	−0.15	−0.15
Friendship	0	0	0	0.17
Future Assignment	0.25	0	0	0.10
Promotion	0	0	0	0.08
Authority	−0.10	−0.20	−0.35**	−0.30
Coercive Power	−0.45**	−0.10	−0.20	−0.25*

Kendall's tau: $^*p < .05$: $^{**}p < .01$.

Note: All associations are measured in Kendall's Tau correlation. The strength of managerial influences is measured as team member perception. Team member support, candor, involvement and innovative performance are measured as managers' perception. Managerial effectiveness is measured as senior managers' perception.

[27]See note 26.

[28]For methods of computing Kendall's Tau rank-order coefficients see S. Siegel, *Nonparametric Statistics or the Behavioral Sciences*, New York: McGraw-Hill, 1956.

management performance: expertise and work challenge. The more expertise, both managerially and technically, managers are perceived as having: (1) the more support they seem to get from their personnel, (2) the better communication seems to be, (3) the more commitment and involvement is generated, and, in the end, (4) the higher is their performance rating by their superiors, The same favorable relationship exists for work challenge. Conversely, engineering and technology managers who are perceived as relying strongly on authority, emphasizing salary, or relying on coercive measures seem to get lower support, less open communication, and lower performance ratings.

5.5.4. 1   *Management Implications*
The popularity of a particular influence base is not necessarily an indication of its effectiveness. Expertise, work challenge, and salary were cited as the three most important reasons for compliance. The performance correlates are somewhat different, however. They indicate that managers who were perceived by their personnel as emphasizing work challenge and expertise not only achieved higher effectiveness ratings on overall project performance, but also tended to foster a climate of better communication and higher involvement among their project personnel. Conversely, the findings suggest that the use of authority, salary, and coercion as an influence method has a negative effect, resulting not only in a lower level of performance but also in less communication and involvement among project personnel. Therefore the more managers rely on expertise and work challenge and the less they emphasize organizationally derived influence bases, such as authority, salary, and penalty, the greater their ability to manage in engineering and technology. One of the most interesting findings is the importance of work challenge as an influence method. Work challenge appears to encompass integrating the personal goals and needs of project personnel with project goals, more than with any other influence methods; work challenge is primarily oriented toward the intrinsic motivation of professional personnel, while other methods are oriented more toward extrinsic rewards without regard to people's preferences and needs. To present the assignments of technology personnel in a professionally interesting and challenging way may indeed have beneficial effects on project performance. In addition, the assignment of challenging work is a variable over which managers may have a great deal of control. Even if the total task structure is fixed, the method by which work is assigned and distributed is discretionary in most cases.

5.6   HOW TO MAKE IT WORK: SUGGESTIONS FOR INCREASING EFFECTIVENESS

The nature of engineering and technology management, the need to elicit support from various organizational units and personnel, the frequently ambiguous authority definition, and the often temporary nature of multidisciplinary technology activities all contribute to the complex operating environment that technology managers experience in the performance of their roles. A number of suggestions may be helpful for increasing managerial effectiveness.

Understanding Motivational Needs. Engineering and technology managers need to understand the interaction of organizational and behavioral elements in order to build an environment conducive to their personnel's motivational needs. This will enhance active participation and minimize dysfunctional conflict. The effective flow of communication is one of the major factors determining the quality of the organizational environment. Because the manager must build task and project teams at various organization levels, it is important that key decisions be communicated properly to all task-related personnel. Regularly scheduled status review meetings can be an important vehicle for communicating and tracking project-related issues.

Adapt Leadership to the Situation. Because their environment is temporary and often untested, engineering and technology managers should seek a leadership style that allows them to adapt to the often conflicting demands existing among their organization, support departments, customers, and senior management. They must learn to "test" the expectations of others by observation and experimentation. Although it is difficult, they must be ready to alter their leadership style as demanded both by the specific tasks and by their participants.

Accommodate Professional Interests. Technology managers should try to accommodate the professional interests and desires of supporting personnel when negotiating their tasks. Task effectiveness depends on how well the manager provides work challenges to motivate those individuals who provide support. Work challenge further helps to unify personal goals with the goals and objectives of the organization. Although the total work of a task group may be fixed, the manager has the flexibility of allocating task assignments among various contributors.

Build Technical Expertise. Technology managers should develop or maintain technical expertise in their fields. Without an understanding of the technology to be managed, they are unable to win the confidence of their team members, to build credibility with the customer community, to participate in the search for solutions, or to lead a unified engineering/technology effort.

Plan Ahead. Effective planning early in the life cycle of a new technology program is highly recommended. Planning is a pervasive activity that leads to personnel involvement, understanding, and commitment. It helps to unify the task team, provides visibility, and minimizes future dysfunctional conflict.

Provide a Role Model. Finally, engineering and technology managers can influence the work climate by their own actions. Their concern for project team members, ability to integrate personal goals and needs of personnel with organizational goals and their ability to create personal enthusiasm for the work itself can foster a climate that is high in motivation, work involvement, open communication, creativity, and engineering/technology performance.

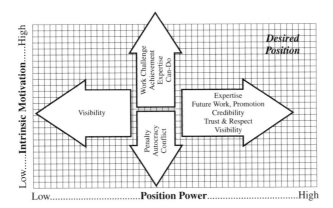

Figure 5.9   Situational effectiveness of environmental factors on motivation and managerial power.

A situational approach to technology management is presented in Figure 5.9, which indicates that the intrinsic motivation of engineering and technology personnel increases with the manager's emphasis on work challenge, their own expertise, and their ability to provide professional growth opportunities. On the other hand, emphasis on penalties and authority and a manager's inability to manage conflict lower personnel motivation. Managers' positional power is further determined by such variables as formal position within the organization, the scope and nature of work, earned authority, and ability to influence promotion and future work assignments. Managers who have strong positional power and can foster a climate of highly motivated personnel not only obtain greater support from their personnel, but also achieve high overall performance ratings from their superiors.

In addition to the six broad suggestions stated above, a number of specific criteria are summarized in this Table 5.3 to help managers in dealing with people and organizational issues effectively. The focus is on technology-based situations that involve the integration of multidisciplinary tasks across functional lines, internal and external to the enterprise.

Table 5.3   Criteria for Managing Effectively in a Technology-Based Organization

Clear Task Assignment. At the outset of any new assignment, managers and project leaders should discuss with their staff/team members the overall task and its scope, timing, resources, deliverables, and objectives.

Table 5.3 (Continued)

Early Project/Mission Involvement and Ownership. A thorough understanding of the task requirements comes usually with intense personal involvement which can be stimulated through participation in project planning, requirements analysis, interface definition, or a producability study. Involvement of the people during the early phases of the assignment, such as bid proposals and project and product planning, can produce great benefits toward plan acceptance, realism, buy-in, personnel matching, and unification of the task team. In addition, any committee-type activity, presentation, or data gathering will help to involve especially new team members and facilitate integration. It also will enable people to better understand their specific tasks and roles in the overall team effort.

Priority Image. Management should clearly articulate the importance of the assignment and its impact on the company and its mission. Senior management can help develop such a "priority image" by their involvement and by effectively communicating key mission parameters. The relationship and contribution of individual work to overall business plans, as well as of individual project objectives and their importance to the organizational mission, must be clear to all personnel.

Team Image. Building a favorable image for an ongoing project and its team in terms of high priority, interesting work, importance to the organization, high visibility, and potential for professional rewards is crucial for attracting and holding high-quality people. Senior management can help develop a "priority image" and communicate the key parameters and management guidelines for specific projects. Moreover, establishing and communicating clear and stable top-down objectives helps in building an image of high visibility, importance, priority, and interesting work. Such a pervasive process fosters a climate of active participation at all levels, helps attract and hold quality people, unifies the team, and minimizes dysfunctional conflict.

Effective Work/Project Planning. Formal planning, using proven tools and techniques, early in the life cycle of a project or specific mission is critical to any mission success. These plans and their methods don't have to be far out, but should be effective. Because engineering/technology managers usually have to integrate various tasks across many functional lines, proper planning requires the participation of the entire multifunctional team, including support departments, subcontractors, and management. Phased project planning (PPP), stage-gate concepts (SGC), and modern project management techniques provide the conceptual framework and tools for effective cross-functional planning and organizing the work toward effective execution.

Process Definition and Team Structure. The proper setup and communication of the operational transfer process, such as concurrent engineering, stage-gate process, CAD/CAE/CAM, and design-build, is important for establishing the cross-functional linkages necessary for innovative engineering performance. Management must also define the basic team structure for each project early in its life cycle. The project plan, task matrix, project charter, and operating procedure are the principal management tools for defining organizational structure and business process.

Work Challenge, Professionally Stimulating Work. Whenever possible, managers should try to accommodate the professional interests and desires of their personnel. Interesting and challenging work is a perception which can be enhanced by the visibility of the work, management attention and support, priority image, and the overlap of personnel values and perceived benefits with organizational objectives. Making work more interesting leads to increased involvement, better communication, lower conflict, higher commitment, stronger work effort, and higher levels of creativity.

Senior Management Support. It is critically important that senior management provide the proper environment for a technology team to function effectively. At the onset of a new development, the responsible manager needs to negotiate the needed resources with the sponsor organization, and obtain commitment from management that these resources will be available. An effective working relationship among resource managers, project leaders, and senior management, critically affects the perceived credibility, visibility and priority of the technology team and its work.

Clear Communication. Poor communication is a major barrier to teamwork and effective technology/engineering performance. In addition to modern technology, such as voice mail, e-mail, electronic bulletin boards, and conferencing, management can facilitate the free flow of information, both horizontally and vertically, by work space design, regular meetings, reviews, and information sessions. Further, well-defined interfaces, task responsibilities, reporting relations, communication channels, and work transfer protocols can greatly enhance communications within the work team and its interfaces, especially in complex organizational settings. The tools for systematically describing the work process and team structure come from the conventional project management system; they include: (1) *project charter*, defining the mission and overall responsibilities of the project organization, including performance measures and key interfaces, (2) *project organization chart,* defining the major reporting and authority relationships, (3) *responsibility matrix* or *task roster,* (4) *project interface chart, such as the N-Squared Chart discussed earlier,* and (5) *job descriptions.*

Commitment. Managers should ensure team member commitment to their project plans, specific objectives, and results. If such commitments appear weak, managers should determine the reason for such lack of commitment of a team member and attempt to modify possible negative views. Because insecurity is often a major reason for low commitment, managers should try to determine why insecurity exists, then work to reduce the team members' fears and anxieties. Conflict with other team members and lack of interest in the project may be other reasons for such lack of commitment.

Leadership Positions. Leadership positions should be carefully defined and staffed for all projects and support functions. Especially critical is the credibility of project leaders among team members, with senior management, and with the program sponsor for the leader's ability to manage multidisciplinary activities effectively across functional lines.

Reward System. Personnel evaluation and reward systems should be designed to reflect the desired power equilibrium and authority/responsibility sharing of an organization. A QFD philosophy helps to focus efforts toward desired results on company internal and external customers, and helps to foster a work environment which is strong on self-direction and self-control.

Problem Avoidance. Engineering and technology managers should focus their efforts on problem avoidance. That is, managers and team leaders, through experience, should recognize potential problems and conflicts at their onset, and deal with them before they become big and their resolution consumes a large amount of time and effort.

Personal Drive and Leadership. Managers can influence the work environment by their own actions. Concern for the team members, the ability to integrate personal needs of their staff with the goals of the organization, and the ability to create personal enthusiasm for a particular project can foster a climate of high motivation, work involvement, open communication, and ultimately high team performance.

5.7 SUMMARY OF KEY POINTS AND CONCLUSIONS

The key points that have been made in this chapter include:

- Technology organizations are seldom structured along traditional functional lines; rather, they operate as matrices or hybrids with a great deal of power and resource sharing.

- Technology managers must be skilled in dealing effectively with technical work, the talent search and management, knowledge, information, communications, collaboration and commitment, direction, and leadership.

- Leaders who succeed within technology organizations must move across various organizational lines to gain services from personnel not reporting directly to them.

- Technical people are a special breed who do not always follow the conventional motivation and leadership models.

- Sixteen professional needs are critical to technology-based performance: (1) interesting and challenging work, (2) a professionally stimulating work environment, (3) professional growth, (4) overall leadership, (5) tangible rewards, (6) technical expertise (7) assisting in problem solving, (8) clearly defined objectives. (9) management control, (10) job security, (11) senior management support, (12) good interpersonal relations, (13) proper planning, (14) clear role definition, (15) open communications, and (16) minimum changes.

- These needs must be satisfied *continuously* before the people can reach and sustain high levels of performance.

- Motivation is an inner drive that transforms activities into desired results.
  "What you expect from yourself or others is most likely to happen" is the essence of the self-fulfilling prophecy.

- Maslow's model suggests that lower-order needs must be fulfilled before the next higher-order need can occur.

- True motivators or satisfiers are derived from the *work content*, while the hygiene factors are related to the *work context* and work environment.

- According to Herzberg's theory, employees are motivated in two stages, by providing (1) hygiene factors and (2) intrinsic motivation.

- Intrinsically motivated people are less concerned about hygiene factors.

- Motivational factors can be enhanced or even created by developing a sense of ownership in the organization, and pride and value for the work itself.

- Vroom's model suggests motivation is based on three conditions: Individuals must perceive/expect that (1) their effort will actually lead to the goal, (2) the goal will actually produce results, and (3) these results are desirable.

- Early leadership concepts focused on personality traits, while contemporary models emphasize the situational nature of leadership, including task, organization, environment, business process, and people relations.

- With increasing task complexity, organizational dynamics, and the evolution of new organizational systems, such as the matrix, a more adaptive and skill-oriented management style evolved.
- The Managerial Grid has become a widely recognized framework for examining leadership styles.
- The Project Life Cycle Model for team leadership provides insight into leadership effectiveness as a function of workgroup maturity.
- Salary is a very important power base, which must be used judiciously but that also must be in line with the employee's output, efforts, and salary expectations.
- The use of authority, salary, and coercion as an influence method has a negative effect, resulting not only in a lower level of performance but also in less communication and involvement among project personnel.
- Technology managers must not only be technically qualified but also be respected, skilled in dealing with people, and comfortable with the latest business practices and administrative techniques.

5.8 CRITICAL THINKING: QUESTIONS FOR DISCUSSION

1. How can motivational theories help in guiding technology managers toward effective leadership of technical professionals?
2. Why are there so may different models explaining motivational behavior? Which model do you like best? Why?
3. Discuss how you could use some of the management models for organizational development or management development in your organization. Consider models such as the Managerial Grid, Project Life Cycle, Inducement-Contribution Model, and Vroom's Expectancy Theory.
4. Define a management development program for a young team leader in preparation for more advanced management assignments.
5. What leadership skills are learnable in the "classroom" and which ones can only be gained experientially?
6. How do you explain the fact that the same manager is being perceived as "autocratic" by some people and "democratic/team-oriented" by others within the same work team?
7. Discuss the issues of self-direction and empowerment. Where is it necessary and where is it risky? What are the managerial limitations and challenges of empowerment? Apply some of the management models in support of your arguments.
8. Discuss the role of salary and financial awards for motivating technology professionals.
9. How can managers and team leaders "earn" their authority, especially when crossing functional lines and dealing with organizations over which they have no formal authority?
10. Discuss the role of trust, respect, and credibility for effective leadership in technology-driven enterprises (relate also to the previous question).

11. What is the role of technical (job-specific) expertise for managerial effectiveness in technology-driven enterprises? Is there a danger in a manager having "too much" technical expertise?

5.9 REFERENCES AND ADDITIONAL READINGS

Barkema, H., Baum, J., and Mannix, E. (2002) "Management challenges in a new time," *Academy of Management Journal*, Vol. 45, No. 5, pp. 916–930.

Barner R. (1997) "The new millennium workplace," *Engineering Management Review (IEEE)*, Vol., 25 No. 3, pp. 114–119.

Blake, R. and McCanse, A. (1991) *Leadership Dilemmas-Grid Solutions*, Houston: Gulf-Publishing.

Blake, R. and Mouton, J. (1964) *The Managerial Grid*, Houston: Gulf Publishing.

Chell, E. and Allman, K. (2003) "Mapping the motivations and intentions of technology orientated entrepreneurs," *R & D Management*, Vol. 33, No. 2 (March), pp. 117–134.

English, K. (2004) "The changing landscape of leadership," *Research Technology Management*, Vol. 47, No. 4 (July/August), p. 9.

Feld, C. (2004) "Getting it right," *Harvard Business Review*, Vol. 82, No. 2 (February), pp. 72–79.

Fiedler, F. (1967) *A Theory of Leadership Effectiveness*, New York: McGraw-Hill.

Foust, J. (2004) "Leading experts: one manager's experience," *Research Technology Management*, Vol. 47, No. 2 (March/April), pp. 12–19.

Gemmill, G. and Thamhain, H. (1974) "Influence styles of project managers: some project performance correlates," *Academy of Management Journal*, Vol. 17, No. 2 (June), pp. 216–224.

Ghoshal, S. (2003) "Miles and snow: enduring insights for managers," *Academy of Management Executive*, Vol. 17, No. 4 (November), pp. 109–114.

Halbesleben, J., Novicevic, M., Harvey, M., and Buckley, M. R. (2003) "Awareness of temporal complexity in leadership of creativity and innovation: a competency-based model," *Leadership Quarterly*, Vol. 14, No. 3 & 4 (August/October), pp. 433–454.

Hersey, P. and Blanchard, K. (1982) *Management of Organizational Behavior and Utilizing Human Resources*, 4th ed., Englewood Cliffs, NJ: Prentice-Hall.

Herzberg, F. (1968) "One more time: how do you motivate workers?" *Harvard Business Review*, Vol. 46, No. 1 (January/February), pp. 53–62

Kahn, P. (2003) "The leadership journey," *Leader to Leader*, Issue 29 (Summer), pp. 7–13.

Kerr, S., House, R., and Filley, A. (1971) "Relation of leader consideration and initiating structure to R&D subordinate satisfaction," *Administrative Science Quarterly*, Vol. 16, No. 1, pp. 19–30.

Kruglianskas, I. and Thamhain, H. (2000) "Managing technology-based projects in multinational environments," *IEEE Transactions on Engineering Management*, Vol. 47, No. 1 (February), pp. 55–64.

Likert, R. (1967) *New Patterns of Management*, New York: McGraw-Hill.

Livingston, S. (1969) "Pygmalion in management," *Harvard Business Review*, Vol. 47, No. 4 (July/August), pp. 81–89.

Maccoby, M. (2003) "Finding the right leader," *Research Technology Management*, Vol. 47, No. 2 (March/April), pp. 60–61.

Martin, R. and Moldoveanu, M. (2003) "Capital versus talent: the battle that's reshaping business," *Harvard Business Review*, Vol. 81, No. 7 (July), pp. 36–41.

Mayo, E. (1933) *The Human Problems of an Industrial Civilization*, New York: Macmillan.

McGregor, D. (1960) *The Human Side of the Enterprise*, New York: McGraw-Hill.

Oakey, R. (2003) "Technical entrepreneurship in high technology small firms: some observations on the implications for management," *Technovation*, Vol. 23, No. 8. (August), p. 679.

Parker, A. (2003) "What creates energy in organizations?" *MIT Sloan Management Review*, Vol. 44, No. 4 (Summer), pp. 51–60.

Richey, J. and Grinnell, M. (2004) "Evolution of roadmapping at Motorola," *Research Technology Management*, Vol. 47, No. 2 (March/April), pp. 37–41.

Roberts, E. (2004) "Linkage, leverage and leadership drive successful technological innovation," *Research Technology Management*, Vol. 47, No. 3 (May/June), pp. 9–11.

Rodriguez, R., Green M., and Ree, M. (2003). "Leading generation X: do the old rules apply?" *Journal of Leadership & Organizational Studies,* Vol. 9, No. 4 (Spring), p. 67.

Senge, P. and Carstedt G. (2001) "Innovating our way to the next industrial revolution," *Sloan Management Review*, Vol. 42, No. 2, pp. 24–38.

Shenhar, Aaron and Thamhain, Hans (1994) "A new mixture of project management skills: meeting the high-technology challenge," *Human Systems Management Journal*, Vol. 13, No. 1 (March), pp. 27–40.

Shim, D., Lee, M. (2001) "Upward influence styles of R&D project leaders." *IEEE Transactions on Engineering Management*, Vol. 48, No. 4, pp. 394–413.

Siegel, S. (1956). *Nonparametric Statistics for the Behavioral Sciences*, New York: McGraw-Hill.

Skinner, B. F. (1953) *Science and Human Behavior,* New York: Macmillan.

Taylor, Frederick (1911) *Principles of Scientific Management*, New York: Harper and Brothers.

Thamhain, H. (1983) "Managing engineers effectively," *IEEE Transactions on Engineering Management*, Vol. 30, No. 4, pp. 231-237.

Thamhain, H. (1990) "Managing technology: the people factor," *Technical & Skill Training Journal* (August/September), pp. 24–31.

Thamhain, H. (1990) "Managing technologically innovative team efforts toward new product success," *Journal of Product Innovation Management*, Vol. 7, No. 1 (March), pp. 5–18.

Thamhain, H. (1996) "Managing self-directed teams toward innovative results," *Engineering Management Journal*, Vol. 8, No. 3, 31–39.

Thamhain, H. (2001) "Team management," Chapter 19 in *Project Management Handbook* (J. Knutson, ed.), New York: Wiley & Sons.

Thamhain, H. (2002) "Criteria for effective leadership in technology-oriented project teams," Chapter 16 in *The Frontiers of Project Management Research* (Slevin, Cleland, and Pinto, eds.), Newton Square, PA: Project Management Institute, pp. 259–270.

Thamhain, H. (2003) "Managing innovative R&D teams," *R&D Management*, Vol. 33, No. 3 (June), pp. 297–312.

Thamhain, H. (2004) "Leading technology-based project teams," *Engineering Management Journal*, Vol. 16, No. 2, pp. 42–51.

Thamhain, Hans and Wilemon, David (1987) "Leadership, conflict, and project management effectiveness," *Executive Bookshelf on Generating Technological Innovations, Sloan Management Review*, Fall 1987, pp. 68–87.

Thamhain, H. and Wilemon, D. (1999) "Building effective teams for complex project environments," *Technology Management*, Vol. 5, No. 2 (May), pp. 203–212.

Tichy, N. and Ulrich, D. (1984) "The leadership challenge: A call for the transformation leader," *Sloan Management Review*, Vol. 26, No. 1 (Fall), pp. 59–68.

Vroom, V. (1964) *Work and Motivation,* New York: Wiley & Sons.

Welch, J. (2003) *Jack: Straight from the Gut*, New York: AOL Time Warner Company.

6

MANAGING
TECHNOLOGY-BASED
PROJECTS

PROJECT MANAGEMENT FOR NASA

When NASA's twin robot geologists, the Mars Exploration Rovers *"Spirit"* and *"Opportunity,"* landed on Mars, January 3 and January 24, 2004, Peter Theisinger and his project team at Jet Propulsion Laboratory were jubilant. And rightly so. Success did not come easy, nor by luck. It was the result of a carefully planned and executed multi-year billion dollar program that culminated in the successful Mars landing, after seven months of space flight. Yet, this is only the beginning of the mission objective.

The Mars Exploration Rover (MER) mission is part of NASA's Mars Exploration Program (MEP), a long-term effort of robotic exploration of the red planet. Primary among the mission's scientific goals is to search for a wide range of rocks and soils that hold clues about the history of our planetary system. To perform the required on-site geological investigations, the rovers must drive up to 40 meters in a single day. Each rover is sort of the mechanical equivalent of a geologist walking the surface of Mars. The mast-mounted cameras are mounted 1.5 meters high and will provide 360-degree, stereoscopic, human-like views of the terrain. The robotic arm will be capable of movement in much the same way as a human arm with an elbow and wrist, and will place instruments directly up against rock and soil targets of interest. In the mechanical "fist" of the arm is a microscopic camera that will serve the same purpose as a geologist's handheld magnifying lens. The Rock Abrasion Tool serves the purpose of a geologist's rock hammer to expose the insides of rocks.

The Challenges of Getting to Mars. Two out of three missions to the red planet have failed. One reason there have been so many losses is that there have been so many attempts. "Mars is a favorite target," says Dr. Firouz Naderi, manager of the Mars Program Office at the Jet Propulsion Laboratory. "We—the United States and former USSR—have been going to Mars for 40 years. The first time we flew by a planet, it was Mars. The first time we orbited a planet, it was Mars. The first time we landed on a planet, it was Mars. However, getting there is hard."

Specifically, *Spirit* and *Opportunity*, the two Mars Exploration Rovers, had to fly through about 300 million miles of deep space and target a very precise spot to land. Adjustments to their flight paths can be made along the way, but a small trajectory error can result in a big detour and or even missing the planet completely.

The space environment isn't friendly. Hazards that range from what engineers call "single event upsets," as when a stray particle of energy passes through a chip in the spacecraft's computer, causing a glitch and possibly corrupting data, to massive solar flares, such as the ones that occurred this fall, can damage or even destroy spacecraft electronics.

A Complex Project. The road to the launch pad is nearly as daunting as the journey to Mars. Even before the trip to Mars can begin, a craft must be built that not only can make the arduous trip but can complete its science mission once it arrives. Nothing less than exceptional technology and planning is required.

If getting to Mars is hard, landing there is even harder. "One colleague describes the entry, descent and landing as six minutes of terror," says Naderi.

Spirit and *Opportunity* entered the Martian space traveling 19,300 kilometers per hour. "During the first four minutes into descent, we used friction with the atmosphere to slow us down considerably," says Naderi. "However, at the end of this phase, we're still traveling at 1,600 kilometers per hour, but now we had only 100 seconds left and were at the altitude that a commercial airliner typically flies. Things need to happen in a hurry. A parachute opened to slow the spacecraft down to 'only' 321 kilometers per hour, but now we had only 6 seconds left and were only 91 meters off the ground. Now, the retro rockets fired to bring the spacecraft down to zero velocity, and we're the height of a four-story building above the surface. Then, the spacecraft freefalls the rest of the way cocooned in airbags to cushion the blow. It hits the ground at 48 kilometers per hour. It bounces as high as a four-story building and continues to bounce afterward, perhaps 30 times all together. What's inside the airbag weighs 453 kilograms (half a ton)."

Mars doesn't exactly put out a welcome mat. Landing is complicated by difficult terrain. The Martian surface is full of obstacles—massive impact craters, cliffs, cracks and jagged boulders. Even the toughest airbag can be punctured if it hits a bad rock. Unpredictable winds can also stir up further complications.

No matter how hard it is, getting to Mars is just the beginning. "The challenge after we land," says Rob Manning, manager of Mars Exploration Rovers entry, descent and landing operations, "is how to get the vehicle out of its cramped cocoon and into a vehicle roving in such a way as to please the scientists."

The rewards are great. "Mars is the most Earth-like of the planets in our solar system," says Naderi. "It has the potential to have been an abode of life."

The risks are also great. "We did everything humanly possible to avoid human mistakes," says Naderi. "That's why we did check, double check, test and test again and then have independent eyes check everything again. Humans, even very smart humans, are fallible, particularly when many thousands of parameters are involved. But even if you have done the best engineering possible, you still don't know what Mars has in store for you on the day you arrive. Mars can get you."

"We are in a tough business," says Naderi. "It is like climbing Mt. Everest. No matter how good you are, you are going to lose your grip sometimes and fall back. Then you have a choice, either retreat to the relative comfort and safety of the base camp, or get up, dust yourself off, get a firmer grip and a surer toehold and head back up for the summit. The space business is not about base camps. It is about summits. And, the exhilaration of discoveries you make once you get there. That is what drives you on."

The Mars Exploration Rover (MER) is a NASA Program, managed by Jet Propulsion Laboratories (JPL), a division of the California Institute of Technology in Pasadena, California.

For additional details and progress updates of NASA's Mars Exploration Program visit the NASA/JPL project web site at http://www.jpl.NASA.gov.

Source: JPL/NASA web site, *http://marsrovers.jpl.nasa.government.*

6.1 MANAGEMENT OF TECHNOLOGY IS PROJECT-ORIENTED

Not only for space exploration, technology-based companies perform a great deal of project work! R&D, new product developments, product enhancements, IT infrastructure upgrades, high-tech construction, and trillions of dollars of aerospace, military, and general government contracts around the world are project-based. Similar to the Mars Exploration Rover (MER), projects are characterized as single, non-repeat undertakings, that have specific results (deliverables) and time and resource constraints. Management of technology and technology-oriented business is project oriented. Therefore, it is not surprising that technology-based companies have long been accustomed to working with projects. In fact, many of the contemporary project management tools and techniques emerged from high-technology work environments. However, producing desirable business results is not a simple matter in today's complex world of business, and it takes a lot more than just implementing the latest project management tools and processes.

6.1.1 Project Management Characteristics in a Changing World

The business environment is quite different from what it used to be. For one thing, new technologies, especially computers and communications, have radically changed the workplace and transformed our global economy, reorienting them toward service and knowledge work, with higher mobility of resources, skills, processes, and technology itself. In a concomitant change, new project management techniques have evolved which are often better integrated with business processes, offering more sophisticated capabilities for project tracking and controlling in culturally diverse environments that contain a broad spectrum of contemporary challenges (Jaafari 2003). Such challenges include time-to-market pressures, accelerating technologies, pressures for innovation, resource limitations, technical complexities, social and ethical issues, operational dynamics, and risk and uncertainty (cf. Bishop 1999, Deschamp & Nayak 1995). As summarized in Table 6.1, facing such a dynamic environment often makes it difficult to manage projects through traditional, linear work processes or top-down controls.

In response to these challenges, many companies and their management have formed new alliances through mergers, acquisitions, and joint ventures, and have explored alternative organizational designs, business processes, and leadership styles, such as concurrent engineering, design-build, and stage-gate protocols (Thamhain 1994, 2004). While these concepts have the potential for organizations to become more agile and responsive, they also require more intense cross-functional teamwork and cooperation, with high levels of resource and power sharing, and complex lines of authority, accountability, and control (cf. Gupta & Govindarajan 2000, Thamhain & Wilemon 1999). As a result, the focus of project management has shifted over the past decades from simply tracking schedule and budget data to the integration of human factors and organizational interfaces into the project management process. The new generation of project leaders must deal effectively with the new challenges and realities of today's business environment, which include highly complex sets of deliverables, as well as demanding timing, environmental, social, political, regulatory, and technological factors. Working effectively in such an intricate environment requires new skills in both project administration and leadership, especially for complex, technology-based and R&D-oriented projects that rely to an increasing extent on innovation, cross-functional teamwork and decision making, intricate multicompany alliances, and highly complex forms of work integration. Project success often depends to a considerable extent on member-generated

Table 6.1 Today's Projects: Characteristics and Challenges

- Changing business models and organizational processes
- Complex joint ventures, alliances, and partnerships
- Complex project performance measurements and data processing requirements
- Complexity of defining project success and deliverables
- Obtaining multifunctional buy-in and commitment
- Global markets
- Integrating broad spectrum of functions and support services
- Integrating project and business processes
- Large groups of stakeholders
- Managing beyond immediate results
- Need for continuous improvement of project operations
- Need for integration across functions, dealing with different organizational cultures and values
- Need for sophisticated human relations skills
- Inevitable organizational conflict, power and politics
- Critical role of organizational members in successful project implementation
- Project complexities, implementation risks, and uncertainties
- Resource constraint, tough performance requirements
- Self-directed teams
- Tight, end-date-driven schedules
- Total project life-cycle considerations
- Virtual organizations, markets, and support systems

performance norms and work processes, rather than supervision, policies, and procedures (cf. Bahrami, 1992; Thamhain & Wilemon, 1999). As a result, self-directed and commitment-based concepts are gradually replacing the traditional, more hierarchically structured project organization.

6.2 MODERN PROJECT MANAGEMENT: A CONTINUOUSLY EVOLVING SYSTEM

In the past, project performance was measured by and large in terms of achieving agreed-on results within given time and resource constraints. Today, these measures have become little more than threshold competencies for companies that espouse project management as a core capability–very important, but unlikely to provide a true competitive advantage. Yet, many firms still measure overall project performance by these threshold factors. Focusing on these factors alone, however, is unlikely to overcome a firm's project management deficiencies. To the contrary, such focus may mask crucial competitive factors such as innovative results, technological

Table 6.2 Change Forces Driving Modern Project Management

Shift from . . .

. . . mostly linear work processes to highly dynamic, organic, and integrated project systems

. . . efficiency toward effectiveness

. . . extensive use of IT to more process-integrated use of IT

. . . information to decision support

. . . project management tools to integrated systems

. . . managerial control to self-direction and accountability

. . . executing projects to enterprise-wide project management

. . . project management as a support system to established standards and professional status

breakthroughs, time-to-market capabilities, flexibility and responsiveness to changing requirements, future business positioning, and client satisfaction.

6.2.1 Forces Driving Modern Project Management

Building on the paradigm shifts discussed in Chapter 2, there are eight major forces that are driving the nature of today's project environment (Table 6.2). These forces must be understood in order to build an effective project management system.

1. *Shift from linear processes to dynamic project systems.* In the past, project management concepts were based on predominately linear models, typically exemplified by production lines, sequential product developments, scheduled services, and discovery-oriented R&D. Managing project in the present environment, in contrast, requires more dynamic and interactive relationships, involving complex sets of interrelated, nonlinear, and often difficult to define processes. These changes have not only increased the complexity of the project environment, they also demand a far more sophisticated management style which relies strongly on group interaction, resource and power sharing, individual accountability, commitment, self-direction, and control. Consequently, many of today's projects and their integration rely to a considerable extent on member-generated performance norms and evaluations, rather than on hierarchical guidelines, policies, and procedures. While this paradigm shift is driven by changing organizational complexities, capabilities, demands, and cultures, it also leads to a radical departure from traditional management philosophy on organizational structure, motivation, leadership, and project control. As a result, traditional "hard-wired" project organizations and processes are being replaced by more flexible and nimble networks. These networks are usually derivatives of the conventional matrix organization, but with more permeable boundaries, more power and resource sharing, and more concurrent project integration.

2. *Shift from efficiency toward effectiveness.* Many companies have broadened their focus from *efficient* execution of projects–emphasizing job skills, teamwork, communications, and resource optimization at the project level–to include *effectiveness* of their project organizations. This shift responds to the need for better integration of project activities into the overall enterprise, making sure the organization is "doing the right thing." Companies are trying to leverage project management as a core competency, integrating closely with other functions, such as marketing, R&D, field services, and strategic business planning. While this shift is enhancing the status and value of project management within the enterprise, it also raises the overall level of responsibility and accountability, placing more demands on project management to perform as a full partner within the integrated enterprise system.

3. *Shift toward more integrated information technology.* The availability and promise of technology have led to an enormous acquisition of IT-based tools and techniques by managers at all levels. These managers are eager to leverage these tools for true value added, increasing the effectiveness of their operations rather than simply generating more data. The challenge for project professionals is to look beyond project management software per se to fully understand and *apply* the technology to the firm's business process, helping project managers solve their problems and increase operating efficiency, rather than just replacing traditional forms of communication, interactions, and problem solving that, in many cases, still play a crucial role in project success.

4. *Shift from information to decision support.* Today's technology provides managers in any part of the enterprise with push-button access to critical information on project status and performance. In addition, IT-based project management tools, in conjunction with well-maintained data bases, offer powerful support for resource estimating, scheduling, and risk analysis. These tools provide better data integrity and decision support, from project initiation to execution, affecting the quality of project control and predictability across the entire enterprise.

5. *Shift from project management tools to integrated systems.* Effective project management today requires far broader skill sets than just dealing with budgeting and scheduling issues. While critically important, these abilities are strongly supported by modern information technology and have become core competencies for project leaders, literal requirements to enter the professional field. Managerial focus has shifted from the mechanics of controlling projects according to established schedules and budgets to optimizing desired results across a wide spectrum of performance measures that integrate with total enterprise performance. An underlying ideal is movement toward developing a true learning organization. The need for understanding project management as an integrated part of the overall business process–its human side and company-external components–all have a profound impact

on the type and scope of project management training, consulting needs, and organizational development required for leveraging project management as an enterprise-wide resource.

6. *Shift from managerial control to self-direction and accountability.* With increasing project complexities, advances in information technology, changing business cultures, and new market structures, companies must look *beyond* traditional managerial control for effective project execution. Especially top-down project controls, based on centralized command and communications, while critically important, are no longer sufficient for generating satisfactory results. Projects rely to an increasing extend on technology, innovation, cross-functional teamwork and decision making, intricate multicompany alliances, and highly complex forms of work integration. The dynamics of these project environments foster to a considerable extent member-generated performance norms and work processes, and a shift toward more team ownership, empowerment, and self-control (cf. Barner 1999, Thamhain & Wilemon 1999, Kruglianskas & Thamhain 2000). All of this has a strong impact on the way project leaders must interface with their teams and support organizations in directing and managing the work effectively. Methods of communication, decision making, soliciting commitment, and risk sharing are shifting constantly away from a centralized, autocratic management style to a team-centered, more self-directed form of project control.

7. *Shift from executing projects to enterprise-wide project management.* Many companies use project management far more extensively than just for implementing projects. A growing number of organizations are relying on project management as a core competency for leveraging their resources, achieving accelerated product developments, capturing higher levels of innovation, ensuring better quality, and, in general, securing better resource utilization. To accomplish this level of competency, companies must integrate their project operations with the strategic goals and business processes across the total enterprise.

8. *Shift from project management as a support function toward full operational responsibility and professional status.* With its own body of knowledge, norms and worldwide standards, professional certification, and formal education programs at the master's and Ph.D. levels, project management has established its professional position over the last two decades. Today, the principles of project management apply across industries and around the globe, virtually in all types of situations. Companies can choose from a growing pool of formally educated project management professionals. They also have access to professional training programs and a broad spectrum of readily available operational tools, techniques, and processes.

6.2.2 New Tools, Techniques, and Management Philosophies

With the increasing complexities of today's projects and their business environments, companies have moved toward more sophisticated tools and techniques for effectively managing their multidisciplinary activities. These tools and techniques range from computer software for sophisticated schedule and budget tracking to intricate organizational process designs, such as concurrent engineering and stage-gate protocols. Even conventional project management tools, such as schedules, budgets, and status reviews, are being continuously upgraded and effectively integrated with modern information technology systems and overall business processes. As part of this evolution, organizations must also shift their focus from simply tracking schedule and budget data to integrating human factors and organizational interfaces into project-control formulae. The new generation of project management tools is designed to deal more effectively with the challenges and realities of today's business environment, which include highly complex sets of deliverables, as well as timing, environmental, social, political, regulatory, and technological factors.

While the shift to more sophisticated project management processes is the result of changing business cultures, project complexities, technological capabilities, and market structures, it also requires radical departures from traditional management philosophy and operating practices on organization, motivation, leadership, and project control. As a result, the traditional management style, designed largely for top-down control, centralized command, and one-way communication, is no longer sufficient for managing effectively in today's team-based environment. The new breed of managers that evolved with these contemporary organizations is often more connected with the organization and its business process, and can deal with a broad spectrum of contemporary challenges, such as time to market, accelerating technologies, innovation, resource limitations, technical complexities, project metrics, operational dynamics, risk, and uncertainty, more effectively than their colleagues in the past. All of this has a profound impact on the way project professionals must manage and lead. It also explains in part why the methods of communication, decision making, soliciting commitment, and risk sharing are continuously shifting away from a centralized, autocratic management style to a team-centered, more self-directed form of project leadership.

6.3   THE FORMAL PROJECT MANAGEMENT SYSTEM

For most large technology undertakings, managers don't have much of a choice but to use formal project management techniques. It just gets too complicated to keep it all in one's head and to communicate it among a large number of people. According to some studies, the threshold between informal and formal planning and control techniques seems to lie at approximately 5 to 10 people working on a single program. However, as shown below, there are many other factors that drive managers toward formal project tools. Specifically, managers are using formal tools so that they can better respond to:

- Specific contract requirements, especially from government clients
- Better performance measurements of work in progress
- Shorter product development cycles

- Organizing multifunctional teams effectively
- Questions of individual accountability
- Changing requirements
- Negotiating resource requirements
- Potential conflict and confusion over plans
- Personnel changes
- Priority shifts
- Subcontractor support requirements
- Geographically dispersed work units
- Language and cultural barriers

All of these reasons provide pressures on technology managers to use a formal planning system with standard project management tools for defining, organizing, and directing their work. In addition, no less significantly, these systems lay the basis for individual accountability, managerial performance appraisal, and rewards. Yet another outfall of this paradigm shift is the increasing number of companies that started to institutionalize their project activities formally within the framework of a *project office.* Such a *project office,* or *PO,* has full responsibilities for the management of all project-related businesses, including the development of the corporate-wide project management system and the integration of the project management platform into the strategic planning process (Kerzner 2003).

6.3.1 Project Management Tools and Techniques

By definition, project management is the application of knowledge, skills, tools, and techniques to project activities in order to meet or exceed stakeholder needs and expectations.[1] This involves balancing the needs and constraints among several competing subsystems of

- Scope, time, cost, and quality
- Stakeholders
- Requirements and expectations

The following tools and management techniques are commonly used for organizing, tracking, and controlling engineering and technology projects. They are classified in this chapter within six principal categories based on their *primary focus* toward specific management applications or project parameters: (1) top-down project definition and integration; (2) project scope management; (3) project time management; (4) project cost management; (5) project human resource management; and (6) tracking and control focus, as summarized in Table 6.3. These categories are also consistent with

[1] Definition of project management according to the Project Management Body of Knowledge (PMBOK), published by the Project Management institute, or PMI. A *Guide to the Proect Management Body of Knowledge,* Newtown Square, PA: Project Management Institute, 2000.

Table 6.3   Principal Project Management Tools and Techniques Grouped by Application Areas

	Primary Tool Application Focus					
	Top-Down	**Work**	**Time**	**Resource**	**Responsibility**	**Tracking & Control**
	(*PMBOK:* Project Integration Management)	(*PMBOK:* Project Scope Management)	(*PMBOK:* Project Time Management)	(*PMBOK:* Project Cost Management)	(*PMBOK:* Project Human Resource Management)	(*PMBOK:* Project Integration Management)
Principal Tools and Techniques	• Work Breakdown Structure • Work Breakdown Structure Dictionary • Project Scope Definition	• Statement of Work • Work Package • Task Authorization	• Milestone Schedule • Bar Graph Schedule (Gantt Chart) • Networks (i.e. PERT, CPM, GERT)	• Resource Plan • Manpower Plan • Budget • Cost Account	• Task Matrix • Task Roster • Project Charter	• Program Evaluation and Review Techniques (PERT) & others • Earned Value System • Variance Analysis • Performance Measurement System.

six of the knowledge areas defined by PMBOK,[2] as shown in Table 6.3. However, a great deal of overlap exists among the various categories and among the application areas of each tool, which provides a broader and more seamless coverage of these tools across the project life cycle. The most common of these project management tools are described below. For the benefit of clarity, the following text organizes the large array of tools into four groups: (1) tools for framing the project, (2) tools for defining the work, (3) tools for defining and managing time and resources, and (4) tools for defining the project organization and its people and structure.

6.4   FRAMING THE PROJECT

Good project management tools are useful across a wide spectrum of the project life cycle. Yet, some tools are especially crucial in framing the overall project at the beginning of a new project, and for getting a quick overview of its scope, status or performance during any point in the project life cycle. These tools are (1) work breakdown structure, (2) statement of scope, (3) milestone schedule, (4) project budget and (5) project charter. They provide the backbone of any project management system. This section discusses only the work breakdown structure. While the other four tools are referenced here as important tools for framing the overall project, they will be discussed in subsequent sections that deal specifically with the project work, timing, resources and organization, respectively.

[2] PMBOK is the *Project Management Body of Knowledge*, published by the Project Management institute (2000).

6.4.1   The Work Breakdown Structure

The work breakdown structure (WBS), sometimes called the *project breakdown structure* (PBS), is a hierarchical family tree of project elements. It is a delivery-oriented grouping of project subsystems and task elements that define the various hardware, software, and service components of a specified project or program. An example is shown in Figure 6.1 for part of an engineering development for a laptop computer. The work breakdown structure (WBS) provides the framework for dividing the total program into manageable subsystems and tasks. It must include all the tasks necessary to conceptualize, design, fabricate, test, and eventually deliver the project as defined under the project scope. The work breakdown structure provides

Table 6.4   WBS Index based on the breakdown shown in Figure 6.1

WBS Level		WBS #	WBS Task Element
0		—	Laptop Computer NPD
I		1	Design/Development
	II	1.1	CPU
	II	1.2	Memory
	II	1.3	Control
	II	1.4	Display
	II	1.5	Keyboard
	II	1.6	Power
	II	1.7	Mechanical
I		2	Production
	II	2.1	Prototype
	II	2.2	Volume
	II	2.3	Logistics
I		3	Marketing
	II	3.1	Task A
	II	3.2	Task B
	II	3.3	Task C
	II	3.4	Task D
	II	3.5	Task E
	II	3.6	Task F
I		4	Field Support
	II	4.1	Task A
	II	4.2	Task B
	II	4.3	Task C
	II	4.4	Task D
	II	4.5	Task E
	II	4.6	Task F
I		5	Project Management
	II	5.1	Reports
	II	5.2	P-Tracking
	II	5.3	P-Reviews
	II	5.4	C-I/F
	II	5.5	Travel
	II	5.6	Websites

the framework for all program planning, budgeting, and scheduling. It becomes the cost model for project budgeting and the framework for the project scope and integration. The WBS is also the central reference point for the project. All other program management tools, such as schedules, task descriptions, and budgets, should clearly correspond to the work breakdown structure.

The Level of Detail Depends on the Project Complexity. Obviously, larger and more complex projects require more depth and detail for their work breakdown. As a guideline, the total project should not be broken down further than the individual task contributor level. Of course, the individual responsible for the task can develop a mini-WBS for his or her own guidance of the work. However, in the interest of saving overhead and clutter, this additional detail does not have to be formally integrated with the overall project plan. Another consideration is the number of levels presented on a single sheet of paper or electronic screen. Normally, three levels of breakdown, as shown in Figure 6.1, is the maximum detail that should be displayed in a single cluster. Anything more detailed clutters the chart and destroys its flow and simplicity. As an example, the work breakdown structure in Figure 6.1 provides sufficient detail to the program manager for integrating the new laptop development. As an example, the program office (level zero) is not really concerned about the specifics of designing the SDRAM unit and integrating it with the computer system. However, someone is. The task leader for the SDRAM subsystem must develop his or her own project plan, starting with a work breakdown structure for the SDRAM subsystem which dovetails into the overall product development plan. This leads to the *modular concept of work breakdown.* That is, specific elements of work are subdivided on a separate sheet (module) into smaller components, until the level of detail, scope, and complexity is sufficiently defined for its proper project execution.

Numbering the Tasks. In order to reference specific tasks consistently throughout all planning and control documents, a decimal system is commonly used. The system is illustrated by listing the Level I and II tasks from the WBS of Figure 6.1, in Table 6.4. Such a listing is called a *WBS index.* If in addition to the index each element is described with a brief statement of work to be performed, the listing in Table 6.4 becomes a *WBS directory.* Lower levels can be included by breaking the decimal system down further. Alternatively, a modular approach to the WBS directory can be chosen by selecting specific elements and providing a detailed work breakdown just for those elements of interest in a separate list.

Grouping of Program Elements. This is a critical step. If the WBS has been structured properly, the various tasks integrate logically along the hierarchical lines of the work breakdown structure. Therefore, the work breakdown structure

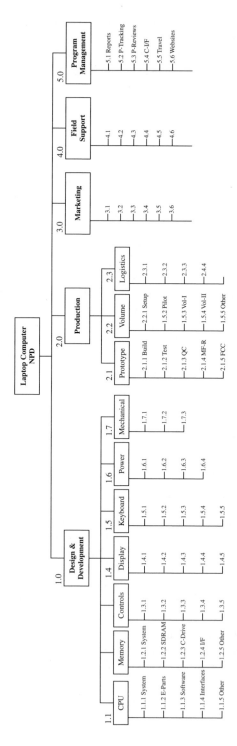

Figure 6.1 Work breakdown structure (WBS) for a laptop computer development project.

should be arranged to satisfy two criteria: (1) it should flow according to the planned integration of the program, and (2) the tasks should be grouped to minimize cross-functional involvement for each subset of activities. Accordingly, the work breakdown structure shown in Figure 6.1 would be most appropriate if the laptop company has organizational resource groups responsible for the activities shown at level I of the WBS, such as *Design & Development, Production, Marketing,* and *Field Services.* A different structure might be advisable if the company is structured along *product group lines* for different laptop applications. In this case, the Level I integration is probably most effectively performed at the product group level. However, it should be emphasized that the work breakdown structure is not an organizational chart. Attempts to force congruency often result in a confused and suboptimal WBS.

Regardless of how carefully the work breakdown structure has been fine-tuned to accommodate the existing organization, specific responsibility and authority relationships must be defined with a task matrix and the statement of work, which are described in the next sections of this chapter. Finally, it should be noted that a specific work breakdown structure may be given by the customer, forcing the program office to organize, integrate, and manage the project along a pre-established structure.

The Work Breakdown Structure as Cost Model. Because the WBS breaks down the overall program into its subsystems, tasks, and deliverable items, it provides an effective and convenient model for estimating the program cost at various task levels. It is equally useful for top-down parametric estimating as well as for bottom-up estimating. As a cost model, the work breakdown structure further provides a management tool for defining the cost drivers, supporting budgetary estimates and economic feasibility studies.

In summary, *the work breakdown structure is the backbone of the project.* It has benefits for both the customer and the contractor because it:

- Defines program building blocks
- Breaks down the program complexity
- Allows further expansion via modules
- Resembles a cost model of the program
- Provides a basis for all program planning and controlling: (1) specifications, (2) responsibility, (3) budget, and (4) schedule

6.5 DEFINING THE WORK

A wide variety of tools and formats exist for defining, communicating and tracking the project work. These tools must be fully integrated with the overall project management system in order to manage the project according to its performance parameters of work quality, timing, and resource utilization. The following tools are most commonly used for defining the project work dimensions: (1) scope statement, (2) statement of work, (3) work authorization, (4) work package, (5) specifications, and (6) deliverables. A brief description of each of these management tools is provided next.

6.5.1 Scope Statement

The scope statement is often the *top-level document for articulating the overall project goals, objectives, and mission-critical parameters*, such as overall timing and resource constraints. The scope statement creates the big picture for the project. It defines the project baseline for all other documentation and management actions. To ensure conciseness and clarity of this top-level document, the scope statement should *not exceed one page*. Typical *categories* within the scope statement include (1) project name and sponsor/customer data, (2) project mission and key objectives, (3) statement of work to be performed, (4) key deliverables, (5) key milestones, (6) key interfaces, and (7) key resources and constraints.

6.5.2 Statement of Work

The statement of work (SOW) is similar in structure and purpose to the scope statement. Its specific focus is on the baseline description of the actual work to be performed. Together with the specification, it forms the contractual basis for most technical projects and programs. While the statement of work can be used to define the project work top-down, it is also a useful format for defining and communicating work at any other project level or work module. The SOW is usually related to a particular work package or task elements in the work breakdown structure, describing precisely what is to be accomplished. Typically, the statement of work includes: (1) The project name and definition of the task module with reference to the corresponding work breakdown structure element, (2) a description of the task, (3) the results and deliverable items to be produced, such as the system concept, hardware, software, tests, documentation, and training, (4) references to specifications, standards, directives, and other documents, and (5) all inputs required from and to other tasks. Top-down, the statement of work is often a key document as part of a customer's request for proposal. Similar to the scope statement, the development of the SOW, especially for large programs, is a project by itself. It is a major effort which requires a considerable amount of interfacing with all stakeholders from the user, customer, and supporting organizations.

6.5.3 Specifications

Specifications (specs) describe the metrics of the project elements to be delivered. Specs form the baseline for developing, producing, and controlling the technical part of the project or program. Good specifications relate to the work breakdown structure. They describe the desired characteristics of the various subsystems in a modular fashion. Depending on the size and complexity of the project, specifications come in different dimensions and modules, such as specification for the overall system, subsystems, components, hardware, software, tests, manufacturing, and quality. In order to define and verify the end product, specifications should be developed with focus on measurability. However, there should also be flexibility to accommodate changes that inevitably evolve, particularly during a development program.

6.5.4 Work Package

A work package is a subset of an overall project. Work packages relate to specific elements of the work breakdown. Referring to the WBS of Figure 6.1, the power module (WBS # 1.6), the prototype module (WBS# 2.1), and the production module (WBS# 2.0) are typical examples of task-sets that could be defined as work packages. In order to provide a useful framework for project management, the work package must define (1) the work to be performed, including reference to the corresponding statements of work and specifications, (2) the responsible organization or individual, (3) the resource requirements, and (4) the schedule. The *task authorization* is the proper tool for summarizing and communicating work package data to the project team.

6.5.5 Task Authorization

All work must be properly defined and authorized by the project manager. This also applies to subcontracted items. The task authorization is a convenient way to summarize the requirements, as well as the budget and schedule constraints, for a particular project subsystem, which is often a work package. As shown in Figure 6.2 the format of a task authorization provides for summary and reference data in four major categories: (1) responsible individual organization, (2) schedule, (3) budget, and (4) work statement. A one-sheet task authorization form is recommended, regardless of the task magnitude. The form is intended as a summary of key data, referencing pertinent documents for details specifications, statement of work, deliverable items, and quality standards.

The task authorization is the written contract between the project manager and the task leader or the performing organization in general. To be meaningful, the task authorization must have been developed together with the key personnel who will perform the work. Moreover, an agreement on the task feasibility, schedule, and budget must exist within the team to have any basis for team commitment to the established objectives.

6.5.6 Deliverables

Deliverables are specific outputs or results from the project activity, such as plans, prototypes, documentations, software, decisions, and approvals. They are tangible, measurable accomplishments, usually associated with a specific milestone. Deliverables can be "hard" or "soft," but they must be "verifiable." An FCC license, a budget decision, and a test sign-off are very specific deliverables that can help in tracking and validating project progression and potential problems. Deliverables are

Task Authorization

Program Name: _____   Program Manager: _____   Date: _____

Customer/Sponsor: _____   Cost Center: _____   Revision: _____

User: _____   Initial Program Value: _____

Task Name: _____   Task Leader: _____   Task Org/Phone: _____
WBS Reference: _____   Resource Manager: _____

1. Overview	Sub-Task Description	Responsible Individual	Cost Account	Budget	Completion Date

2. Schedule	Key Milestones	Completion Date		Comments

3. Budget	Total Approved Budget: _____ Contractor/Vendor Budget: _____	Expenditure Profile: Date: _____ $: _____ Date: _____ $: _____ Date: _____ $: _____ Date: _____ $: _____

4. Statement of Work (SOW)	Brief description of work to be performed under the Task Authorization, including key deliverables (give references to applicable documents such as contract, specifications and project plan):

Figure 6.2   Task authorization form.

also useful measures in determining progress payments, and in providing visibility and recognition for team accomplishments. Deliverable items should be defined for each major milestone. The responsibility for defining deliverable items rests with the project leader. Close cooperation with the performing organizations is necessary in order to ensure meaningful deliverables with measurable parameters. A *list of key deliverables* is often part of a project summary plan or contract document, which can provide useful focus for detailed project planning, tracking, and management.

6.6   MANAGING TIME AND RESOURCES

Time and resources are the two major constraints of the project. They define the project boundaries and directly affect its scope and quality. The old saying "any project is possible given unlimited time and money" reminds us of the luxury that we don't have in today's business environment. Most projects are end-date-driven and resource limited. Defining and effectively documenting the time and resource requirements, consistent with the project scope and its objectives, is one of the most challenging responsibilities of the project leader. Managers have available today a wide variety of tools and techniques for dealing with time and resource issues. These tools must be fully integrated with the overall project management system in order to manage the project within its set of performance parameters of work quality, timing, and resource utilization. The following tools are most commonly used for defining, communicating, tracking, and controlling project time and resource parameters: (1) milestone chart, (2) bar graph, (3) network diagram/technique, (4) budget, and (5) cost account. A brief description of each of these management tools is provided next, grouped into two categories: (a) schedules and networks and (b) budgets and cost accounts.

6.6.1   Schedules and Networks

Schedules are the cornerstones in any project planning and control system. They present a time-phased picture of the activities to be performed and highlight the major milestones to be tracked throughout the program. Although schedules come in many forms and levels of detail, they should be related to a master schedule and their activity structure should be consistent with that of other project planning and control documents, particularly the work breakdown structure. For larger programs, a modular arrangement is a necessity to avoid cluttering, as suggested below for the laptop project discussed earlier

Level-I:	Master Project Schedule			
Level-II:	>Design Schedule	>Production Schedule	>Marketing Schedule	>Field Supt Schedule
Level-III:	CPU Schedules	Prototype Schedule	Mkt Research Schedule	Parts Schedule
	Memory Schedules	Volume Pro Schedule	Mkt Planning Schedule	Service Schedule
	Control Schedules	Logistics Schedule	Distribution Schedule	Training
	Display Schedules		Promotion Schedule	Schedule
	Keyboard Schedules			Support
				Schedule
	Power Schedules			Remote
	Mechanical Schedules			Diagnostics
				Schedule

Schedules are working tools for project/program planning, evaluation, and control. They are developed via many iterations with project team members and the sponsor. Schedules should remain dynamic throughout the program life cycle. Every program has unique management requirements. The most comprehensive schedule is not necessarily the best choice for all programs. Selecting the right schedule is important. There are three principal schedule types that are most commonly used in project management: milestone charts, Gantt charts (bar graphs), and networks.

6.6.2   Milestone Chart

A good way to start any schedule development is to define the key milestones for the work to be performed. Once agreed upon, the milestone chart becomes the backbone for the master schedule and subsequent subsystem schedules.

A key milestone is defined as an important event in the project life cycle, such as the start of a new project phase, a status review, a test, or the first shipment of a deliverable. Ideally, the completion of a key milestone should be easily verifiable. In reality, however, most milestones, no matter how crucial, are not easily verifiable. "System design completed," "first article test," and "final design review" are examples of typical key milestones that must be defined in specific detail to be measurable and useful for subsequent project control.

Key milestones should be defined for all major project phases prior to start-up. The type and number of these milestones must be carefully determined to ensure meaningful tracking of the project/program. If the milestones are spread too far apart, continuity problems in program tracking and control can arise. On the other hand, too many milestones can result in unnecessary busywork, confusion, inappropriate controls, and increased overhead cost. As a guideline *for* multiyear programs, four key milestones per year seem to provide sufficient inputs for detailed project tracking without overburdening the system.

The program office typically has the responsibility for defining key milestones, in close cooperation with the customer and the supporting resource groups.

Selecting the right *type* of milestone is critical. Every key milestone should represent a checkpoint for the completion of a cluster of activities. Ideally, major milestones are located at strategic time points of a project, encompassing a significant program segment with well-defined boundaries. Examples are listed below:

- Project kickoff
- Requirement analysis complete
- Preliminary design review
- Critical design review
- Prototype fabricated
- Integration and testing
- Value engineering review
- Start volume production
- Promotional program defined
- First shipment
- Customer acceptance test complete

6.6.3 The Bar Graph

The most widely used management tool for project schedule planning and control is the bar graph. Its development dates back to the work by Henry L. Gantt during World War I, which is the reason why bar graphs are often referred to as Gantt charts. Figure 6.3 illustrates the basic features of the bar graph by showing a partial master schedule for the minicomputer development of the work breakdown structure of Figure 6.1. The tasks on the left-hand side of the schedule should correspond directly

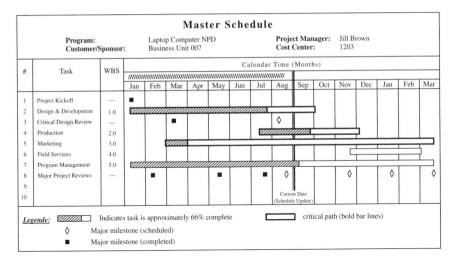

Figure 6.3   Bar graph schedule (Gantt chart) for laptop computer development project.

to the work breakdown structure and its numbering system. In fact, there should be no tasks on the schedule which cannot be found in the work breakdown structure.

If it becomes necessary to introduce a new task on the schedule, the work breakdown structure should be revised accordingly, which is part of the iterative nature of project planning. However, it is common practice and appropriate to list milestones such as project kickoff and critical design review on the bar graph schedule along with the tasks.

Bar graphs are simple to generate and easy to understand. They show the schedule start and finish of the tasks to be managed. Also, bar graphs can be modified to indicate project status, most critical activity, etc. The hatched areas in the example of Figure 6.3 indicate the approximate *percentile of project completion*, while the bold bar lines indicate that task items 2 (design), 4 (production), and 5 (marketing) are *most critical,* that is, they are on the longest time path through the project (see also next section on network techniques).

Bar graphs can be further modified to show budget status by adding a column that lists planned and actual expenditures for each new task. Many variations of the original bar graph have been developed to provide more detailed information to project managers. One commonly used method is to replace the bars with lines and triangular markers at the end-points. By using different symbols at the end-points, one can indicate the original time line and its revisions and thus trace schedule changes through the project life cycle.

The problem with these additional features is that they take away from the clarity and simplicity of the original bar graph, and often cause confusion interpreting the data. However, in many cases the additional information helps in communicating, project status reporting, and subsequent control and is considered of benefit in spite of the "fog factor" it carries.

The major limitation of the bar graph schedule is its inability to show task interdependence and time-resource trade-offs. Network techniques, which usually work together with computer data processing, have been developed especially for larger projects. These are powerful but also expensive techniques, which help project managers to plan, track, and control their larger, more complex projects effectively.

6.6.4 Network Techniques

Several techniques evolved in the late 1950s to support planning and tracking of projects with large numbers of interdependent activities. Best known today are PERT (Program Evaluation and Review Technique) and CPM (Critical Path Method). PERT was developed by the U.S. Navy in the late 1950s to aid in the management of the *Polaris* missile program, while CPM was jointly developed by DuPont and Remington Rand, also in the 1950s. CPM originally had an additional feature in comparison to PERT: it could track resource requirements. Originally, each technique had its own unique features, but today's commercially available project tracking software combines both features, often referred to as PERT/CPM. The concept of PERT/CPM and its features are discussed as part of an example derived from our laptop development. The network diagram shown in Figure 6.4 represents the flow of major activities, *similar* to our laptop development (cf. WBS of Figure 6.1 and bar graph schedule of Figure 6.3).

It is clear that the network diagram is more powerful than the bar graph in describing the flow of project phases through time and showing task interdependencies. The network provides a dynamic picture of the events and activities and their interrelationships. Yet, this is only the tip of its capability. The main value of PERT/CPM is in its ability to track the timing and cost parameters of the project, and to draw specific conclusions regarding project status and performance.

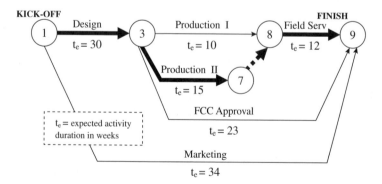

Figure 6.4   Network diagram for laptop development project similar to the WBS in Figure 6.1 and the schedule in Figure 6.3.

6.6.4.1   PERT/CPM Example

PERT/CPM is an excellent system for keeping track of all activities, especially in large projects where there are thousands of interdependent tasks. Moreover, PERT/CPM is a powerful tool for quick impact assessment of what-if scenarios and contingencies during project execution. To explain the principal quantitative methods and features of PERT /CPM we use an example, similar to the laptop development discussed earlier (see Figures 6.1 and 6.3). *The methodology* for setting up the PERT or CPM system is as follows:

1. *Define the Work.* The work breakdown structure, such as shown in Figure 6.1, usually provides a good framework for delineating the work or project subsystems and for deciding the level of detail for the PERT/CPM data base. Let us assume that we want to prepare the network on level I of the work breakdown. We choose the activities that should be tracked with the PERT/CPM system and list them in the PERT-table (left column) of Table 6.5.

2. *Define the Resources and Time Estimates.* At the chosen level of detail (i.e. WBS level I), we have to list all the activities and develop the labor budget (e.g. in staff-hours, staff-weeks, or staff-months). Then, based on the available personnel, we have to calculate a time estimate, t_e, for each activity duration and record the time estimate in column five). Many of the commercially available PERT software packages allow the user to calculate an expected time duration for each activity, based on the weighted average of three estimates: *most optimistic, most pessimistic,* and *most likely.* For simplicity of the example we work here only with *one* time estimate, t_e. I also like to point out that using a range of estimates, while providing data for interesting statistics, doesn't make the resource estimate any more accurate, nor does it help in getting commitment to a realistic or stretch target date.

3. *Establish the Skeleton Diagram (Network).* The activities from Table 6.5 must be organized manually into a skeleton diagram or network to reflect

Table 6.5   Summary of PERT Calculations for PERT/CPM Example

Activity (n)	WBS Reference (#)	Resource Requirement (staff-weeks)	Available Personnel (P)	Expected Duration (t_e weeks)	Preceding Event (PE)	Succeeding Event (SE)	Earliest Start (ES)	Earliest Finish (EF)	Latest Start (LS)	Latest Finish (LF)	Total Slack (TS)	Free Slack (FS)
Design	1.0	2000	100	30	1	3	1	30	-4	25	-5	0
Production-I	2.0	500	60	10	3	8	31	40	31	40	0	+5
Production-II	2.2	250	20	15	3	7	31	45	26	40	-5	0
FCC Approval	3.0	100	5	23	3	9	31	53	30	52	-1	+4
Marketing	4.0	100	5	34	1	9	1	34	19	52	+18	+23
Field Services	6.0	50	6	12	8	9	46	57	41	52	-5	0
Dummy	-	0	0	0	7	8	45	45	40	40	-5	0

their interdependence and resource constraints. Structuring the network diagram is an art that requires a great deal of work process knowledge and judgment, and the unified input of all resource groups across the organization. The result is a PERT/CPM network diagram such as shown in Figure 6.4. Please note that in our example the circles are *events* and the lines are *activities*. Therefore, this is an Activity-on-Arrow (AOA) diagram. While this is the most common form used in industry, PERT/CPM software programs often use an Activity-on-Node connotation. The skeleton diagram becomes the basis for setting up the PERT/CPM table.

Important Rule. When using AOA, all events must be uniquely numbered, but the numbers do not have to be sequential. Each activity must be uniquely defined by a *preceding* and a *succeeding event.* For example, the *production-I* activity is uniquely defined by events 3 and 8. In order to define the production-II activity (executed in parallel to production-I) uniquely, an additional *event 7* must be inserted, resulting in the dummy activity 7-8, which is shown as a dashed line in Figure 6.4.

4. *Define Schedule Constraints.* Let us assume that the project kickoff cannot take place before January 1 and the total project should be completed by December 31. Therefore, the earliest start (ES) = week 1, and the latest finish (LF) = week 52. These are the schedule limitations or constraints that must be specified, together with the start and finish events (or event numbers).

Table 6.6   Formulas Used in PERT Analysis (Table 6.5)

Legend:		*Formulas:*
ES	= Earliest Start	$ES = EF_{n-1} + 1$
EF	= Earliest Finish	$EF_n = ES_n + t_e - 1$
LS	= Latest Start	$LS_n = LF_n - t_e + 1$
LF	= Latest Finish	$LF_n = LS_{n=1} - 1$
TS	= Total Slack	$TS_n = LF - EF = LS - ES$
FS	= Free Slack	$FS_n = TS_n - TS_{MCP}$
CP	= Critical Path	
MCP	= Most Critical Path, this is the longest network path	
PE	= Preceding event	
SE	= Succeeding event	
P	= Number of full-time personnel available	
SW	= Work Effort in staff-weeks	
TS_{MCP}	= Total slack of activities on the most critical (longest) path in the network	Within the string of networked activities, n – 1 represents the preceding activity, and n + 1 the
t_e	= Expected duration (in weeks)	succeeding activity.

5. *Generate Computer Inputs and Process Data.* The computer must be instructed regarding (1) how the network is constructed and (2) the time duration for each activity. The network topology is defined by the preceding and succeeding event numbers that tell the computer which activities are executed sequentially and which are in parallel. Technically, the PERT computer software needs only data from columns 5, 6, and 7 to calculate all PERT/CPM results. However, to generate more user-friendly outputs, such as shown in Table 6.5, the first seven columns of Table 6.5 should be uploaded to the PERT/CPM software. The computer can now generate a PERT-table, as shown in Table 6.5, based on formulas and calculations summarized in Table 6.6.

6. *Interpretation of Results.* Looking at the earliest start (ES) and earliest finish (EF) figures will identify the specific week when we plan to start the project and will actually finish. The latest start (LS) and latest finish (LF) figures identify the specific week when we *should* start the project in order to finish as we want to by the end of week 52. Specifically, the following conclusions can be drawn from the computer-generated Table 6.5:

PROJECT DURATION: 57 weeks. We find this result by examining the earliest finishing (EF) times. The largest number (week 57) indicates that the field services activity is the last one to finish at the end of week 57. This is 5 weeks later than we want to finish the program.

PROJECT START & FINISH: If we start the project at the beginning of week 1 (as planned), we will finish at the end of week 57 (as discussed above). However, if we want to finish at the end of week 52 (planned), we will have to start the project 5 weeks earlier, at week −4 (please note that "week 0" is the week prior to "week −1)." This often causes confusion for manual calculations, but is necessary to make the formulas consistent and workable for both positive and negative dates.

TOTAL SLACK (TS): The TS-column indicates the amount of *extra time* we have for each activity, before the project exceeds the schedule constraint (becomes late). For example, TS = −5 indicates that the design activity is already 5 weeks late, while production-I is just on time (TS = 0) and marketing has 18 weeks of extra time on hand.

FREE SLACK (FS): The FS-column (also called *float*) indicates the amount of *extra time* we have for each activity, before it affects the actual project finish date. For example, let's take a look at the FCC approval activity. While this activity is actually 1 week late, according to the desired project finish time, the field service activity (that enters the same event) finishes even later, exactly 5 weeks late (TS = −5). Therefore, we could spend an additional 4 weeks on the FCC approval before it impacts the overall project schedule.

CRITICAL PATH (CP): As a common definition, any activity with a *total slack of less than zero* is *critical*. That is, activities on the critical path will directly impact the planned finish time of the project, if they experience any delay. Without any other critical path information given, this definition applies. However, project managers can set their own level of criticality, such as TS < 3 weeks.

MOST CRITICAL PATH (MCP): This is the longest activity path through the network. We can quickly identify the MCP by examining the total slack (TS) or free slack (FS) columns. Under total slack, all activities with the lowest TS (e.g., most negative) are on the most critical path. Alternatively, all activities with *zero free slack* are on the most critical path. By definition, any delay of activities on the most critical path will directly delay the actual finish time of the project.

6.6.4.2   The Benefits of PERT/CPM

The full benefits of PERT/CPM and other network techniques are only realized on larger and more enterprise-integrated programs that can also utilize the PERT/CPM system for overall contract management, customer reporting, procurement, and contractor management. In our small example, the information gleaned from the PERT-table might not look like a big deal and help to project managers. However, when we have to track thousands of activities, this type of bookkeeping is very helpful and leads to other types of automatic recordings, checks, and balances.

The more powerful indicators are the slack figures. *Total slack* indicates the number of time units that we can spare and still meet the schedule constraints. In our .example, a TS of –5 for the design activity indicates that we will be 5 weeks behind schedule with this activity and any other activity in the same path, unless we shorten the expected duration of t_e. Therefore, scanning the TS column assists in identifying which activities need help and where we have the luxury of slack, and therefore could "borrow" resources. Any activity with *negative* total slack is termed a *critical activity*. The activities on the longest path, that is, those activities that are most negative or least positive. are called *most critical.*

Free slack indicates the amount of additional time we have available without impacting the schedule outcome. For example, the FCC activity will be late by one week, as indicated by TS = –1. However, the field service activity will be late by five weeks (TS = –5). Therefore, the FCC activity really has another 4 weeks before an impact on the delivery date of the overall project is being felt.

6.6.4.3   Computer-Supported Project Management

Sophisticated computer software, high-speed data lines, and Internet technology provide the tools and support systems for collecting, processing, storing, and distributing the information needed for effective project management. These systems are a vital part of the project management infrastructure, supporting the whole spectrum of project administration and project leadership. These systems have the capacity to integrate scope, resource, and timing information during the project planning stage, to aid in decision making, track project status, and perform in-depth analysis of project data, far beyond of what was possible only a few years ago.

Yet, in spite of these powerful and sophisticated techniques, it is left to the project manager and the key stakeholders to define the principal framework of the project and its organizational environment, including scope, work flow, team organization, and interfaces. In this context, the fundamental tools of modern project management, such as work breakdown structure, task matrix, bar graph, network diagram, resource budget, and statement of work, are indispensable. Computer, modern telecommunications, and Internet technology are essential to support project management and to provide relevant and timely information to all project stakeholders.

6.6.4.4 *Government Reporting Requirements*

Government contracts, such as those procured by the Department of Defense, the Department of Energy, and the National Aeronautics and Space Administration, require particular project planning and reporting formats for their larger programs. The requirements vary with contract type and size. However, contracts over $1 million are expected to have some formal reporting requirements specified by the contractor. The PERT-COST requirements imposed during the 1960s have been eased in favor of less costly methods. Currently nearly all engineering and technology projects contracted by the U.S. government require detailed scheduling, budgeting, and reporting, however. Many specialized systems have been devised by various government agencies. Most widely used is the performance measurement system (PMS), also known as cost/schedule control system criteria (C/SCSC). Although these systems are based on conventional tools and techniques, such as work breakdown structure, task matrix, network techniques, activity cost accounting, earned-value concepts, and variance analysis, their understanding and implementation at the operational level often require considerable effort, cost, and time. It is not unusual for a company with government contracts to maintain a special support department for establishing and administering these reporting requirements, and for training project personnel in the use of these techniques. Special publications are available from the procurement offices of government agencies, which describe the specific requirements and techniques in detail.[3]

6.6.5 Project Budgets and Cost Accounts

Project budgets are important management tools for defining resource requirements and profit expectations. Budgets are based on cost estimates. In many cases, the first basic cost estimate is needed to support a feasibility analysis, long before the project baseline is fully defined. Such early cost estimates are referred to as *budgetary estimates* or *rough order of magnitude (ROM) estimates*. Later on, during the project definition phase, when the project requirements are known in sufficient detail, a more accurate estimate (often called a *bottom-up estimate)* can be developed. Project budgets provide a critical input to management decisions on bidding, pricing, new product development, strategic business plans, and the project management process.

> *Cost-Estimating Philosophy.* In spite of the sophisticated tools and computer-aided models available today, estimating the resource requirements for a technology development remains an art, at least to some degree. The driving forces

---

[3] Listings of specific procurement related publications are available from the Government Printing Office, Washington, D.C., 20548.

toward a quality estimate are (1) knowledge of the project baseline, (2) historic cost availability, (3) competence of the estimator, and (4) willingness of the estimator to provide a realistic estimate. In addition, it is important to involve the whole project team in the cost estimate. Without such an involvement, it is difficult to develop an integrated resource plan, and to obtain commitment to any budget or cost objective or design-to-cost approach.

Who Should Perform the Estimate? Ultimately it is the project manager who is responsible to management for the cost estimate. However, many staff groups, such as cost engineering or cost accounting, can help to support the effort by developing the cost model and coordinating and integrating the many activities that come into play during a major cost estimating effort. The critical question is, Who should make the estimate? And at what level of detail should it be made? Much of the answer depends on the level of project detail that is available at the time of the estimate.

During the early stages of the project formation, when little detail is known, the estimate should be performed on a global level by a few senior people who, based on their experience with similar projects, can *judge* the effort involved. This is often referred to as *parametric estimating*. The best-qualified person to make the estimate may not be the task specialist most familiar with the work but someone with a broad perspective, who can judge multidisciplinary integration efforts as well as task redundancies, and, above all, is motivated to provide a realistic resource estimate. Breaking the overall project into its major subsystems, such as defined on Level I of the WBS, may serve as a sufficient cost model.

Later, a detailed cost estimate is needed as a basis for contract pricing, operating budgets, and cost control. The level of estimating detailed depends on the level at which the project cost is to be controlled. As a ground rule, the estimate should be performed at one level of greater detail below each cost account. The cost account becomes the level at which the cost is tracked and controlled.

Unit Measurements. Eventually, the question of unit measurement must be addressed. Budgets are usually established and controlled in dollars or some other local currency. This is convenient because it normalizes labor, material, travel, overhead, and other cost components. Furthermore, money becomes the ultimate measure of financial performance. However, when estimating the labor component of a technology or engineering project, it is advisable to estimate the actual effort in staff-hours, staff-weeks, and staff-months. Later on, a cost accounting specialist can translate these figures into dollars by properly considering labor rate, overhead, and cost escalators. The cost-estimating form provides a convenient format for compiling labor time units associated with their job classification and the distribution of effort over time periods. For larger projects, the aid of a computer for number crunching and formatting becomes essential. In our relatively simple laptop example of Figure 6.1, already over 75 elements of cost must be calculated and summarized. Each task might require the efforts of several people, with different labor rates and overheads.

Moreover, cost estimating often requires many iterations. Once the cost model has been set up on the computer, changes in the original estimate can be handled relatively quickly. An additional advantage of the computer is its ability to perform parametric comparisons, relative cost distributions, ranking of cost drivers, and other checks and balances. As a rule of thumb, computers should be used to prepare cost estimates for projects in excess of 10 cost accounts. However, once established, many companies find computer-aided estimating beneficial and economical for projects of all sizes. An even stronger argument for computer-aided cost estimating can be made for organizations that use project management software for developing their project plans and tracking their implementation. Most of these project-specific software packages collect and process resource information as an ongoing part of the project plan development.

Design-to-Cost Approaches. Budgetary constraints and design-to-cost are realities in today's world. The feasibility of a new project often hinges on its affordability. Therefore, the traditional process of conceptualizing a system based on customer requirements and then estimating its cost is often reversed. That is, the project baseline may have to be designed to a given budget. The challenge here is to come up with a design that is acceptable to the sponsor without exceeding the pre-established timing and target cost. An additional challenge exists if the project is a prep-production development that must be designed to meet a specified unit production cost. Such design-to-unit-production-cost (DTUPC) efforts are major undertakings, quite common for new product developments or specialty projects across all industries.

6.6.6 The Budgeting Process

The budgeting process is filled emotions. Each group tries to negotiate for enough money to perform the work comfortably, maybe even to fund some other developments or professional activities. At the same time, the project manager is resource-constrained from the sponsor to perform against a fixed budget. Moreover, project personnel may not see any incentive for a low cost estimate which often leads to inflated resource estimates that might kill projects already at the proposal stage. All of this adds up to great challenges for realistic, effective, and competitive budgeting. It is up to the project manager to make the project desirable to the team and to convince the people that getting the project started will depend on the team's ability to come up with an affordable budget. Second, the project manager should emphasize, and obtain upper management's endorsement, that team performance is not measured against arbitrary cost over- and underruns, but is measured by the team's (1) willingness and effort to design to established cost targets, (2) contribution to winning a contract, (3) innovative methods for cost savings, (4) ability to handle risks and contingencies, (5) support of overall team efforts, and (6) effort in supporting customer or sponsor relations. Moreover, field research shows that with a project manager who can foster a work environment that focuses on interesting, challenging work, recognition for achievements, and professional rewards, but downplays penalties for cost overruns, the team is willing to cost-estimate more aggressively and make an effort to define a project baseline that is within an affordable budget range.

Techniques for designing a baseline concept to a given resource constraint rely on the design-to-cost approach. That is, cost targets are established for all major subsystems, and cost-performance trade-offs are permitted at various phases of the systems design and project implementation. Both cost and performance parameters are frequently reviewed during the conceptual phases of the project, keeping management options open for renegotiating both cost and performance targets. *The key steps of a typical design-to-cost process are listed below:*

1. Define the target cost for the total project.
2. Develop the cost model, such as the work breakdown structure.
3. Establish the target budgets for labor, materials, travel, etc.
4. Allocate relative (percent) efforts to all project subsystems at level I of the work breakdown (the total project must add up to 100%), based on parametric estimates, such as experiences and historic cost data from similar previous projects.
5. Establish staff-hour target budgets for each subsystem, based on the percentage allocation of the total budget.
6. Distribute all target cost information to all project team members.
7. Ask functional resource managers ultimately responsible for each subsystem to estimate, together with their task teams, the resource needs and other costs. For the first round, the estimate should be performed at a high project level, such as work breakdown level I. It also should be performed top-down, based on past experiences and parametric comparisons. After the initial estimates begin to converge toward the target costs, more detailed estimates are recommended, which eventually will be one level below each cost account.
8. The estimators provide four sets of data: (1) best estimate of effort, (2) cost drivers, (3) cost-performance trade-off analysis, and (4) alternative baseline for given target budget.
9. Analyze cost-performance trade-offs. Select acceptable approaches for negotiations with the project sponsor.
10. Recalculate the relative (percent) effort for all subsystems and trade-offs.
11. Identify the cost drivers (subsystems) of the project. Try to work out alternative solutions. Consider the total project, its underlying concepts and subsystems and estimates. The functional resource personnel have to work as a team together with the project office and the sponsor or customer community.
12. Go back to step 6 and repeat the process until a satisfactory solution has been worked out to all parties.

6.6.7   Cost Accounts

The cost account is the lowest level at which program performance is measured and reported. It represents a specific task, which is identifiable on the work breakdown structure; it contains its relevant schedule, budget, task specifications, and responsible individual and organization. Budgets for each cost account are time-phased over the program life cycle. Especially for larger technology programs, cost accounts become the focal point for project budget planning, cost control, and reporting.

A simple cost- tracking mechanism is the conventional *time card*, or its electronic equivalent, that exists for record keeping in most organizations. If project personnel, or their supervisors, are required to record on these "time cards" the number of hours worked on each task, which is referenced by a cost account, then all cost information is available for data processing. Cost reports can be generated which feature various summaries for convenient cost tracking and ultimate cost control. Examples are listed below:

- Actual total project cost versus planned cost, both in weekly increments and cumulatively
- Variance reports
- Expenditures by task categories and project subsystems
- Listing of project personnel together with their task and time allocations.

If established within the proper organizational support system, cost accounts, together with their reporting systems, are powerful tools for the project manager in measuring and controlling project cost. The cost account is also the lowest level at which actual cost budgets, schedules, and technical performance measures can be integrated.

6.7 DEFINING THE PROJECT ORGANIZATION

People must understand where they fit in the organization. While the traditional organizational chart is a common and useful device to define the command channels of conventional superior-subordinate relations, additional tools are needed to describe the specific authorities, responsibilities, and interfaces associated with each key position within the project organization. Project charters, task rosters, task matrices, and job descriptions are common devices for delineating the roles of key project personnel and for communicating them to all team members. In addition, companies may use management directives to clarify definitions, organizational processes, and relations.

The project charter is a particularly useful and powerful tool for identifying and communicating the top-down mission, authority structure, and key parameters of a project (Anderson 2003). Eventually, these project charters include the names of key personnel, while at the early stages of a project definition the charter serves as a top-down planning document. It is also very useful in identifying and negotiation key team members.

Policies and procedures are yet another set of management tools for defining the inner working of a project organization. They provide operating guidelines for running the business in a proven, standard way. The depth of these documents can range from simply acknowledging project activities to detailed procedural guidelines which carry the project team through the complete project life cycle. Unless there are extenuating circumstances, project personnel must follow these policies and procedures. At the same time, these documents must be written broadly enough to leave operating leeway to management and to accommodate the variety of projects that run through the organization. Further, to ensure proper integration of these policies and procedures with the enterprise system, they should be tested and fine-tuned before being formally issued.

Project managers who find themselves without adequate guidelines often develop and issue their own procedures at the project level. This is an excellent way to communicate the established operating standards. Further, these task-level procedures often become "test cases" for more formal, higher-level directives.

6.7.1   Task Matrix and Task Roster

The task matrix is a simple but powerful tool to define responsibility relationships among the various program tasks and the performing organizations. The task matrix, also known as responsibility matrix or linear responsibility chart, evolved out of frustrations over conventional organization charts, which do not show relationships "within" and "between" organizational subsystems—that is, they do not show who is responsible for what at the task level. Equally important, the task matrix has a great amount of flexibility and does not restrain the system by emphasizing status and position, which is an awkward limitation of conventional organizational charts. The task matrix is derived from the work breakdown structure. Therefore, all task descriptions should correspond to the WBS. Figure 6.5 illustrates this management tool by defining the task responsibilities for the laptop development which we discussed earlier based on the work breakdown in Figure 6.1.

The task descriptions on the left-hand side of the matrix are derived directly from the work breakdown structure and should correspond exactly to its reference numbers and structured hierarchy as shown in the WBS of Figure 6.1. The right-hand side shows either the responsible organizations with their section heads or individual task managers. This structure should correspond to the actual performing organizations, which might be submerged within a functional, projectized, or matrix framework. The relationship between tasks and responsibilities is indicated by a symbol in the matrix. It is common to use P for *"prime responsibility"* and S for *"supporting responsibility."* However, quite elaborate schemes for identifying various degrees of responsibility and participation can be devised if needed.

During the program planning phase, the task matrix can be used effectively as a cost model for budgeting. Cost estimating data are collected by recording the working hours, dollar estimates or budgets needed to perform the task at the crossing points of the task and responsibility. Adding up the rows and columns provides a convenient summary of the estimated cost for all task and performing organizations.

In summary, the task matrix provides a single framework for planning and negotiating project assignments and resources throughout the enterprise, including support departments, subcontractors, and the customer. Furthermore, the task matrix provides

Task Matrix

Program:	Laptop Computer NPD		Project Manager:	Jill Brown		Legend:	P = Prime Responsibility
Customer/Sponsor:	Business Unit 007		Cost Center:	1203			S = Supporting Responsibility

Description	WBS Reference	R&D	System Design	Engg Section I	Engg Section II	Testing	Quality Assur.	Manufacturing	Marketing	Etc.
Design & Development	**1.0**		P	S	S		S		S	
CPU	1.1	S	S	P	S		S		S	
System Design	1.1.1	S		P	S	S	S			
E-Parts	1.1.2		S		P		S			
Software	1.1.3		P		S					
Interfaces	1.1.4	S	P	S	S			S		
Other	1.1.5			S	S	P				
Memory	1.2	P	S							
System Design	1.2.1	P	S		S					
SDRAM	1.2.2	P	S	S	S					
C-Drive	1.2.3	P	S	S	S					
I/F	1.2.4	P	S							
Other	1.2.5		P							
Controls	1.3	S		S	S	P				
Task A	1.3.1	S	S	S	P					
Task B	1.3.2	S	S		P		S			
Task C	1.3.3	S			P		S			
Task D	1.3.4	S		P			S			
Displays	1.4			S	P		S			
Task A	1.4.1			S	P		S	S	S	
Task B	1.4.2			S	P	S				
...	...									
...	...									
Production	**2.0**				S	S	S	P		
Prototype	2.1			S	S	S	S	P		
Volume Production	2.2			S	S	S	S	P		
Logistics	2.3			S		S	S	P		
Marketing	**3.0**		S		S				P	
Task A	3.1		S	S					P	
Task B	3.2		P			S			S	
Task C	3.3				S				P	
Task D	3.4				S				P	
...	...									
...	...									

Figure 6.5 Segment of task matrix based on work breakdown structure (WBS) in Figure 6.1.

excellent visibility of who is responsible for what, throughout the organizations. It also provides some information on organizational interfaces for integrating particular subtasks.

6.7.1.1 The Task Roster

An alternative to the task matrix is the *task roster*. Its clear and simple format makes it especially attractive for smaller projects. As shown in Figure 6.6, the task roster is a listing of the project team members, including their organizational affiliation and telephone number, next to their task responsibilities with reference to the WBS. Although the task roster is less sophisticated, it is often preferred over the task matrix because of its clarity and change flexibility. The task roster is, furthermore, a team-building tool. It summarizes in a simple format, usually on one sheet of paper, *who is responsible for what*. It recognizes the project contributors individually and as a team, regardless of their status and position within the company, and their duration on the project. More than any other management tool, the task roster fosters a sense of belonging, pride, ownership, and commitment to the project. Because of these motivational benefits, the task roster is often used in large projects in addition to the task matrix.

Task Roster
Laptop Computer NPD

Customer/Sponsor: Business Unit 007

Project Manager: Jill Brown
Cost Center: 1203 WBS Reference

Responsible Individual	Organization/ Telephone		Task	WBS Ref
Al	Syst Des	x364	Syst Des	1
Beth	End-I	x733	E-Parts	1.1.1
Charlie	Mfgg	x445	Vol Prod	2.2
Don	Quality Ass	x233	QC	2
Eric	Testing	x521	Prototype	2.1
Fran	Testingx	x633	Test+Integr	2.1
George	Eng-II	x375	Software Des	1.1.3
Helena	Eng-II	x387	Controls	1.3
Ian	Syst Des	x745	Sys D&D	1.2.1
Jeff	Mfgg	x488	Logistics	2.3
Karen	Marketing	x333	Prod Plan	3.1
...	...		...	...
...	...		...	...
...	...		...	...

Figure 6. 6 Task roster (based on laptop example).

6.8 USING PROJECT MANAGEMENT TOOLS PROPERLY

Fundamentally, project management tools are communication devices. They have been designed to *define* and *communicate* the requirements to all parties involved, to *measure* performance, and finally to *direct* and control the effort toward the pre-established requirements. Effective project management tools have the following characteristics:

- They assure measurability of all parameters used.
- They use standard formats for all tools throughout the organization.
- They use a uniform number system for project elements, consistent with all tools (e.g., a project subsystem number should be the same in the work breakdown structure as it is in the statement of work or cost account).
- They do not clutter the manager's tools with too much detail. They use modular concepts.
- They keep the manger's documents current.

The following suggestions should help in utilizing project management tools effectively and in building a high-performing project organization:

1. An agreed-upon program plan is absolutely essential. The plan defines the requirements in measurable steps and is the key to successful project performance. Measurability of technical status is crucial for engineering projects. If we cannot measure, we cannot control.

2. The use of standard formats for all tools is recommended. Nothing can be more confusing than looking at two different schedule or budget forms for the same project. It is the responsibility of the project manager to issue a standard set of forms if the organization does not have the planning process already proceduralized.

3. Number systems are often another area of unnecessary conflict and confusion. Many of the tools, such as schedules, work breakdown structures, and task matrices, label each task with a number. There is no need to label a particular task differently in the work breakdown structure than in the schedule, for example. All tasks should be easily traceable throughout the project plan.

4. Use modular concepts to break down the complexity of the plans. For example, rather than showing all activities in a schedule in detail, partition the schedule in to a master schedule, which provides just an overview, a subsystem A schedule, a subsystem B schedule, and so on. Or, alternately, use time phasing to break down the complexity, issuing a system design schedule, prototype design schedule, prototype fabrication schedule, and so on.

5. Keep your documents current. Nothing outlives its usefulness faster than an outdated document. Review your plans regularly and make revisions as needed. A document control system should be maintained by an assigned individual such as a secretary, who makes sure that agreed-upon changes get properly recorded and distributed.

6.9   A MODEL FOR PROJECT PERFORMANCE

Based on the preceding discussion, the issues affecting project performance can be analyzed in terms of eight principal categories summarized in Figure 6.7: (1) the *people* on the project team and its support organizations, (2) *leadership*, (3) *project tools and techniques*, and (4) *business processes* that power and support the project activities; these four categories are overlapping and intricately affected by (5) *the organizational infrastructure and support systems*, (6) *managerial support*, (7) *project complexity*, and (8) the *overall business environment*. These categories not only determine project performance, but also hold the DNA for the type of consulting services that are best suited for improving specific project management situations.

The model emphasizes the human side as critically important to successful project execution. The project team attitude, effort, and commitment all are influenced by managerial leadership and the work environment, which in turn influence team effectiveness and overall project performance. Research shows consistently and measurably that the strongest drivers toward project performance are *derived from the work itself*, including *personal interest, pride and satisfaction with the work,*

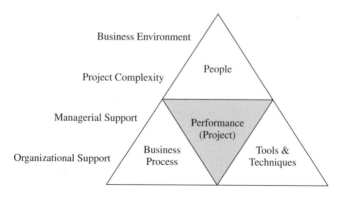

Figure 6.7 Influence on project performance.

professional work challenge, accomplishments and recognition (cf. Thamhain 1999, 2002, 2004). Other important influences include effective communications among team members and support units across organizational lines, good team spirit, mutual trust and respect, low interpersonal conflict, opportunities for career development and advancement, and, to some lesser degree, job security. All of these factors help in building a unified project team that can leverage the organizational strengths and competencies.

Other critical factors relate to the organizational structure and business process, including the technology transfer system, which by and large relies on the tools and techniques of modern project management. These tools are the backbone of any project management system. They must be carefully defined, developed, and integrated with the enterprise and its culture and value system.

6.10 SUMMARY OF KEY POINTS AND CONCLUSIONS

The key points that have been made in this chapter include:

- The management of technology and engineering projects is supported by a highly diverse set of sophisticated tools and techniques that need to be carefully fine-tuned and integrated with the enterprise and its culture.
- There is a built-in relationship among project management tools that helps to connect the vital project parameters of work quality, timing, and resources into an integrated tool kit for project planning, tracking, and control. However, to fully utilize these linkages, project management tools must be custom-fitted to the organizational environment and the specific projects to be managed.
- The work breakdown structure is the backbone of any project plan and its execution.

Project planning and tracking systems must be able to measure work quality, timing, and resources as an integrated set. For example, the project leader must be able to determine at any given time (1) the amount of work that has been completed, (2) the actual cost of the work completed against budgeted cost, and (3) the schedule of actual work completed against the original schedule.

- Deliverables are specific outputs or results of the project activity, useful for determining progress and for providing visibility and recognition for team accomplishments. Deliverables must be tangible and measurable, and should be associated with specific milestones.

- The human side is critical to project success. Team involvement in scope definition, resource estimating, and scheduling is crucial to the development of realistic project plans, team ownership, and commitment.

- Project plans are living documents that must be reviewed on a regular basis and updated to reflect changing requirements, objectives, and conditions in the project environment.

- The project information system provides an important tool set for collecting, processing, storing, and distributing the information needed for effective project management. Yet, it is the project manager, not the computer, who by skill, experience, and team leadership defines the principal framework of the project and its environment, including scope, work flow, team organization, and interfaces.

- Team resistance to the introduction of new project management tools is natural. It can be minimized by involving the people who will be affected by the new tools in the selection, testing, and fine-tuning of these tools.

- The shift to more sophisticated project management processes requires a radical departure from traditional management practices and leadership styles. The methods of communication, decision making, soliciting commitment, and risk sharing are continuously shifting away from a centralized, autocratic management style to a team-centered, more self-directed form of project leadership.

- The strongest drivers toward project performance are derived from the work itself, including personal interest, pride and satisfaction with the work, professional work challenge, accomplishments, and recognition.

6.11   CRITICAL THINKING: QUESTIONS FOR DISCUSSION

1. Why is the work bretctwn structure called "the backbone of the project plan"?
2. What are the role and significance of the WBS numbering system?
3. Why is the WBS called a "cost model"?
4. Describe the drivers toward realistic cost estimates.
5. Define a methodology for estimating the development cost for a major technology project.

6. Discuss the challenges of design-to-cost approaches, and make suggestions for building an environment conducive to DTC.

7. Compare the three popular scheduling tools (1) milestone chart, (2) Gantt chart, and (3) PERT/CPM network diagram regarding their strengths, weaknesses, and limitations.

8. What type of tool (or tools) would you suggest for summarizing a $100M project on one page?

9. Why is modularity within a project plan an important feature?

10. What benefits do you gain from establishing a project management information system? What costs?

11. How does modern information technology affect managerial power and leadership style?

12. How do you overcome team member resistance to accept or use a new project management tool, such as following a new design review process or using a new method for reporting progress or changes?

6.12 REFERENCES AND ADDITIONAL READINGS

Anderson, E. (2003) "Understanding your project organization's charter," *Project Management Journal,* Vol. 34, No. 4, pp. 4–11.

Badir, Y., Founou, F., Sticker, C. and Bourquin, V. (2003). "Management of global large-scale projects through a federation of multiple web-based workflow management systems," *Project Management Journal,* Vol. 34, No. 3, pp. 40–47.

Bahrami, H. (1992). "The emerging flexible organization: perspectives from Silicon Valley," *California Management Review,* Vol. 34, No. 4, pp. 33–52.

Bishop, S. K. (1999). "Cross-functional project teams in functionally aligned organizations," *Project Management Journal,* Vol. 30, No. 3, pp. 6–12.

Cleland, D. & Itreland, L. (2002). *Project Management: Strategic Design and Implementation,* New York: McGraw-Hill.

Deschamps, J. and Nayak, P. R. (1995). "Implementing world-class process," *Product Juggernauts* (Chapter 5, pp. 175–214). Cambridge, MA: HBS Press.

Gupta, A. and Govindarajan, V. (2000) "Knowledge management's social dimension," *Sloan Management Review,* Vol. 42, No. 1 (Fall), pp. 71–80.

Iansiti, M. & MacCormack, A. (1997). "Developing product on Internet time," *Harvard Business Review,* Vol. 75, No. 5, pp. 108–117.

Jaafari, A. (2003). "Project management in the age of complexity and change," *Project Management Journal,* Vol. 34, No. 2, pp. 13–25.

Kerzner, H. (2003). "Strategic planning for a project office," *Project Management Journal,* Vol. 34, No. 4, pp. 47–57.

Kerzner, H. (2003). *Project Management: A Systems Approach to Planning, Scheduling and Controlling,* New York: Wiley and Sons.

Kruglianskas, I. & Thamhain, H. (2000). "Managing technology-based projects in multinational environments," *IEEE Transactions on Engineering Management,* Vol. 47, No. 1, pp. 55–64.

Thamhain, Hans (1991). "Skill developments for project managers," *Project Management Journal,* Vol. 22, No. 3 (September), pp. 39–45.

Thamhain, H. (1994) "Designing project management systems for a radically changing world," *Project Management Journal,* Vol. 25, No. 4, pp. 6–7.

Thamhain, H. (1999) "Controlling projects according to plan," *Essentials of Project Control* (J. Pinto and J. Tailor, eds), Philadelphia: PMI Press.

Thamhain, H. J. (2002). "Criteria for effective leadership in technology-oriented project teams," Chapter 16 in *The Frontiers of Project Management Research* (Slevin, Cleland and Pinto, eds.), Newton Square, PA: Project Management Institute, pp. 259–270.

Thamhain, H. (2004). "Concurrent engineering for integrated product development," Chapter 19 in *Wiley Guide to Project Management,* (Morris & Pinto, editors), New York: Wiley and Sons.

Thamhain, H. (2004). "Leading technology-based project teams," *Engineering Management Journal,* Vol. 16, No. 2, pp. 42–51.

Thamhain, H. and Wilemon, D. (1999) "Building effective teams for complex project environments," *Technology Management*, Vol. 5, No. 2, pp. 203–212.

Thomas, J., Delisle, C., Jugdev, K. and Buckle, P. (2001). "Selling project management to senior executives," *pmNetwork*, Vol. 15, No. 1, pp. 59–62.

7

MEASURING AND CONTROLLING THE WORK

PROJECT MANAGEMENT PRINCIPLES KEPT TRANSPORTATION SECURITY PROJECT ON TRACK

The United States Transportation Security Administration (TSA) had only three months and less than $20 million to build phase I of what is expected to be one of the nation's most critical command, control, and communications facilities, the Transportation Security Coordination Center (TSCC).

The TSCC is charged with gathering intelligence and recommending security risk mitigation strategies. Its daunting duties include monitoring the activities of security screeners and federal air marshals at 429 commercial airports nationwide. The new facility is also responsible for processing and integrating security information from other security systems that monitor truck and rail transport.

In addition to the TSCC facilities and their IT infrastructure, the contract called for strong security measures during project execution which stressed the "no-slack" around-the-clock schedule even further. "We had an extremely tight budget that could not be exceeded by a single penny and an end-date-driven schedule that had to be met," says TSCC Director Curt Powell. "One of the things we had to do was accelerate the cycle for government procurement and reduce approval times for permits and other sign-offs." TSA gives credit to the project leaders for their highly organized and disciplined approach to the project execution, which included a dedicated project management office

(PMO) for resource coordination and integration. In addition to modern project-tracking and control techniques, the project team used three project specialists from a consulting firm to oversee the master schedule, track the budget, and expedite procurements. "Without this specialized help we could never have done this in 97 days," says Powell.

While accurate tracking of project progress against the plan was extremely important so that the project office knew how things are moving along and to catch problems in their early stages, it was the dedication of the team and their commitment to meeting the established milestones that ultimately brought the project in on time and 3 percent under budget. "Once the people on the team understood the importance of this project and bought into their roles, everything was flowing. People reached out to solve problems and find innovative solutions," says Bill Eaton the project manager. "To give you an example, we had up to 300 people working simultaneously on different components of the project. Not all tasks can be planned for optimal integration ahead of time. But, by talking to different task groups, people figured out how to execute work packages in parallel. People would install IT equipment ahead of schedule while the facility group was still finishing up its part. It was wonderful to see the enthusiasm and dedication that this project team brought to the job," says Eaton. "Without it, we could not have done it!"

Source: Transportation Security Administration press releases published in 2004 at http://www.tsa.gov/public/display?theme=44 and Statement by Under Secretary Hutchinson before the U.S. House of Representatives, Committee on Appropriations, Subcommittee on Homeland Security, on March 11, 2004 (see http://appropriations.house.gov/_files/asahutchinsontestimony.pdf).

7.1 THE CHALLENGES OF MANAGERIAL CONTROL IN TECHNOLOGY

The challenges presented in the above scenario are typical for today's technology undertakings. End-date-driven schedules, resource limitation, changing requirements, and complex project situations are quite common together with pressures for innovation, quality, and implementation flexibility. It is also no coincidence that the human side was a critical factor in the TSA success story. Even the most sophisticated control process and the latest technology are no substitute for the unified buy-in and commitment of team personnel to a project plan and its mission objectives. Project plans and tracking systems are important prerequisites for managerial control. Yet, as shown in the lead-in scenario, without the cooperation, dedication, and innovative engagement of the team members, it is difficult, if not impossible, to maintain control, especially in complex and dynamic mission environments. Let's start by defining managerial control.

A Definition.   As a working definition, managerial control is the process of taking the necessary actions for implementing a plan and reaching its objectives, in spite of changing and often unpredictable conditions. This statement implies many assumptions. First, we must have a plan. Second, we must be able to measure current status and performance in order to determine possible deviations from our plan and to assess their impact. Third, we assume that the root cause of the plan deviation can actually be determined and corrective measures can be found. Finally, there is a whole spectrum of assumptions that relate to the organization and its culture, values, and people.

Table 7.1   Challenges of Managerial Control in High Technology

- Contingencies, risks, problems
- Ambiguous progress and performance measures
- Determining causes (not symptoms)
- Learning curve; uncertain estimates
- Changes, interference
- Constraints: time, budget, quality, features
- Complex organizational interactions
- Tough enterprise mandates
- Complex solutions
- Unfavorable perceptions: anxieties, fear, confusion, fairness
- Conflict, power, politics.

Why Is It So Difficult?   Few managers would argue with the need to control projects and missions according to established plans. Even fewer would think of it as a trivial matter. The challenge is to apply the available tools and techniques effectively. Even with the best preparation, detailed planning, and team involvement, managers find it difficult to control these projects to achieve desired results because of work complexities, changes, and contingencies. As shown in Table 7.1, the challenges of managerial control range across the whole spectrum of business operations. Moreover, these challenges are increasing with the complexity of the project and its business environment. Appendix A lists some of the more common parameters for characterizing projects and their organizational complexities.[1] These complexity factors include project size, duration, and complexity; team size and skill level; budget; number of subsystems to be integrated; technology advances; evolving solutions; risk factors; and external communities. Appendix A can be used as a tool to determine project complexity as a comparative measure.

Why Technology Projects Fail.   The challenges in Table 7.1 are also the factors that are often blamed for project failures as listed in the first part of Table 7.2.

---

[1]These parameters have been identified in a survey of project managers (Thamhain 2002) as indicators of project complexity. As a profile these parameters can also help in establishing a framework for classifying and comparing projects.

However, it is interesting to note that managers blame project performance problems and failures predominately on situations outside their sphere of control, such as scope changes, market shifts, and project support problems, while senior management points at the project leaders. Senior management argues that projects fail by and large because of insufficient planning, tracking, and control; poor communication: and weak leadership. Yet, a third perspective can be obtained from independent observations during field research, which trace the root causes of many of the project performance problems and failures to the broader issues and difficulties of understanding and communicating the complexities of the project, its applications, and the support environment. These field investigations produced a longer list of more subtle reasons for failures, including unrealistic expectations for scope, schedule, and budget, underfunding, unclear requirements, and weak sponsor commitment, as summarized in part three of Table 7.2.

Table 7.2   Top Reasons for Project Failure

Project Manager's Perspective:

• Too many changes and contingencies (scope, $, priorities, technology, market, sponsor ...)

• Lacking project support (inside and contractors)

• Suboptimal resources (competencies)

• Underestimating project complexity

• Cascading effects

• Interfering administrative processes and requirements

Senior Management Perspective:

• Inaccurate/insufficient planning

• Insufficient performance measurements

• Insufficient communication/escalation of problems

• Lacking change control

• Weak project leadership

Outsider Perspective (Field Research)

Difficulties in understanding/communicating complexities of project, its applications, and the support environment

• Unrealistic expectations (scope, $)

• Underfunding

• Underestimating complexities

• Unclear requirements

• Weak sponsor commitment

7.2 WHAT WE KNOW ABOUT MANAGERIAL CONTROLS IN COMPLEX WORK ENVIRONMENTS

The concepts and methods of managerial control have been known for a long time (Randolph and Posner 1988). Yet, the applications of managerial control have expanded continuously to respond to the ever-increasing challenges and complexities of our business environment. In addition, both competitive pressures to transform ideas into products quickly and economically and changes in our technological, social, political, and economic environment have forced companies to rethink their management concepts and business practices. As a result, new organizational forms and special processes, such as concurrent engineering, CAD, CAM, IRM, MRP and DFM/A,[2] intricate multicompany alliances, and highly complex forms of work integration across wide geographic and multinational regions emerged and gradually replaced traditional, more hierarchically structured organizations and their linear work processes. This shift also significantly affects management styles and the tools and techniques for orchestrating and controlling complex projects,[3] regardless of their product, process, or service orientation.

7.2.1 Evolution of Control Tools and Techniques

Specifically, in the past, management concepts were based predominately on linear models, typically exemplified by production lines, sequential product developments, scheduled services, and discovery-oriented R&D. Today's management has to operate in a much more dynamic and interactive way, involving complex sets of interrelations of nonlinear and often messy processes (Senge et al 2004). As a result, companies have moved toward more sophisticated tools and techniques for effectively managing multidisciplinary activities (Rigby 1995; Thamhain 1996c, 1999). These tools and techniques range from computer software for sophisticated schedule and budget tracking to intricate organizational process designs, such as concurrent engineering and Stage-Gate protocols (Thamhain, 1996a). Even conventional project management tools, such as schedules, budgets, and status reviews, are continuously upgraded and effectively integrated with modern information technology systems and overall business processes. During this evolution, the focus of project management has shifted from simple tracking of schedule and budget data to the integration of human factors and organizational interfaces into the project control formula, as illustrated in Figure 7.1. The new generation of project management tools is designed to deal more effectively with the new challenges and realities of today's business environment, which include highly complex sets of deliverables, as well as timing, environmental, social, political, regulatory, and technological factors. All of these changes not only increase the complexity of the project environment, but

---

[2]Acronyms: CAD, computer-aided design; CAM, computer-aided manufacturing; IRM, information resource management; MRP, manufacturing resource planning; and DFM/A, design for manufacture and assembly.

[3]Consistent with today's emphasis on teamwork, technology-intensive applications have a project focus, which is also reflected in this chapter. Of cause, the concepts discussed here are also applicable to other result-driven operations, missions, processes, and ventures.

also demand a more sophisticated management style, which relies strongly on group interaction, resource and power sharing, individual accountability, commitment, self-direction, and control. Consequently, many of today's technology projects and their integration rely to a considerable extent on member-generated performance norms and evaluations, rather than on hierarchical guidelines, policies, and procedures. While this paradigm shift is driven by changing organizational complexities, capabilities, demands, and cultures, it also leads to a radical departure from traditional management philosophy on organizational structure, motivation, leadership, and project control. As a result, traditional management tools, designed largely for top-down centralized command, control, and communication, are often no longer sufficient for managing projects effectively.

7.2.2   Control Tools and Techniques Available Today

Because of their importance to business performance, managerial controls have been in use for a long time. Their methods and practices have been studied and extensively discussed in the literature (Christensen 1994, Cleland 2004, Dinsmore 2005, Kerzner 2003, Meredith and Mantel 2000, Rigby 1995). It is not surprising, however, that field research confirms what managers have been saying for some time: the conventional tools and methods of project control by themselves seldom produce satisfactory results in today's work environment. Especially for more complex and technology-intensive projects, which require innovative solutions to challenging problems and flexible change-oriented implementation of the project plan (see Tables 7.1 and 7.2), the tools and techniques in use today have changed considerably over the last two decades.

As illustrated in Figure 7.1, up to the 1980s, project-oriented management focused, by and large, on tracking of schedule, resource data, and deliverables, without too much concern about true project performance, flexibility, and responsiveness to dynamically changing business environments. This tracking of basic project metrics

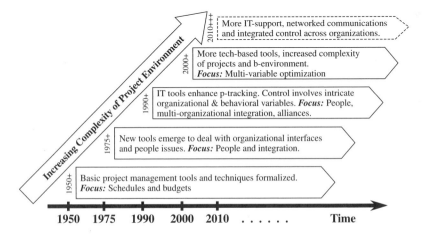

Figure 7.1   Evolution of managerial control tools and techniques.

is still an important part of project management today. However, the business environment is quite different from what it used to be. For one thing, new technologies, especially computers and communications, have radically changed the workplace and transformed our global economy more and more to one of service and knowledge work, with a high mobility of resources, skills, processes, and technology itself. New project management techniques have evolved that are often more integrated with the business process, offering more sophisticated capabilities for project tracking and control, in an environment that not only is different in culture, but also has to deal with a broad spectrum of contemporary challenges, such as time to market, accelerating technologies, innovation, resource limitations, technical complexities, social and ethical issues, operational dynamics, risk, and uncertainty. Faced with such contemporary measures of project performance as those summarized in Table 7.3, traditional methods of strict schedule and budget control are often useless and sometimes even counterproductive to overall project performance.

Table 7.3   Contemporary Measures of Project Performance

• Producing agreed-on results (deliverables) on time and within budget

• Innovative, value-added project implementation

• Time-to-market acceleration

• Technological advances and acceleration

• Capturing opportunities for performance enhancements and time and cost savings

• Quality concerns and enhancements of deliverables

• Ability to work toward objectives and target results (e.g., evolving solutions, R&D, etc.)

• Ability to deal with risks and contingencies

• Ability to deal with conflict and politics

• Reaching stretch goals

• Recognizing changing environment and proactive response

• Recognizing new market opportunities or technological applications

• Flexibility toward customer requirements (i.e., change orientation)

• Recognizing and capturing follow-on business

• Minimizing organizational disruptions and interferences

• Creating synergism among enterprise units (i.e., leveraging resources)

• Aligning project work with enterprise strategy

• Concern for natural environment

• Capturing experiences and organizational learning

• Training of personnel

7.2.3 How Do Companies Cope?

Driven by business pressures and advances in information technology, many companies have invested heavily in *new management tools and techniques*, promising more effective alternatives and enhancements to traditional forms of project control. Table 7.4 summarizes the most popular project management tools used today.[4] For convenience of discussion, these tools are grouped into three categories: (1) analytical tools and techniques, (2) procedural tools and techniques, and (3) people-oriented tools and techniques. This grouping also provides descriptive focus and a convenient way to catalog these control tools (Thamhain 1996b). However, it should be recognized that applications overlap a great deal among these categories.

In spite of the long list of project management methods, tools, and techniques, there has been no universal evidence on the effectiveness of these tools, especially for the more contemporary ones. Further, management tools do not come with a user's manual. Few guidelines have been published in the literature on how and where to use these tools and techniques most appropriately. Perhaps one *of the greatest challenges* for management is to *seek out* management tools and techniques that meet the triple constraint of being, (1) compatible with the business environment, processes, cultures and values, (2) conducive to specific problem solving, which usually involves a whole spectrum of factors from innovation to decision making, cross-functional communications, and dealing with risks and uncertainty, and (3) useful for tacking and controlling the project according to established plans. The *second major challenge* is for management to *implement* selected project control tools and techniques as part of the business process. This involves careful integration of these tools with the various physical, informational, managerial, and psychological subsystems of the enterprise to be helpful in guiding the work and to minimize tool rejection. The *third major challenge* is for management to *create and facilitate a learning process* for these tools and techniques so that they become institutionalized and therefore are used by the people in the organization, because these tools help in getting projects done more effectively and create visibility and recognition for their work.

Many of these contemporary tools also require new administrative skills and a more sophisticated management style. All of this has a profound impact on the way that project leaders must manage toward desired results. The methods of communication, decision making, soliciting commitment, and risk sharing are shifting constantly away from a centralized, autocratic management style to a team-centered, more self-directed form of project control. Equally important, project control has radically departed from its narrow focus on satisfying schedule and budget constraints to a much broader and more balanced managerial approach, which focuses on the effective search for solutions to complex problems, as reflected by the performance metrics in Table 7.4. This requires trade-offs among many parameters, such as creativity, change-orientation, quality, *and* traditional schedule and budget constraints. Control also requires accountability and commitment from the team members toward the project objectives.

---

[4]The project management tools summarized in Table 7.4 were originally compiled as part of a field study by Hans Thamhain, (1996b). It is estimated that this listing in Table 7.4 includes about 90 percent of all project management tools in use today.

Table 7.4a Analytical Management Techniques for Project Control

Analytical Management Technique	Description	Elements of Control	Conditions for Successful Control
Action Item/Report	A memo or report defining specific action items agreed upon with the resolver, necessary to move the project forward or to correct a deficiency.	Responsibility identfICat'n, personal commitment, peer pressure.	Individual commitment, management support, and incentives.
Computer Software	Computer software to support project planning, tracking and control. Provides various reports of project status and performance analysis, and documentation.	Schedule, budget, pert/cpm, resource leveling scheduling, $-t trade-off.	Ability to measure status. Valid input data. Willingness to correct deviations. Leadership.
Critical Path Analysis	Analysis of the longest paths within a network schedule with the objective of (1) determining the impact of task delays, problems, contingencies, and organizational dependencies, (2) finding solutions, and (3) optimizing schedule performance.	Schedule, budget, deliverables. Cost-time trade-off.	Accurate estimates of effort, cost, and duration.
Budget Tracking	Analysis of planned versus actual budget expenditures relative to work performed. The objective is to detect and correct project performance problems and to deal with projected cost variances in these early developments	Cost, budget, deliverables, project status.	Accurate estimates of effort, cost, and duration.
Deficiency Report	Description of an emerging deficiency (work, timing, or budget), including impact analysis and recommended resolution.	Schedule, costs, configuration mgmt, impact analysis.	Candor, commitment to plan. Management direction.
Earned Value Analysis	Comparison of project completion status to budget expenditure. The regular calculation and analysis of earned value and performance index allows projections of cost variances and schedule slips and serves as an early warning system of project performance problems.	Schedule, budget, deliverables. Cost-time trade-off.	Measurable milestones. Ability to estimate cost and time-to-complete. Trust. Risk sharing. Ownership.

Interface Chart	A chart of N × N elements defining the inputs, outputs, and timing to and from N interfacing work groups. Chart can also be used as part of QFD to define and manage the "customers" of the business process.	Task leaders, cross-functional communications, QFD framework.	Established cross-functional linkages. Management support and leadership.
PERT/CPM	Time-activity network showing task flow, interfaces, and dependencies. Used for comprehensive analysis of project schedules and schedule changes.	Schedule, budget, deliverables. Cost-time trade-off.	Accurate cost, time, and technical performance data. Measurable milestones.
Schedule Compression Analysis	Graphical technique for showing compression of overlapping activities due to slippage of earlier or preceding milestones. Serves as early warning system for runaway schedules and costs.	Milestones. Deliverables.	Accurate cost, time, and technical performance data. Measurable milestones.
Schedule Tracking	Incremental tracking of activities through time by measuring predefined partial results against plan.	Measurable milestone. Deliverables. Micro schedule.	Accurate cost, time, and technical performance data.
Simulation	Simulation of a technical, business, or project situation based on some form of a model. Applications range from a simple test to computer-assisted analysis of complex business scenarios.	Advanced results, feasibility, technology transfer.	Relevant input data and appropriate model. Meaningful interpretation.
Status Assessment	Systematic comparison of technical progress with project schedule and budget data. Analysis of status against plan and possible revision of plan, scope, and business strategy.	Valid project plan. Review process. Earned value. Variance analysis.	Accurate cost, time, and technical performance data. Measurable milestones.
Variance Analysis	Analysis of causes of cost and schedule variances, cost-at-completion, earned value, percent of project completion, and performance index. Applied to project status assessment, reporting, and control.	Schedule, costs, configuration mgmt, impact analysis. Management.	Accurate cost, time, and technical performance data. Measurable milestones.

Table 7.4b   Process-Oriented Management Techniques for Project Control

Process-Oriented Management Technique	Description	Elements of Control	Conditions for Successful Control
Concurrent Engineering	In-parallel/concurrent execution of project phases; "seamless product development." Objectives: reducing project cost and cycle time, increasing responsiveness to customer/market dynamics. Also an effective tool for multifunctional integration and technology transfer.	Input-output definition. Interface definition. QFD. DFM/A. DICE. Rapid prototyping. Structured analysis.	Org interface agreement. Personal commitment. Effective communications and organizational linkages.
Benchmarking	Comparing one system, process, or practice to another one (usually best-in-class) with the objective to improve performance.	Performance metrics. Business process.	Measurability of comparative metrics. Ability to diagnose/ analyze the cause of differences. Ability to adopt tool or method.
Design Review	Review of the project baseline at various stages of development, such as preliminary, critical, or final design review. Objective: examine and predict functionality of the deliverable system early in the project cycle.	Baseline. Design parameters. Documentation. Multifunctional reviewers. Agenda.	
Out-of-Bounds Review	Critical review and impact assessment of a situation declared as *out-of-bounds*.	Review. Analysis. Corrective action plan. Visibility. Peer pressure. Management control.	
Project Definition	Front-end planning of a project and its resource and timing requirements. Objective: project definition/ organization, task delegation, project tracking, and control.	Schedule. Budget. Task matrix. Statement of work. Task authorization. Resource leveling.	Multidisciplinary preparation/ homework. Presence of all technology transfer agents. Competence and willingness to analyze implementation and transferability. Mutual trust and respect among team members. Power sharing among managers and with team.

Project Review	Technical and contractional review of project status against established plans.	Professional review. C/SCSC.	
Prototyping	Advanced build of a design for the purpose of testing functionality and performance prior to production or deployment.	Design process. Simulation. CAD/CAM. Project management.	Team involvement. Desire to participate. Risk sharing and power sharing among managers and with team. Senior management involvement and support.
Quality Function Deployment (QFD)	TQM concept, known as House of Quality. Used for mapping the technology transfer flow throughout an organization and its markets, identifying for each organizational unit, inputs, outputs, and specific "internal customers" and their requirements.	Organizational interface and input-output definition. Internal and external customer focus.	
Requirements Analysis	Distinct project planning phase which defines the specific technical, resource, market, and timing requirements for the project baseline. Often coupled with Voice of the Customer and strategic assessment.	Needs assessment. Voice of the customer. Project planning.	Relevant baseline. Effective prototype testing and evaluation.
Stage-Gate Process	Phased approach to project planing and management. Defines "gates" for consecutive stages which check the feasibility, and provide implementation focus and control.	Phased planning. Modular work plan. Deliverables. Check points. Sign-offs. Gate reviews. Process action teams. Focus groups.	Team involvement. Desire to participate. Risk sharing and power sharing among managers and with team. Senior management involvement and support.
Stage-Gate Review	Specific review at the end of a project stage with deliverables and checkpoints, predefined as part of the stage-gate process. Can incorporate other project or design reviews	Multidisciplinary teams. Management.	
Voice of the Customer	Distinct project planning phase which defines the specific technical, resource, market, and timing requirements for the project baseline, with focus on the specific customer needs.	Needs assessment. Market analysis. Focus group. Survey. Customer feedback.	

Table 7.4c People-Oriented Management Techniques for Project Control

Analytical Management Technique	Description	Elements of Control	Conditions for Successful Control
Core Team	A team of resource managers responsible for planning, organization, and execution of many projects of a similar nature.	Dedicated functional management team. Minimum cross-functional impedance.	Proper team design and charter. Effective communication channels, internal and external. Competent team members and desire to participate. Risk sharing and power sharing among managers and within team. Autonomy. Senior management involvement and support. Effective conflict management system. Team-based evaluations and awards. Minimum fear. Reasonable job security. Managerial checks and balances. Team leadership.
Design/Build	Typically used in product development projects; new team members are integrated into the project team as it moves from the product design to the product build stages, while retaining key members from the earlier project stages.	Shared multifunctional experience. Project/team ownership. High technical competence of team.	

Focus Group	A group of "stakeholders" within a project organization or its support functions, engaging in self-study and analysis of the project management system or the business process with the objective to improve it.	Problem ownership. Pride. Personal/professional needs. Will to change status quo.
Joint Performance Evaluation	Both project performance and individual performance are defined in terms of end-objectives, hence including multifunctional measures. The objective is to enhance cross-functional cooperation and team integration.	Stakeholder/ownership. Mutual dependency. Risk sharing. Internal customer orientation.
Self-Directed Team	Individual team members and the project teams as a whole are given high levels of autonomy and accountability (empowerment) toward plan implementation. This forces higher degrees of multidisciplinary decision making and work integration at the operational level.	Stakeholder/ownership. Mutual dependency. Risk sharing. Internal customer orientation. Personal commitment and drive. Team communication and decision making.

7.3 CHARACTERISTICS OF EFFECTIVE CONTROLS

Managerial controls have many dimensions and formats. Traditionally, budgets, schedules, and performance appraisals were primarily used in addition to reports, procedures, and directives to control the implementation of managerial plans. In spite of the limitations just discussed, these tools are still very important in today's world of technology. However, given the complexities, dynamics, risks, and uncertainties of technology-intensive work, these conventional tools need to be "upgraded" to respond effectively to the challenges of our contemporary business environment. While the conventional tools provide the foundation for managerial control, they must be augmented and expanded into the various subsystems of the enterprise to have access to the widest possible range of information and resources for dealing with the complexities and uncertainties of a given mission, project, or venture. As shown with the TSA situation, especially important is the human side of managerial control, which is being emphasized in this chapter.

Every manager has his or her own set of criteria for effective controls. Yet, these controls have some common characteristics, which are summarized below.

- *Realistic Plans.* Plans are implemented by people. It is difficult, if not impossible, to control activities leading toward objectives if these objectives are perceived as unrealistic. People who are ultimately held accountable for achieving these results should be involved in the planning. This personal involvement leads to a better understanding of the requirements, and hopefully to a commitment to the plan and its objectives. It also results in a more realistic plan, which is validated by the implementer.

- *Commitment.* Control of complex work, such as technology-based projects, depends a great deal on self-imposed control. Personal commitment is a strong intrinsic driver toward self-direction and control. Without this commitment, controls will focus more on the cause and perpetrators of a problem than on its correction. Commitment is also expected from senior management regarding resources, priorities, and contingency support.

- *Competence.* The people implementing the plan must be professionally competent. The best plan and strongest commitment are meaningless unless the people have the capacity to perform. Attracting and holding high-quality people is a primary mission of management. Furthermore, the sign-on process for assigning personnel to a new task or project can help considerably in matching personal skills, capabilities, and interests with the project and its requirements.

- *Measurability.* If you can't measure, you can't control. Any assignment of an activity, project, milestone, or mission must have measurable results. These deliverables should be compared against the plan, including its resources and timing, on an ongoing basis.

- *Appropriate Controls.* Management controls must be meaningful and appropriate for the type of mission. This includes both managerial action and timing. For example, close supervision and incentive bonuses might be appropriate for a beta site test operation, but of much less value in R&D. Many controls work only in combination with other methods, such as review meetings, focus groups, reports, and management actions.
- *Focus on Key Objectives.* According to Peter Drucker, "control must follow strategy." Managerial control should be related to key objectives. Trivia should not be measured or controlled. It only dilutes the principal objectives and leads into an "activity trap"! That is, we are busy earning Brownie points, following procedures, and writing progress reports, rather than dealing with risks, implementation challenges, and contingencies, which is the essence of managerial control toward desired results.
- *Simplicity and Adaptability.* Management controls must be simple. Their purpose is to identify problems early and to take corrective actions. Good controls produce a quick reaction to potential problems. They have a feedback mechanism. They also have some built-in automatic problem detection system that provides early warnings and adapts dynamically to changing situations and emerging problems. Examples of such in-process controls are *process action teams* and *rapid prototyping.*
- *Early Problem Detection.* Controls should focus on problem prevention and early detection. Direct management involvement and review meetings are more likely to have this desired focus than reports which concentrate more on the documentation and justification of the problem, rather than its resolution.
- *Controlling Authority.* Management controls should be in the hands of the people who are accountable for results. As we move up the organization, senior management is responsible for more global business results, and therefore should exercise more global controls rather than micromanage.

7.4 HOW DO MANAGERS CONTROL TECHNOLOGY-INTENSIVE WORK?

Cultural and philosophical differences obviously influence the way companies manage their projects and set up control systems. Even within one organization, methods may differ from project to project. However, recent field studies of best practices[5]

---

[5]This section reports the results of an ongoing field study into the *best practices* of controlling technology-based projects. Two hundred eighty projects and their management processes were examined and analyzed between 2001 and 2004. For specific method and results please see Thamhain 2001, 2002, 2004.

show that a general agreement seems to exist among project leaders regarding the types of tools and techniques that find general application in today's technology-intensive project environment. The results of these field studies will be discussed in three parts. First, the popularity and value of various project control techniques will be summarized. Second, the skills needed for effective application will be analyzed. Finally, the managerial implications and recommendations derived from the study will be presented.

7.4.1   Popularity and Value of Project Control Techniques

Effective project management involves a great variety of tools and methods as discussed earlier (see Table 7.4). These tools are listed again in Table 7.5, rank-ordered by popularity as found during the field study. It is interesting to note that in spite of their limitations, traditional project management tools and techniques such as budget and schedule tracking and project reviews are considerably more popular among professionals than contemporary techniques such as schedule compression analysis, simulation, and out-of-bounds reviews. In part, this lower popularity is related to the finding that both project managers and team members are less familiar with contemporary techniques, which explains 82 percent of the variance between using and not using a particular technique. That is, on average 82 percent of project leaders who choose not to use a particular technique also are not familiar with its application. *The significance* of this finding is related to judging a technique's *value*. Clearly, it is inappropriate to determine the value of a particular technique solely based on its popularity. From our interviews and observations, we can argue even further that it is very difficult and risky to assess the perceived value of a management technique from the population that rejected it, even if they claimed familiarity with it.. In fact, follow-up interviews showed that some of the most frustrated and adversely positioned managers are those who have tried a particular technique with marginal results or great disappointment. Yet, in most cases researchers were unable to determine whether the dismal experience was the result of poor familiarity with the technique, improper integration with the business process, insufficient senior management support, poor matching of the tool with the situation, or a combination of all of these factors. While this is clearly an area of further research, it also suggests that *project management tools are underutilized because project managers have insufficient familiarity with them.*

Because of these complexities, the value or usefulness of specific project management techniques has been measured as a perception of those managers who actually use them. The results show that the users perceive 85 percent the tools summarized in Table 7.5 as being at least of "significant value and important for managing their projects effectively." It is also interesting to note that the traditional project management techniques are not only among the most popular ones, but also among the most valued techniques. Rank-ordered by value, the 12 most important techniques include:

Table 7.5 **Management Techniques: Popularity, Skill Level, and Value**

Management Technique	Popularity	Skill Type & Level Needed					Value
MT	%	AN	BA	TB	SM	L	V
Schedule Tracking	99	2.00	1.00	3.00	2.00	3.50	3.25
Project Definition	98	3.00	2.75	2.00	2.25	3.00	3.75
Project Review	93	2.75	3.25	3.25	3.00	2.75	3.15
Budget Tracking	92	2.00	1.50	2.50	2.00	1.50	3.25
Design Review	87	3.00	2.00	2.75	2.25	3.00	3.50
Prototyping	82	3.50	1.00	1.50	1.00	1.50	3.25
Status Assessment	82	3.25	2.50	3.00	2.00	3.00	3.75
Computer Software	71	3.75	2.25	2.00	1.50	1.75	2.50
Deficiency Report	68	1.50	0.75	0.50	0.50	1.00	2.50
Action Item Report	65	1.00	1.00	0.50	0.50	0.50	3.00
Requirements Analysis	52	3.25	2.75	2.00	1.50	1.50	3.20
Benchmarking	52	3.25	2.50	1.00	1.75	1.75	1.50
PERT/CPM	42	3.00	1.75	2.00	0.75	1.75	1.50
Earned Value Analysis	41	2.50	0.50	2.75	0.75	2.00	1.75
Stage-Gate Process	40	3.50	3.25	3.25	3.50	3.25	3.25
Variance Analysis	39	2.50	1.50	1.50	0.75	1.75	1.50
Core Team	38	2.25	2.75	3.25	3.75	3.25	2.35
Interface Chart	38	2.00	1.50	2.00	2.00	2.00	2.00
Stage-Gate Review	35	3.25	3.25	3.00	3.25	3.00	3.25
Critical Path Analysis	32	1.75	0.50	0.25	0.50	0.00	2.00
Concurrent Engineering	32	3.25	3.25	3.50	3.25	3.75	2.00
Quality Function Deployment	28	3.25	3.50	3.25	3.25	3.25	2.00
Focus Group	28	2.50	1.25	1.50	1.25	1.75	2.00
Voice of the Customer	25	2.75	3.00	1.75	2.25	2.25	2.75
Self-Directed Team	23	2.25	2.50	3.50	3.25	3.50	2.50
Design/Build	18	2.75	3.25	3.25	3.50	3.50	3.00
Schedule Compression Analysis	18	1.75	0.25	0.25	0.50	0.25	1.00
Joint Performance Evaluation	15	3.00	3.25	2.75	3.75	3.00	2.25
Out-of-Bounds Review	12	1.50	1.00	2.50	2.75	2.25	2.25

Legend: AN Analytical skills SM Senior management support
BA Business administrative skills TB Team building skills
L Leadership skills % Popularity of technique (percentile of actual users)

All skill levels and values are perceptions of project leaders and have been measured on a five-point scale, 0-4: 0 = little or no skills or value, 1 = some, 2 = considerable, 3 = high and significant, 4 = very high and critical skill level or value.

The percentile figures (%) in the popularity column refer to the number of project professionals actually using the technique.

Project definition
Status assessment
Design review
Project review
Schedule tracking
Budget tracking
Prototyping
Stage-gate process
Stage-gate review
Requirements analysis
Action item report
Design/build

All 12 techniques were rated on average as "significant" and "important" to project success (average score = 3.3), with the top half of the techniques receiving ratings of 3.5 or better; that is close to the rating of "very highly significant and crucial to achieving planned project performance."

Many of these tools, such as design reviews and schedule and budget tracking, have been around for a long time. However, in recent years virtually all of the traditional project management tools have been redefined and enhanced with information technology and sophisticated computer software to fit the contemporary challenges of today's business environment more appropriately, and to work flexibly and effectively with complex processes, human factors, and cross-functional alliances.

Further, we find that for project control to work effectively, it must be congruent with the business process and the human side of management, a conclusion that is being supported by other field research (Weisinger and Traut 2003; Davenport 1998; Schultz, Slevin, and Pinto 1987). That is, managerial techniques work only if (1) the inputs, such as time and resource estimates, percent complete, and skill levels, are accurate, and (2) the people responsible for implementation will use these tools and analytical results to correct problems and deficiencies in the plan. Therefore, it is not surprising that many of the control techniques depend on a whole spectrum of people issues: clear direction and guidance; ability to plan and elicit commitments; communication skills; assistance in problem solving; ability to deal effectively with managers and support personnel across functional lines, often with only little or no formal authority; information-processing skills, the ability to collect and filter relevant data valid for decision making in a dynamic environment; and the ability to integrate individual demands, requirements, and limitations into decisions that benefit the overall project. It further involves the project manager's ability to resolve intergroup conflicts and to build multifunctional teams. These findings are further supported by other research (Doolan and Hacker 2003, McDonough and Leifer 1986). They consistently emphasize the important role of organizational and cultural factors, including mutual trust, respect, candor, risk sharing, and the ability to fail safely. These factors involve the people side of project control. This complicates the otherwise relatively straightforward analytical processes. In response to these challenges, an increasing number of companies are focusing on self-forcing control, with work challenge, personal pride, and commitment as the most critical elements of managerial control. Equally important, the work environment must foster effective communication, cross-functional linkages, and business processes that interconnect people, activities, and support functions. Taken together, *effective control of technology-intensive work, such as projects, requires:*

- Understanding of modern management tool and applications
- Management involvement and support
- Project team ownership of control techniques
- Controls integrated with project management process
- Controls integrated with the human resource system and personal rewards
- Effective cross-functional communications
- Managers who share power and control with their project teams
- Competent, motivated project teams with ownership of project objectives
- Project teams involved in tool development and implementation
- Provisions to deal with emerging conflicts
- Control process integration with functional support groups
- Mutual trust and respect among team members
- Risk sharing and fail-safe measures
- Project team structure conducive to the use of tools and techniques
- People able and willing to work in unstructured environment
- Motivation and direction toward achieving results
- Managerial leadership

7.4.2   Satisfaction with Project Management Techniques

One wonders why many of the managerial tools that were designed to improve project performance and came highly recommended by the management literature have not been more widely adopted. Popularity of management tools and techniques in the literature and their actual field applications are two different things. Few companies go into a major restructuring of their business processes lightly. At best, the introduction of a new project control technique is painful, costly, and disruptive to ongoing operations. At worst, it can destroy existing managerial controls. It can lead to mistrust between the team and management, game playing, power struggles, conflict, and misleading information. It can also lead to a transfer of accountability away from team members and project leaders to the control tool mechanics and their administrators.

7.4.2.1 *Anxieties Caused by New Tools and Techniques*

In fact, for many managers, the risks of introducing a new project control tool are so substantial that they are willing to live with an inefficient system rather than going through the trouble and uncertainties of changing it. Most skeptical of course are managers who have tried a specific tool and obtained disappointing results or outright failures. Such negative impressions are often most intense for complex process-oriented controls such as Stage-Gate techniques, which rely on intricate and often fuzzy performance measures. Table 7.6 summarizes the typical reasons for rejecting or underutilizing project controls. The listing is rank-ordered by frequency as given by project leaders. Specifically, Table 7.6 summarizes the responses to a field survey,[6] asking project leaders, team members and resource managers the question "Why are the project control tools and techniques used in your organization underused or rejected?" The reasons for underusing or rejecting project controls can be divided into four classes[7]:

- Lack of confidence that tools will produce benefits
- Anxieties over the potentially harmful side effects
- Conflict among users over the method or results
- Method is too difficult and burdensome, or interferes with the work process

Interestingly, most of the perceived problems are based on "feelings" and "assumptions" rather than on specific measures, comparisons, and analytical facts.

---

[6]This question was a follow-on to a series of questions probing into the type and effectiveness of control tools and techniques in use at a particular organization. Questionnaires and interviews were used to collect the data; 1,350 responses were summarized into the composite shown in Table 7.4.

[7]Although resource managers, project leaders, and team members responded differently regarding the type of reason, frequency, and situational impact, these differences were not statistically significant enough to separate these populations. In fact, agreement among the three populations was verified at a 95 percent confidence level, using a Kruskal-Wallis analysis of variance by ranks, indicating that the concerns and reasons for underusing or rejecting tools are shared across different levels of managerial responsibilities.

Table 7.6   Perception of New Project Management Techniques

Why Project Management Tools and Techniques Are Rejected ?

[Typical reasons listed in decreasing order of frequency perceived by project leaders:]

1. Lack of understanding on how to use tools properly.

2. General anxiety over methods and information use and misuse.

3. Use of tool requires too much work, is too time-consuming, and requires too much paperwork.

4. Tools reduce personal drive and willingness to fix ad hoc problems and contingencies.

5. Not consistent with already established project management procedures and business processes.

6. Control method is threatening regarding performance assessment, personal freedom, or autonomy.

7. Conflicting points of view among team members regarding the tool value or appropriate use.

8. "Not invented here" syndrome exists.

9. Conflict among managers or project team members about value of tool or application method.

10. Cost of acquisition and implementation is too high.

11. Tools focus on project management metrics, neglecting the importance of teamwork and cooperation.

12. Purpose, objectives, benefits, and value of tool unclear.

13. Tool leads to unwanted additional policies and procedures.

14. Too busy to learn new tool or technique.

15. Uncomfortable with new or unfamiliar methods.

16. Disagreement over application method or use of data.

17. Tools don't help in control, but help to maintain status quo when project performance deteriorates.

18. Stifle multifunctional communication and complex decision making.

19. Reduce face-to-face communication and multidisciplinary problem solving.

20. Stifle technical innovation and search for solutions to complex problems.

21. Tools are seen as a substitute for management support and decision making.

22. Tools isolate team members and their leaders.

23. Prior bad experience with tool or technique.

24. Tools weaken managerial power.

25. Not consistent with a self-directed team concept.

Only approximately 20 percent of the reasons given could be supported with examples from personal experience. This suggests a potentially negative bias toward new and untried tools and techniques. Of cause, basic psychology tells us that this behavior is quite normal and should be anticipated. Management must recognize the potential barriers toward project controls, which should be expected based on anxieties, misunderstandings, unpleasant experiences, and other unfavorable perceptions. Management must deal effectively with these perceptions and develop a positive attitude among project team members toward these new tools to avoid game playing or outright rejection. Specifically, the findings emphasize the important role of behavioral variables. Good leadership, personal interest, commitment to the project, potential for recognition, management involvement, and project importance and priority all have a favorable impact on the team members' willingness to work toward the established objectives, and to engage proactively in the work process.

7.5 RECOMMENDATIONS FOR USING MANAGEMENT CONTROLS EFFECTIVELY

The pressures for improving business performance, combined with the emergence of new tools and techniques which promise more effective project control, have prompted management to cautiously shift their focus away from operational processes that narrowly emphasize schedule/budget performance measures. They had no choice but to explore new concepts.

To be successful in implementing managerial controls, business leaders must pay attention to three important criteria that are often overlooked: *First*, tools and techniques must be properly integrated with the business process. *Second*, the impact on intrinsic project performance must be considered. That is, the impact of a new control process on innovation, creativity, quality, customer relations, and the ability to cope with changing requirements must be evaluated and factored into the design of new controls and their fine-tuning. *Third*, the human side of organizational change must be carefully considered. Introducing new managerial controls, as minor as a new weekly review procedure or as massive as changing to concurrent engineering, involves organizational change, a process that is often poorly managed or outright ignored. These changes are associated with hopes, fears, anxieties, and likes and dislikes, plus a good dose of organizational politics. One of the strongest messages from field research points at the fact that people are still the most vital part of any managerial control system. A new system has at least a chance for consideration if the project team believes in the value of a tool for making their job easier and more effective, helping to produce desired results, visibility, and recognition, while minimizing anxieties and fears over administrative burden, restrictions of freedom and autonomy, and any negative impact on personal growth and security. Unfortunately, research shows that few companies have systematically attacked these softer, more subtle issues of project control and tool implementation, which are crucial for overall business performance, especially in technology-based environments. These issues often need special attention and resources to ensure effective integration of these tools into the business process and effective use by project team members.

A number of recommendations for effectively implementing and using managerial controls for technical projects have been derived from the field study. These recommendations should help both project leaders and their managers to understand the complex interaction of organizational and behavioral variables that affect implementing appropriate managerial tools and techniques to control technology-based projects. These findings should increase the awareness of what works and what doesn't, and help project management fine-tune control systems for high performance. Finally, the findings should help management scholars in better understanding the complexities involved in project control, and hence provide building blocks for further research.

Cross-Reference to Chapter 4. I would like to make reference to the recommendations toward the end of Chapter 4 that address similar issues of organizational change and development. While both sets of recommendations overlap in tone and emphasis, they are meant to complement each other. The recommendations for controlling technology-intensive work are summarized below.

Involve the Team. Make your people part of the tool selection and implementation process. Both the project team and the project manager should be involved in assessing the project situation and evaluating new control tools. *Critical factor analysis, focus groups*, and *process action teams* are good vehicles for such team involvement and collective decision making. These actions lead to ownership, greater acceptance of the selected tool, and a willingness to work toward continuous improvement and effective use.

Make Tools Consistent with the Work Process. Management controls should be an integral part of the business process. Particular attention should be paid to the workability of the tools and techniques for integrating tasks and transferring technology across organizational lines.

Build on Existing Tools and Systems. The highest levels of acceptance and success are found in areas where new management tools are added incrementally to already existing control systems. These situations should be identified and addressed first.

Connect with Established Management Practices. Team members feel more comfortable with management procedures that they have familiarity with and can relate to their work. To minimize anxieties, suspicions, and the risk of rejection, management should refer to established project management practices when introducing new tools or techniques. For example, an organization implementing a Stage-Gate process should make an effort to integrate it into the process already established and proven procedures for project definition, documentation, status reports, reviews, and sign-offs. This will make the new management process look evolutionary rather than like a radical change. Thus, if done right, management can use the existing project management system to build upon, and incrementally enhance and test, new managerial tools and techniques.

Make Tools User-Friendly. New project management tools or techniques are more likely to be accepted if they are easy to use and produce results that are helpful to the users and their work, including the organization's senior management.

Anticipate Anxieties and Conflicts. Especially when introducing new tools and techniques (i.e. a new concurrent engineering system discussed in Chapter 4), project leaders should anticipate anxieties and conflict among their team members. These negative biases come from uncertainties associated with new working conditions and requirements. They range from personal discomfort with skill requirements to anxieties over the impact of the tool on the work process and personal performance evaluation. These problems should be anticipated and dealt with in a straightforward manner as early as possible.

Ensure That There Is No Threat. Management must foster a project team environment of mutual trust and cooperation, an environment that is low on personal conflict, power struggles, surprises, unrealistic demands, and threats to personal and professional integrity. Cooperation with a new (or existing) tool or technique, and commitment to it, can be expected only if its use is relatively risk-free to the user. Unnecessary references to performance appraisals, tight supervision, reduced personal freedom and autonomy, and overhead requirements should be avoided, and any concerns dealt with promptly on a personal level.

Foster a Challenging Work Environment. Professionally interesting and stimulating work appears to be one of the strongest drivers toward desired results. Verified by several field studies, we find consistently that the degree of interest and excitement derived from work is directly related to personal effort, the level of team involvement, cross-functional communication, commitment toward established plans, and creativity. Work challenge also produces higher levels of cooperation and some tolerance for risk and conflict. Taken together, work challenge seems to foster a desired behavior conducive to exploring new methods and innovatively applying them to project situations. Further, people who are strongly engaged with their work have a more positive attitude toward change. Therefore, work challenge seems to be a catalyst for integrating team members' personal goals with project objectives and organizational goals. All of this fosters a favorable climate toward acceptance and effective use of managerial controls. Project leaders should try to accommodate the professional interests and desires of their personnel whenever possible. One of the best ways to ensure that the work is interesting to team members is to match carefully their personal interests with the scope and needs of the tasks when "signing on" team personnel. In addition, managers should build a project image of importance and high visibility, which can elevate the desirability of participation and contribution.

Pretest New Tools and Techniques. Preferably, new concepts should be tried first with a small project and an experienced, high-performance project team. Asking such a team to test, evaluate, and fine-tune the new tool for the company is often seen an honor and a professional challenge. Further, it will start the implementation with a positive attitude and can create an environment of open communication.

Continuous Improvement. Project management tools and techniques are part of the continuously changing business process. Provisions must be made for updating and fine-tuning these tools on an ongoing basis to ensure relevancy to today's project management challenges.

Senior Management Support. Management tools require top-down support to succeed. Through its involvement and communications, management can stress the importance of these tools to the organization, span organizational and cultural boundaries, and unify objectives.

Ensure Proper Direction and Leadership. Throughout the implementation phase of a new management tool or technique, managers can influence the attitude and commitment of their people toward a new concept by their own actions. Concern for project team members, assistance with the use of the tool, and enthusiasm for the project and its administrative support systems can foster a climate of high motivation, involvement with the project and its management, open communications, and willingness to cooperate with the new requirements and use them effectively.

7.6 CONCLUSION

The proper implementation and use of an effective management control system, including its tools and techniques, can critically determine the success of any project, especially for technology-based undertakings. Successful application of these management controls involves a complex set of variables. The tools must be consistent with the business process and be an integral part of the existing managerial control and personal reward system. Most importantly, managers must pay attention to human factors. To enhance cooperation with the evaluation, implementation, and effective use of project management controls, management and team leaders must foster a work environment where people find the controls useful, or at least not threatening or interfering with the work process. Further, professionally stimulating work, refueled by visibility and recognition, is conducive to change and cooperation. Such a professionally stimulating environment seems to lower anxieties over managerial controls, reduce communications barriers and conflict, and enhance the desire of personnel to cooperate and to succeed. It also seems to enhance organizational awareness of the surrounding business environment and increases the ability to respond to these challenges effectively by using modern project management techniques. Further, effective use of modern project control techniques requires administrative skills for planning and defining project efforts properly and realistically, as a prerequisite for tracking the project through its life cycle. Effective project leaders understand the interaction of organizational and behavioral variables and can foster a climate of active participation and minimal dysfunctional conflict. They also build alliances with support organizations and upper management to ensure organizational visibility, priority, resource availability, and overall support for the multifunctional activities of the project throughout its life cycle.

In the decades ahead, the ability to effectively manage complex projects, for both internal and external clients, will play a critical role in separating the winning companies from losers. Largely because of new technology in computers and communications, project management tools will further proliferate and find applications across a wide range of business situations. The sophistication and effectiveness with which these tools are used will profoundly influence the way companies (1) do business, (2) utilize their resources, (3) handle project complexities, (4) achieve quality, and (5) respond to market requirements with speed and accuracy. These challenges will be true for any enterprise, but they are expected to be amplified in technology-based business environments.

7.7   SUMMARY OF KEY POINTS AND CONCLUSIONS

The key points that have been made in this chapter include:

- Managerial control is the process of implementing a plan and reaching its objectives in spite of changing conditions.
- Effective controls have some common characteristics, such as (1) realistic plans, (2) commitment, (3) competence, (4) measurability, (5) meaningful tracking systems, (6) focus on key objectives, (7) simplicity and adaptability, (8) early problem detection, and (9) controlling authority.
- Management control of technology-based work is project focused.
- New organizational forms such as concurrent engineering, special processes, and self-directed teams have emerged, gradually replacing traditional, more hierarchically structured organizations and their linear work processes.
- Today's technology projects rely to a considerable extent on member-generated performance norms and evaluations, rather than hierarchical guidelines, policies, and procedures.
- Traditional management tools, designed largely for top-down centralized command, control, and communications, are often no longer sufficient for managing projects effectively.
- Conventional tools and methods of project control by themselves seldom produce satisfactory results in today's work environment.

- New project management techniques are often more integrated with the business process, offering more sophisticated capabilities for project tracking and control.
- Driven by business pressures and advances in information technology, many companies have invested heavily in new management tools and techniques, promising more effective alternatives and enhancements to traditional forms of project control.
- One of the greatest challenges for management is to seek out tools and techniques that meet the triple constraint of being (1) compatible with the business environment, processes, culture, and values, (2) conducive to specific problem solving, and (3) useful for tackling and controlling the project according to established plans.
- Project control has radically departed from its narrow focus on satisfying schedule and budget constraints to a much broader and more balanced managerial approach which focuses on the effective search for solutions to complex problems.
- Project management tools are underutilized because of insufficient familiarity on the part of project leaders.
- Project control must be congruent with the business process and the human side of management to work effectively.
- Field research consistently emphasizes the importance of organizational and cultural factors for managerial control, including mutual trust, respect, candor, risk sharing, and the ability to fail safely.

  Self-forcing or self-directed control uses work challenge, personal pride, and commitment as motivators for team self-governance and accountability.
- Project control requires work processes with effective communication channels and cross-functional linkages, interconnecting people, activities, and support functions.
- Few companies go into a major restructuring of their business processes lightly. At best, the introduction of a new project control technique is painful, costly, and disruptive to ongoing operations. At worst, it can destroy existing managerial controls.
- For many managers, the risks of introducing a new project control tool are so substantial that they are willing to live with an inefficient system rather than going through the trouble and uncertainties of changing it.
- The reasons for underusing or rejecting project controls can be divided into four classes: (1) lack of confidence that tools will produce benefits, (2) anxieties over potentially harmful side effects, (3) conflict among users over method or results, and (4) the method is too difficult or interferes with the work process.
- People are still the most vital part of any managerial control system.
- Effective project leaders understand the interaction of organizational and behavioral variables and can foster a climate of active participation and minimal dysfunctional conflict.

7.8   CRITICAL THINKING: QUESTIONS FOR DISCUSSION

1. Why is it so difficult to control technology-intensive activities and projects according to established plans?

2. Define the most basic components of managerial control for high-technology projects.

3. Why can a narrow focus on budget and schedule parameters be counterproductive to project performance, especially in high-tech work environments?

4. How does a professionally stimulating work environment enhance team self-direction and control life cycle established project objectives?

5. Discuss the organizational environment and management style most conducive to self-controlled teamwork.

6. What conditions in the work environment are most conducive to personal commitment to established project or mission goals?

7. Define some guidelines for an R&D team for "self-managing" a one-year project.

8. Why are project team members often reluctant to cooperate with a new project control technique, such as a new method for budget or schedule tracking?

9. Discuss the challenges of implementing a new gate review process for a given work environment (select a work environment that you know or can identify with). Could you build on existing tools and make the process congruent with the "old" work environment?

10. How can you make a process for budget or schedule tracking more user-friendly?

11. Discuss "risk sharing." What does risk sharing among members of a new product development team mean? How can you encourage and enhance risk sharing among these team members?

12. Discuss methods for pretesting a new design review process.

13. Discuss ways to encourage and engage team members to make continuous improvements in their project tracking and control system.

7.9   REFERENCES AND ADDITIONAL READINGS

Abdel-Hamid, T. and Madnick, S. (1990) "The elusive silver lining: how we fail to learn from software development failures," *Sloan Management Review*, Vol. 32, No. 1 (Fall), pp. 39–48.

Adler, P., McDonald, D., and MacDonald, F. (1992) "Strategic management of strategic functions," *Sloan Management Review*, Vol. 33, No. 2 (Winter), pp. 19–37.

Anbari, F. (1985) "A systems approach to project evaluation," *Project Management Journal*, Vol. 16 (August), pp. 21–26.

Archibald, R. (2001) *Managing High-Technology Programs and Projects*, New York: Wiley & Sons.

Badaway, M. (1982) *Developing Managerial Skills in Engineers and Scientists*, New York: Van Nostrand Reinhold.

Bahrami, H. (1992) "The emerging flexible organization: perspectives from Silicon Valley," *California Management Review*, Vol. 34, No. 4 (Summer), pp. 33–52.

Bailetti, A., Callahan, J. and DiPietro, P. (1994) "A coordination structure approach to the management of projects," *IEEE Transactions on Engineering Management*, Vol. 41, No. 4 (November), pp. 394–403.

Beacon, G., et al. (1994) "Managing product definition in high-technology industries: a pilot study," *California Management Review*, Vol. 36, No. 3 (Spring), pp. 32–56.

Berman, E. and Vasconcellos, E. (1994) "The future of technology management," *Organization Dynamics*, Winter 1994, and *IEEE Engineering Management Review*, Vol. 22, No. 3 (Fall), pp. 13–19.

Bohn, R. (1994) "Measuring and managing technological knowledge," *Sloan Management Review*, Vol. 36, No. 1 (Spring), pp. 61–73.

Cash, C. and Fox, R. (1992) "Elements of successful project management," *Project Management Journal*, Vol. 23, No. 2 (June), pp. 43–47.

Christensen, D. (1994) "A review of the cost/schedule control systems criteria literature," *Project Management Journal*, Vol. 25, No. 3 (September), pp. 32–39.

Clark, K. and Wheelwright, S. (1992) "Creating product plans to focus product development," *Harvard Business Review*, Vol. 70, No. 2 (March/April), pp. 70–82.

Cleland, D. (2004) *Project Management: Strategic Design and Implementation*, New York: McGraw-Hill.

Cordero, R. (1999) "Developing the knowledge and skills for R&D professionals to achieve process outcomes in cross-functional teams," *Journal of High Technology Management Research*, Vol. 10, No. 1, pp. 61–78.

Corero, R. and Ferris, G. (2004) "Supervisors in R&D laboratories: using technical, people and administrative skills," *IEEE Transactions on Engineering Management*, Vol. 51, No. 1 (February), pp. 19–30.

Davenport, T. (1998) "Putting the enterprise into the enterprise system," *Harvard Business Review*, Vol. 76, No. 4 (July/August), pp. 121–131.

Deschamps, J., Nayak, P., and Ranganath, R. (1995) "Implementing world-class process," Chapter 5 in *Product Juggernauts*, Cambridge, MA: Harvard Press.

Dinsmore, P. (2005) *Handbook of Project Management*, New York: AMACOM.

Doolan, T. and Hacker, M. (2003) "The importance of situating culture in cross-cultural IT management," *IEEE Transactions on Engineering Management*, Vol. 50, No. 3 (August), pp. 285–296.

Editorial. (1995) "1995 project management software survey," *Project Management Network,* Vol. 9, No. 7 (July), pp. 35–44.

Gido, J. and Clements, J. (2003) *Project Management*, Cincinnati, OH: Southwestern Publishing Company.

Gobeli, D. and Brown, D. (1993) "Improving the process of product innovation," *Research-Technology Management*, Vol. 32, Vol. 2 (March/April), pp. 38–44.

Gopalakrishnan, S. and Santoro, M. (2004). "Distinguishing between knowledge transfer and technology transfer activities: the role of key organizational factors," *IEEE Transactions on Engineering Management*, Vol. 51, No. 1 (February), pp. 57–69.

Hatfield, M. (1995) "Managing to the corner cube: three-dimensional management in a three-dimensional world," *Project Management Journal*, Vol. 26 (March), pp. 13–20.

Kerzner, H. (2003) *Project Management: A Systems Approach*, New York: Wiley & Sons.

Knutson, J. (1996) "A socio-technical model of project management," *Project Management Network*, Vol. 10, No. 8 (August), p. 507.

McDonough, E. and Leifer, R. (1986) "Effective control of new product projects: the interaction of organizational culture and project leadership," *Journal of Product Innovation Management*, Vol. 3, No. 2, pp. 149–157.

Meridith, J. and Mantel, J. (2000) *Project Management*, New York: Wiley & Sons.

Moores, T. and Gregory, F. (2000) "Cultural problems in applying SSM for IS development," *Journal of Global Information Management*, Vol. 8, No. 1, pp. 14–19.

Ramabadran, R., Dean, J., Evans, J., and Raturi, A. (2004) "Testing the relationship between team and partner characteristics and cooperative benchmarking outcomes," *IEEE Transactions on Engineering Management*, Vol. 51, No. 2 (May), pp. 208–225.

Ramabadron, R., Evans, J., and Dean, J. (1997) "Benchmarking and project management: a review and organizational model," *International Journal of Benchmarking for Quality Management Technology*, Vol. 4, No. 1, pp. 34–46.

Randolph, W. and Posner, B. (1988). "What every manager needs to know about project management," *Sloan Management Review*, Vol. 29, No. 4 (Summer), pp. 65–73.

Rigby, D. (1995) "Managing the management tools," *Engineering Management Review (IEEE)*, Vol. 23, No.1 (Spring), pp. 88–92.

Schultz, R., Slevin, D., and Pinto, J. (1987) "Strategy and tactics in a process model of project implementation," *Interfaces*, Vol. 3, No. 3 (May/June), pp. 34–46.

Senge, P., Scharmer, C., Jaworski, J., and Flowers, B. (2004). *Presence: Human Purpose and the Field of the Future,* Cambridge, MA: SOL Publication.

Thamhain, H. and Wilemon, D. (1986) "Criteria for controlling projects according to plan," *Project Management Journal* (June), pp. 75–81.

Thamhain, H. (1989) "Validating project plans," *Project Management Journal*, Vol. 20, No. 4 (October), pp. 60–68.

Thamhain, H. (1994) "A manager's guide to effective concurrent project management," *Project Management Network*, Vol. 8, No.11 (November), pp. 6–10.

Thamhain, H. (1994). "Designing project management systems for a radically changing world," *Project Management Journal*, Vol. 25, No. 4 (December), pp. 6–7.

Thamhain, H. (1996a) "Managing technology-based innovation," Chapter 9 in *Handbook of Technology Management* (G. Gayner, ed.), New York: McGraw-Hill.

Thamhain, H. (1996c). "Applying stage-gate reviews to accelerated product developments," Proceedings, *PMI-'96 Annual Symposium of the Project Management Institute*, Boston, October 4–10.

Thamhain, H. (1996b). "Best practices for controlling technology-based projects," *Project Management Journal*, Vol. 27, No. 4 (December), pp. 37-48.

Thamhain, H. (1999) "Best practices for controlling technology-based projects," in *Essentials of Project Control* (Pinto and Tailor, eds), Newtown Square, PA: Project Management Institute Press, pp. 7–28.

Turtle, Q. (1994) *Implementing Concurrent Project Management*, Englewood Cliffs, NJ: Prentice-Hall.

Weisinger, J. and Traut, E. (2003) "The impact on organizational context on work team effectiveness," *IEEE Transactions on Engineering Management*, Vol. 50, No. 1 (February), pp. 26–30.

Appendix A: Project Complexity Metrics

	Parameter	Complexity Measure & Multiplier (0...3)				Σ	Weight	Total Complexity Index	Comments
		Low	Medium	High	Very High				
		(0)	(1)	(2)	(3)	0, 1, 2 or 3		(FxG)	
	(A)	(B)	(C)	(D)	(E)	(F)	(G)	(H)	(I)
1	Number of team members	< 10	10-30	30-50	> 50		4		
2	% Masters & PhDs on team	< 10%	10-30%	30-50%	> 50%		2		
3	Number of functions on team	< 5	5-10	10-40	> 40		3		
4	Number contractors	< 5	5-20	20-50	> 50		3		
5	Project budget (total)	< 100K	$.1-2M	$2-10M	> $10M		3		
6	Project duration	< 1 yr	1-4 yrs	4-7 yrs	> 7 yrs		3		
7	Ratio of budget/duration	< 100K/yr	5-10M/yr	≤ 100M/yr	> 100M/yr		2		
8	Similar to previous work	Yes	Somewhat	Little	First time		2		
9	New technology application	Low	Medium	High	Very high		2		
10	R&D content	Low	Medium	High	Very high		2		
11	Risk of failure	Low	Medium	High	Very high		1		
12	Joint venture/collaboration (#)	< 3	3-7	8-15	> 15		2		
13	No. of end user communities	< 3	3-7	8-15	> 15		3		
14	Classification of deliverable	Component	Subsystem	System	Array		5		
15	Other:								
16	Other:								
17	Other:								
18	Other:								
19	Other:								
						Grand Total.....			

Process of Determining Project Complexity:

1. *Judge the complexity* of each project parameter listed in column (A).
2. For each parameter, circle the corresponding *Complexity Level*, either column (B=0), (C=1), (D=2) or (E=3).
3. For each parameter, record the *Complexity Measure*, 0, 1, 2 or 3, in column (F).
4. For each parameter, multiply Complexity Measure in column (F) with Weight in column (G), then record in column (H).
5. Calculate the Grand Total Complexity Index by adding all numbers of column (H).

Interpretation of Numbers:

Given the set of 14 parameters, project complexity can range between 0 and 111.

For the purpose of classification and comparison, the following Complexity Levels have been defined:

Low Project Complexity	< 25 points
Medium Project Complexity	25–50 points
High Project Complexity	51–75 points
Very High Project Complexity	> 75 points

8

PROJECT EVALUATION
AND SELECTION

director of business strategy for IBM's server group. "Expertise in enterprise computing is not in Dell's skill set." Mike Winkler, HP chief marketing officer, predicts that his company's archrival has finally overextended. "The closest analogy," he says, "is Napoleon's invasion of Russia."

Michael Dell has heard all that before. Competitors once dismissed him as an underage computer geek who would never amount to anything against computer giants like Compaq or IBM, and who for sure could never move into the complex world of corporate computing. Both predictions were wrong. However, Dell now wants a much larger share of the industrial-strength hardware market, at the heart of corporate systems where the stakes and profit margins are higher. These are the engines that either drive or crash entire businesses.

Dell wants to leverage its manufacturing efficiency into these product lines. "When it comes to assembling servers, we enjoy all the same advantages from our procurement, logistics, and manufacturing capabilities as we do from PC making," says Randy Groves, vice president and general manager of Dell's enterprise systems group. Yet, there are more factors than just low price in determining customer preference and market success. Many of the servers that Dell tries to replace run on proprietary technology, such as IBM's Power4 microprocessors or various flavors of UNIX operating systems. Selling into these established markets is a problem for Dell, which opted for de facto industry standards, such as Intel microprocessors and Windows or Linux operating systems, which don't lock users into any particular hardware.

Dell bet on the notion that in a world dominated by standard platforms, hardware running the platform eventually becomes a commodity, and the most important criterion for choosing a vendor is price. And when it comes to price Dell can compete with anyone.

So far, Dell has been right, with success in many of its new business ventures. After passing HP-Compaq as the top provider of Intel-based servers, with more than one-third market share, the next logical step was to venture into storage systems that houses an enterprise's most crucial data. However, in contrast to PCs and servers, there are no standard storage technologies. Dell tried first unsuccessfully to develop a system in-house. The company quickly realized that it didn't have the expertise. In a rare acquisition, Dell purchased network storage specialist ConvergeNet Technologies for $332 million a few years ago, only to find that ConvergeNet's elegant but complex technology made a poor fit with Dell's commodity-producer business model. The investment had to be written off as a loss.

Dell had to compromise on its low-cost commodity production philosophy that served the company so well for two decades. Dell accepted the closest thing to an industry standard, the storage systems built by industry leader EMC. Dell eventually agreed to co-market its midrange Clarion storage system as a joint-venture until 2006. While Dell ponders the lessons of jumping into a new product sector, unprepared for some surprises and market realities, the ECM-Dell joint-venture seems to be working. The deal brought a new partner to ECM in a sector where it had been hard pressed by the competition, while Dell got into a key enterprise market without additional development expenditures. Yet, technology is not standing still. New technologies that link small computing devices to storage networks are beginning to replace proprietary systems such as EMC's. If cheap networked storage devices become the standard, Dell will be the logical winner.

Regardless of technology changes, Dell is not standing still either. In a networked world, a natural extension of Dell's products and services is into the routers and switches of corporate networks. Here Dell is moving more cautiously. Its sole offering in this category is its PowerConnect switch family, a lower-cost version of dominant products by Cisco, 3Com, Enterasys, and Nortel. "To challenge Cisco in higher-end equipment, Dell would have to build its own," says Jim Slaby, an analyst at the Giga Information Group. "You would have to duplicate 15 years of software development and thousands of worker-years of R&D. It would be like creating a word processor to compete with Microsoft Word."

Source. Kathryn Jones, "The Dell Way," *Business 2.0*, Vol. 4, No. 1 (February 2003), pp. 60–66. Excerpts reprinted with permission from *Business 2.0*.

For additional information visit *www.dell.com*.

8.1 MAKING THE RIGHT DECISIONS UP FRONT

Few decisions are more fundamental to business viability than resource allocations for new ventures. Every organization must continuously assess its options for survival and growth, and select projects for implementation, ranging from product development to organizational improvements, and from customer contracts to R&D and bid proposals. The Dell story is typical of the complexities, uncertainties, and risks involved in selecting winning projects for implementation. In Dell's case, many of the projects focused on new products and services, R&D, and promotional strategies. They also involved standards, technologies, system acquisitions, and joint ventures. Having the best business model and the fact that management has always made the right decisions on similar projects in the past do not guarantee success in the future. The surprises and challenges that Dell experienced with the network storage systems might be typical of the things that can go wrong with projects in spite of their careful assessment, evaluation, and planning for implementation.

Pursuing the "wrong" project not only (1) drains company resources, but also causes the enterprise to (2) miss alternative opportunities, (3) operate less flexibly and responsively in the marketplace, and (4) miss opportunities for leveraging core competencies. Project opportunities must be analyzed relative to their potential value, strength, and importance to the enterprise. Four major dimensions should be considered: (1) the added value of the new project, (2) the cost of the project, (3) the readiness of the enterprise to execute the project, and (4) managerial desire. A well-organized *project evaluation and selection process* provides the framework for systematic data gathering and informed decision making toward resource allocation. Typically, these decisions can be broken into four principal categories:

1. *Deciding Initial Feasibility.* Screening and filtering and making a quick decision on the viability of an emerging project for further evaluation
2. *Deciding Strategic Value to Enterprise.* Identifying alternatives and options to the proposed project
3. *Deciding Detailed Feasibility.* Determining the chances of success for a proposed project
4. *Deciding Project Go/No-Go.* Committing resources for a project implementation

While these decisions look logical and straightforward, developing meaningful support data is a complex process. It is also expensive, time-consuming, and often highly eclectic. Typically, decision making requires the following inputs:

i. Specific resource requirements
ii. Specific implementation risks
iii. Specific benefits (economics, technology, markets, etc.)
iv. Benchmarking and comparative analysis
v. Strategic perspective.

While there are plenty of challenges in evaluating project opportunities in terms of cost, time, risks, and benefits such as those shown in Table 8.1, it is relatively straightforward in comparison to *predicting project success.* The difficulty is in defining a meaningful *aggregate measure for rating project value and success.* Methods range from purely intuitive to highly analytical. No method is seen as truly reliable in predicting success, especially for more complex types of projects. Yet, some companies have a better track record in selecting "winning" projects than others, which seems to be related to the ability to create an integrated picture of the potential benefits, costs, and risks for the proposed project relative to the company's strength and strategic objectives. Producing such a composite is both a science and an art. Traditionally, managers have used predominately *rational selection processes* to support project selections. However, purely rational-analytical processes apply only to a limited number of business situations. Many of today's complex project scenarios require the integration of both analytical and judgmental techniques to evaluate projects in a meaningful way toward conclusions on *"right, successful, or best choice."*

Yet, in spite of the intricacies involved in making project selections, systematic information gathering and standardized methods are at the heart of any project evaluation process, and provide the best assurance for reliably predicting project outcome and repeatability of the decision process. Approaches to project evaluation and selection fall into one of three principal classes:

- Primarily *quantitative* and *rational* approaches
- Primarily *qualitative* and *intuitive* approaches
- *Mixed approaches*, combining both quantitative and qualitative methods

Because of the interdisciplinary complexities involved, analyzing a new project opportunity is a highly interactive effort among the various resource groups of the enterprise and its partners. Often, many meetings are needed before (1) a clear picture emerges of potential benefits, costs, and risks involved in the project, and (2) data emerge that are useful for the project evaluation and selection process, regardless of its quantitative, qualitative, or combined nature.

Table 8.1   Typical Criteria for Project Evaluation and Selection

---

The criteria relevant to the evaluation and selection of a particular project depend on the specific project type and business situation, such as project development, custom project, process development, industry, and market. Typically, evaluation procedures include the following criteria:

- Development cost
- Development time
- Technical complexity
- Risk
- Return on investment
- Cost-benefit
- Product life cycle
- Sales volume
- Market share
- Project business follow-on
- Organizational readiness and strength
- Consistency with business plan
- Resource availability
- Cash flow, revenue, and profit
- Impact on other business activities

8.2 QUANTITATIVE APPROACHES TO PROJECT EVALUATION AND SELECTION

Quantitative approaches are often favored to support project evaluation and selections if the decisions require economic justification. They are also commonly used to support judgment-based project selections. One of the features of quantitative approaches is the generation of numeric measures for simple and effective comparison, ranking, and selection. These approaches also help to establish quantifiable norms and standards, and lead to repeatable processes. Yet, the ultimate usefulness of these methods depends on the assumption that the decision parameters, such as cash flow, risks, and the underlying economic, social, political, and market factors, can actually be quantified and reliably estimated over the project life cycle. Therefore, quantitative techniques are effective and powerful decision-support tools, if meaningful estimates of cost-benefits, such as capital expenditures and future revenues, can be obtained and converted into net present values for comparison. As an example, Table 8.2 describes four project options that will be evaluated in this chapter, using various quantitative methods, with the results summarized in Table 8.3.

8.2.1   Net Present Value (NPV) Comparison

This method uses discounted cash flow as the basis for comparing the relative merits of alternative project opportunities. It assumes that all investment costs and revenues are known, and that economic analysis is a valid basis for project selection.

We can determine the *net present value (NPV)* of a single revenue, or stream of future revenues, or costs expected in the future. Two types of presentations are common: (1) present worth and (2) net present value.

Table 8.2   Description of Four Project Proposals

Project Option P1. Do not accept any new project proposal, hence does not require any investment capital, nor generate any revenue.

Project Option P2. This opportunity requires a $1,000 investment at the beginning of the first year and generates a $200 revenue at the end of *each* of the following *five* years.

Project Option P3. This opportunity requires a $2,000 investment at the beginning of the first year, and generates a variable stream of net revenues at the end of *each* of the next *five* years as follows: $1,500; $1,000; $800; $900; $1,200.

Project Option P4. This opportunity requires a $5,000 investment at the beginning of the first year and generates a variable stream of net revenues at the end of *each* of the next *five* years as follows: $1,000; $1,500; $2,000; $3,000; $4,000.

Table 8.3   Cash Flow of Four Project Options or Proposals, Assuming a MARR of $i = 10\%$

End of Year	Do-Nothing Option P1	Project Option P2	Project Option P3	Project Option P4
0	0	−1,000	−2,000	−5,000
1	0	200	1,500	1,000
2	0	200	1,000	1,500
3	0	200	800	2,000
4	0	200	900	3,000
5	0	200	1,200	4,000
Net Cash Flow	0	0	+3,400	+7,500
$NPV\mid_{N=5}$	0	−242	+2,153	+3,192
$NPV\mid_{N=\infty}$	0	+1,000	+9,904	+28,030
$ROI\mid_{N=5}$	0	20%	54%	46%
$CB = ROI_{NPV}\mid_{N=5}$	0	76%	108%	164%
$N_{PBP}\mid_{i=0}$	0	5	1.5	3.3
$N_{NPV}\mid_{i=10\%}$	0	7.3	5	3.8

Present Worth (PW). This is the single revenue or cost (also called annuity A) that occurs at the end of a period n, subject to the prevailing interest rate i. Depending on the management philosophy and enterprise policies, this interest rate can be (1) the internal rate of return (IRR) realized by the company on similar investments, (2) the minimum attractive rate of return (MARR) acceptable to company management, or the prevailing discount rate. The present worth is calculated as:

$$PW\left(A\mid i, n\right) = PW = A\frac{1}{\left(1 + i\right)^n}$$

For the examples used in this chapter, we consider the *internal rate of return (IRR)* (the average return realized on similar investments) as the prevailing interest rate.

Net Present Value (NPV). The net present value is defined a series of revenues or costs, A_n, over N periods of time, at a prevailing interest rate i:

$$NPV\left(A_n\mid i, N\right) = \sum_{n=1}^{N} A_n\frac{1}{\left(1 + i\right)^n} = \sum_{n=1}^{N} PW_n$$

Three special cases exist for the net present value calculation: (1) *for a uniform series of revenues or costs* over N periods: $NPV(A_n\mid i, N) = A[(1 + i)^{N-1}]/i(1 + i)^N$;

(2) for an annuity or interest rate i approaching zero: NPV = A * N; *and (3) for the revenue or cost series to continue forever:* NPV = A/i.

Table 8.3 applies these formulas to four project alternatives described in Table 8.2, showing the most favorable five-year net present value of $3,192 for project option P3.

8.2.2 Return on Investment (ROI) Comparison

Perhaps one of the most popular measures for project evaluation is the *return on investment, ROI:*

$$ROI = \frac{Revenue\,(R) - Cost\,(C)}{Investment\,(I)}$$

ROI calculates the ratio of net revenue over investment. In its simplest form the stream of cash flow is *not* discounted. One can look at the revenue on a year-by-year basis, relative to the initial investment. For example, project option 1 in Table 8.3 would produce a 20 percent ROI each year, while project option 2 would produce a 75 percent ROI during the first year, 50 percent during the second year, and so on. Although this is a popular measure, it does not permit a meaningful comparative analysis of alternative projects with fluctuating costs and revenues. Furthermore, it does not consider the time value of money. In a more sophisticated way, we can calculate the *average ROI per year:*

$$\overline{ROI} = \left(A_n, I_n | N \right) = \left[\sum_{n=1}^{N} \tfrac{A_n}{I_n} \right] | N$$

We can then *compare the average ROI to the minimum attractive rate of return, MARR.* Given a MARR of 10 percent for our project environment, all three project options—P1, P2, and P3—compare favorably, with project P2 yielding the highest average return on investment of 54 percent.

8.2.3   Cost-Benefit (CB)

Alternatively, we can calculate the *net present value* of the total ROI over the project's life cycle. This measure, known as the *cost-benefit (CB)*, is calculated as the present-value stream of net revenues divided by the present-value stream of investments. It is an effective measure for comparing project alternatives with fluctuating cash flows:

$$CB = ROI_{NPV} \left(A_n, I_n | i, N \right) = \left[\sum_{n=1}^{N} NPV \left(A_n, I_n | i, N \right) \right] / \left[\sum_{n=1}^{N} NPV \left(I_n | i, N \right) \right]$$

In our example of four project options (see Table 8.3), project proposal P2 produces the highest cost-benefit of 206 percent under the given assumption of $i = MARR = 10\%$.

8.2.4   Payback Period (PBP) Comparison

Another popular figure of merit for comparing project alternatives is the *payback period (PBP)*. It indicates the time period of net revenues required to return the capital investment made on the project. For simplicity, *undiscounted* cash flows are often used to calculate a quick figure for comparison, which is quite meaningful if we deal with an initial investment and a steady stream of net revenue. However, for fluctuating revenue and/or cost steams, the net present value must be *calculated for each period individually* and cumulatively added up to the "breakeven point" in time, N_{PBP}, when the net present value of revenue equals the investment. Mathematically,

$$N_{PBP} \dots \triangleright\triangleright\triangleright \; when \sum_{n=1}^{N} NPV\left(A_n|i\right) \geq \sum_{n=1}^{N} NPV\left(I_n|i\right)$$

In our example of four project options (Table 8.3), project proposal P2 produces the shortest, most favorable payback period of 1.9 years under the given assumption of i = MARR = 10%.

8.2.5   Pacifico and Sobelman Project Ratings

The previously discussed methods of evaluating projects rely heavily on the assumption that technical and commercial success is ensured and that all costs and revenues are predictable. Because these assumptions do not always hold, many companies have developed their own special procedures and formulas for comparing project alternatives. Two examples illustrate this special category of project evaluation metrics. The *project rating factor (PR)* was originally developed by Carl Pacifico for assessing chemical products and predicting commercial success:

$$PR = \frac{pT * pC * R}{TC}$$

Pacifico's formula is in essence an ROI calculation adjusted for risk. It includes probability of technical success [.1 < pT <1.0:], probability of commercial success [.1 < pC <1.0:], total net revenue over project life cycle *[R:]*, and total capital investment for product development, manufacturing setup, marketing, and related overheads *[TC:]*.

The second example shows a formula developed by Sobelman:

$$z = \left(P * T_{LC}\right) - \left(C * T_D\right)$$

It represents a modified cost-benefit measure that takes into account both the development time and the commercial life cycle of the product. It also includes average profit per year *[P]*, estimated product life cycle *[T_{LC}]*, average development cost per year *[C]*, and years of development *[T_D]*.

8.2.6 Going Beyond Simple Formulas

While quantitative methods of project evaluation have the benefit of producing relatively quickly a measure of merit for simple comparison and ranking, they also have many limitations, as summarized in Table 8.4. Because of these limitations, alternatives to these strictly quantitative methods have been developed that use a broader

Table 8.4 Comparison of Quantitative and Qualitative Approaches to Project Evaluation

Quantitative Methods	Qualitative Methods
Benefits:	*Benefits:*
Simple comparison, ranking, selection	Search for meaningful evaluation metrics
Repeatable process	Broad-based organizational involvement
Encourages data gathering and measurability	Understanding of problems, benefits, opportunities
Benchmarking opportunities	Problem solving as part of selection process
Programmable	Broad knowledge base
Useful input to sensitivity analysis and simulation	Multiple solutions and alternatives Multifunctional involvement leads to buy-in Risk sharing
Limitations:	*Limitations:*
Many success factors are nonquantifiable	Complex, time-consuming process
Probabilities and weights may change	Biases via power and politics
True measures do not exist	Difficult to proceduralize or repeat
Analyses and conclusions are often misleading	Conflict and energy intensive
Methods mask unique problems and opportunities	Do not fit conventional decision-making processes
Stifle innovative decision making	Intuition and emotion dominate over facts
Lack people involvement, buy-in, commitment	Justify wants over needs
Do not deal well with multifunctional issues and dynamic situations	Lead to more fact finding than decision making
May mask hidden costs and benefits	
Pressure to act quickly and possibly prematurely	

set of measures in determining the long-range cost and benefits of a project proposal to the enterprise. These methods rely to a large degree on *qualitative, judgmental decision making*. They cast a broad data gathering net and consider a wide spectrum of factors that are often difficult to quantify or even to describe. Yet, in spite of the limitations of quantitative evaluation and the increased use of qualitative approaches, virtually every organization supports its project selections with some form of quantitative measures—the most popular ones are ROI, cost-benefit, and payback period.

8.3 QUALITATIVE APPROACHES TO PROJECT EVALUATION AND SELECTION

Especially for project evaluations that involve complex sets of business criteria, the narrowly focused quantitative methods must often be supplemented by broad-scanning, intuitive processes and collective, multifunctional decision making such as *Delphi, nominal group technology, brainstorming, focus groups, sensitivity analysis,* and *benchmarking*. Each of these techniques can be used by itself to determine the "*best, most successful, or most valuable*" option. Or these techniques can be integrated into an analytical framework for *collective multifunctional decision making*, which is discussed next.

8.3.1 Collective, Multifunctional Evaluations

This process relies on subject experts from various functional areas to collectively define and evaluate broad project success criteria, employing both quantitative and qualitative methods. *The first step* is to define the specific organizational areas critical to project success and to assign expert evaluators. For a typical product development project, these organizations may include R&D, engineering, testing, manufacturing, marketing, product assurance, and customer services. These functional experts should be given the time necessary for the evaluation. They also should have the commitment from senior management for full organizational support. Ideally, these evaluators should have the responsibility for ultimate project implementation should the project be selected

Evaluation Factors. The next step is for the evaluation team to define the factors that appear critical to the ultimate success of the projects under evaluation and arrange them into a concise list which includes both quantitative and qualitative factors. A mutually acceptable scale must be worked out for scoring the evaluation criteria. Studies of collective multifunctional assessment practices show that simplicity of scales is crucial to a workable team solution. Three types of scales have produced the most favorable results in field studies: (1) a 10-point scale, ranging from +5 = most favorable to –5 = most unfavorable, (2) a 3-point scale, +1 = favorable, 0 = neutral or can't judge, –1 = unfavorable, and (3) a 5-point scale, A = highly favorable, B = favorable, C = marginally favorable, D = most likely unfavorable, F = definitely unfavorable. Weighing of criteria is not recommended for most applications because it complicates and often distorts the collective evaluation.

Evaluators score first individually all of the factors that they feel qualified to make an expert judgement on. Collective discussions follow. Initial discussions of the project alternatives, the markets, business opportunities, and the technologies involved are usually beneficial but not necessary for the first round of the evaluation process. The objective of this first round of expert judgments is to calibrate the scale of the opportunities and challenges presented. Further, each evaluator has the opportunity to recommend (1) actions needed for better assessment of the project, (2) additional data needed, and (3) suggestions that would enhance project success and the evaluation score. Before meeting for the next group session, agreed-on action items and activities for improving the decision process should be completed. With each iteration, the functional expert meetings are enhanced with more refined project data. Typically, between three and five iterations are required before a project selection can be finalized.

8.4 RECOMMENDATIONS FOR EFFECTIVE PROJECT EVALUATION AND SELECTION

Effective evaluation and selection of project opportunities involve many variables in the organizational and technological environment, often reaching far beyond cost and revenue measures. While economic models provide an important dimension of the project selection process, most situations are too complex to use simple quantitative methods as the sole basis for decision making. Many of today's project evaluation procedures include a broad spectrum of variables and rely on a combination of rational and intuitive processes for defining the value of a new project venture to the enterprise. The better a firm understands its business processes, markets, customers, and technologies, the better it will be able to evaluate the value of a new project venture. Further, manageability of the evaluation process is critical to its results, especially in complex situations. The process must have a certain degree of structure, discipline, and measurability to be conducive to the intricate multivariable analysis. One method of achieving structure and manageability of the process calls for grouping the evaluation variables into four categories: (1) degree of consistency and strength of the project as it relates to the business mission, strategy, and plan, (2) multifunctional ability to produce the project results, including technical, cost, and time factors, (3) success in the customer environment, and (4) economics, including profitability. Modern phase management and Stage-Gate processes provide managers with the tools for organizing and conducting project evaluations in a systematic way. The following section summarizes suggestions that can help managers in effectively evaluating projects for successful implementation.

Seek Out Relevant Information. Meaningful project evaluations require relevant quality information. The four categories of variables, identified above, can provide a framework for establishing the proper metrics and detailed data gathering.

Ensure Competence and Relevancy. Ensure that the right people become involved in the data collection and judgment processes.

Take Top-Down Look First, Detail Comes Later. Detail is less important than information relevancy and evaluator expertise. Don't get hung up on lack of data during the early phases of the project evaluation. Evaluation processes should be iterative. It does not make sense to spend a lot of time and resources in gathering perfect data to justify a "no-go" decision.

Select and Match the Right People. Whether the project evaluation consists of a simple economic analysis or a complex multifunctional assessment, competent people from functions critical to the overall success of the project should be involved.

Define Success Criteria. Whether deciding on a single project or choosing among alternatives, evaluation criteria must be defined. They can be quantitative, such as ROI, or qualitative, such as the chances of winning a contract. In either case, these evaluation criteria should cover the true spectrum of factors affecting success and failure of the project(s). Only the functional experts are qualified to identify these success criteria. Often, people from outside the company, such as vendors, subcontractors, or customers, must be included in this expert group.

Strictly Quantitative Criteria Can Be Misleading. Be aware of evaluation procedures based on quantitative criteria only (ROI, cost, market share, MARR, etc.). The input data used to calculate these criteria are likely based on rough estimates and are often unreliable. Evaluations based on predominately quantitative criteria should at least be augmented with some expert judgment as a "sanity check."

Condense Criteria List. Combine evaluation criteria, especially among the judgment categories, to keep the list manageable. As a goal, try to stay within 12 criteria for each category.

Gain a Broad Perspective. The inputs to the project selection process should include the broadest possible spectrum of data from the business environment that affect the success, failure, and limitations of the new project opportunity. Assumptions should be carefully examined.

Communicate. Facilitate communication among evaluators and functional support groups. Define the process for organizing the team and conducting the evaluation and selection process.

Ensure Cross-Functional Representation and Cooperation. People on the evaluation team must share a strategic vision across organizational lines. They also must sense the desire of their host organizations to support the project if it is selected for implementation. The purpose, goals, objectives, and relationships of the project to the business mission should be clear to all parties involved in the evaluation/selection process.

Don't Lose Sight of the Big Picture. As discussions go into detail during the evaluation, the team should maintain a broad perspective. Two global judgment factors can help to focus on the big picture of project success: (1) overall benefit-to-cost perception and (2) overall risk of failure perception. These factors can be recorded on a 10-point scale, –5 to +5. This also leads to an effective two-dimensional graphic display of competing project proposals.

Do Your Homework between Iterations. Because project evaluations are most likely conducted progressively, action items necessary for obtaining more information, clarification, and further analysis surface. These action items should be properly assigned and followed up, thereby enhancing the quality of the evaluation with each consecutive iteration.

Take a Project-Oriented Approach. Plan, organize, and manage your project evaluation/selection process as a *project.*

Resource Availability and Timing. Don't forget to include in your selection criteria the availability and timing of resources. Many otherwise successful projects fail because they cannot be completed within a required time period.

Use Red-Team Reviews. Set up a special review team of senior personnel. This is especially useful for large and complex projects with a major impact on overall business performance. This review team examines the decision parameters, qualitative measures, and assumptions used in the evaluation process. Limitations, biases, and misinterpretations that may otherwise remain hidden can often be identified and dealt with.

Stimulate Creativity and Candor. Senior management should foster an innovative ambience for the evaluation team. Evaluating complex project situations for potential success or failure involves intricate sets of variables, linked among organization, technology, and business environment. It also involves dealing with risks and uncertainty. Innovative approaches are required to evaluate the true potential of success for these projects. Risk sharing by senior management, recognition, visibility, and a favorable image in terms of high priority, interesting work, and importance of the project to the organization have been found to be strong drivers toward attracting and holding high-quality people on the evaluation team, and to gaining their active and innovative participation in the process.

Manage and Lead. The evaluation team should be chaired by someone who has trust, respect, and leadership credibility with the team members. Senior management can positively influence the work environment and the process by providing guidelines, charters, visibility, resources, and active support to the project evaluation team.

In summary, effective project evaluation and selection require a broad-ranging process across all segments of the enterprise and its environment to deal with the risks, uncertainties, ambiguities, and imperfections of data available for assessing the value of a new project venture relative to other opportunities. No single set of broad guidelines exists that guarantees the selection of successful projects. However, the process is not random! A better understanding of the organizational dynamics that affect project performance and the factors that drive cost, revenue, and other benefits can help in gaining a better, more meaningful insight into the future value of a prospective new project. Seeking out both quantitative and qualitative measures and incorporating them into a combined rational-judgmental evaluation process often yield the most reliable predictor of future project value and desirability. Equally important, the process requires managerial leadership and skills in planning, organizing, and communicating. Above all, leaders of the project evaluation team must be social architects, who can unify the multifunctional process and its people. They must share risks and foster an environment that is professionally stimulating and strongly linked with the support organizations eventually needed for project implementation. This is an environment that is conducive to cross-functional communication, cooperation, and integration of the intricate variables needed for effective project evaluation and selection.

8.5 SUMMARY OF KEY POINTS AND CONCLUSIONS

The key points that have been made in this chapter include:

- The project evaluation and selection process provides the framework for systematic data gathering and informed decision making for resource allocation.
- Decision making on new projects can be broken into four principal categories: (1) deciding initial feasibility, (2) screening and filtering for a quick viability decision on an emerging project, (3) deciding on the strategic value alternatives and options, and (4) deciding on the detailed feasibility and project go/no-go.
- Defining a meaningful aggregate measure for rating project value and success is difficult.
- Many of today's complex project scenarios require the integration of both analytical and judgmental techniques to evaluate their cost-benefits and feasibilities.
- Quantitative approaches to project evaluation and selection are often favored if the decisions require economic justification.
- The usefulness of quantitative methods depends on the assumption that the decision parameters, such as cash flow, risks, and economic factors, can actually be quantified.
- The most common quantitative methods of project evaluation and selection are (1) net present value (NPV) comparison, (2) return on investment (ROI) comparison, and (3) payback period (PBP) comparison.
- Project evaluations that involve complex criteria are often supplemented by broad-ranging, intuitive processes and collective, multifunctional decision making such

as. *Delphi, nominal group technology, brainstorming, focus groups, sensitivity analysis, and benchmarking.*

- *Collective, multifunctional evaluations* rely on experts from various functional areas for collectively defining and evaluating broad project success criteria, employing both quantitative and qualitative methods.
- While economic models provide an important dimension of the project selection process, most situations are too complex to use simple quantitative methods as the sole basis for decision making.
- The better a firm understands its business processes, markets, customers, and technologies, the better it will be able to evaluate the value of a new project venture.
- Understanding the organizational dynamics and the factors that drive cost, revenue, and other benefits can help one to gain a better, more meaningful insight into the future value of a prospective project.
- Any project evaluation and selection process must have structure, discipline, and measurability.
- One method of achieving structure and manageability is grouping the evaluation variables into four categories. (1) degree of consistency and strength between the project and the business mission, strategy, and plan, (2) ability to produce the project results, including technical, cost, and time factors, (3) success in the customer environment, and (4), economics, including profitability. Modern phase management concepts provide the tools for systematic decision making.
- Effective project evaluation and selection require a broad-ranging process across all segments of the enterprise and its environment to deal with the risks, uncertainties, ambiguities, and data imperfections.

8.6   CRITICAL THINKING: QUESTIONS FOR DISCUSSION

1. Discuss the strength and weaknesses of (a) quantitative and (b) qualitative methods of project evaluation and selection.
2. While this chapter suggests that collective, multifunctional evaluations are most powerful and reliable, do you see any limitations and managerial challenges in applying this method? Discuss.
3. Discuss project evaluation and selection methods in addition to those discussed in this chapter. Look what is being discussed in the literature and what is being used in your company. Discuss differences similarities, benefits, challenges, and limitations.
4. How does modern information technology and computer software help in the project evaluation and selection process?
5. What are the human factors (i.e., biases, trust, competence, and conflict) in properly evaluating and selecting new projects?
6. Develop a project evaluation and section procedure for your work environment.

8.7 REFERENCES AND ADDITIONAL READINGS

Balachandra, R. and Friar, J. (1997) "Factors for success in R&D projects and new product innovation: a contextual framework," *IEEE Transactions on Engineering Management*, Vol. 44, No. 3 (August), pp. 276–287.

Brenner, M. (1994) "Practical R&D project prioritization," *Research-Technology Management*, Vol. 37, No. 5, pp. 38–42.

Bulick, W. (1993) "Project evaluation procedures," *Cost Engineering*, Vol. 35, No. 10, pp. 27–32.

Campbell, K. and Helleloid, D. (2002). "An exercise to explore the future impact of new technologies," *Journal of Product Innovation Management*, Vol. 19, No. 1 (January), pp. 69–80.

Debruyne, M., Moenaertm, R., Griffen, A., Hart, S., Hultink, J., and Robben, H. (2002) "The impact of new product launch strategies on competitive reaction in industrial markets," *Journal of Product Innovation Management*, Vol. 19, No. 3 (May), pp. 159–170.

Hultink, J. and Langerak, F. (2002) "Launch decisions and competitive reactions: an exploratory market signaling study," *Journal of Product Innovation Management*, Vol. 19, No. 3 (May), pp. 199–212.

Kahn, K. (2002). "An exploratory investigation of new product forecasting practices," *Journal of Product Innovation Management*, Vol. 19, No. 2 (March), pp. 133–143.

Khurana, A. and Rosenthal, S. (1997). "Integrating the fuzzy front end of new product development," *Sloan Management Review*, Vol. 38, No. 2, pp. 103–120.

Kim, J. and Wilemon, D. (2002) "Focusing on the fuzzy front end in new product development," *R&D Management*, Vol. 32, No. 4, pp. 269–279.

Lin, C. and Chen, C. (2004). "New product go/no-go evaluation at the front end: a fuzzy linguistic approach," *IEEE Transactions on Engineering Management*, Vol. 51, No. 2 (May), pp. 197–207.

MacCormack, A. and Verganti, R. (2003) "Managing the sources of uncertainty: matching process and context in software development," *Journal of Product Innovation Management*, Vol. 20, No. 3 (May), pp. 217–232.

Menke, M. (1994) "Improving R&D decisions and execution," *Research-Technology Management,* Vol. 37, No. 5, pp. 25–32.

Obradovitch, M. and Stephanou, S. (1990). *Project Management: Risk and Productivity.* Bend, OR: Daniel Spencer Publishers.

O'Conner, V. and Veryzer, G. (2001) "The nature of market visioning for technology-based radical innovation," *Journal of Product Innovation Management*, Vol. 18, No. 4 (July), pp. 231–246.

Remer, D., Stokdyk, S., and Van Driel, M. (1993). "Survey of project evaluation techniques currently used in industry," *International Journal of Production Economics*, Vol. 32, No. 1, pp. 103–115.

Schmidt, R. (1993) "A model for R&D project selection," *IEEE Transactions on EM*, Vol. 40, No. 4, pp. 403–410.

Shtub, A., Bard, J., and Globerson, S. (1994) *Project Management: Engineering, Technology, and Implementation.* Engelwood Cliffs, NJ: Prentice-Hall.

Skelton, M. and Thamhain, H. (1993) "Concurrent project management: a tool for technology transfer," *Project Management Journal*, Vol. 26, No. 4, pp. 41–48.

Song, M. and Pary, M. (2002) "The determinants of Japanese new product success," *Journal of Marketing Research*, Vol. 34, No. 1 (May), pp. 64–73.

Thamhain, H. (2003) "Project evaluation and selection," *The Engineering Handbook* (Richard C. Dorf, ed.), Boca Raton, FL: CRC Press.

Ward, T. (1994) "Which product is BE$T?" *Chemical Engineering*, Vol. 101, No. 1, pp. 102–107.

Ziamou, P. (2002) "Commercializing new technologies: consumers' response to a new interface," *Journal of Product Innovation Management*, Vol. 19, No. 5 (September), pp. 365–374.

Resources for Further Information

The following journals are good sources of additional information on "project evaluation and selection" (the professional associations that publish these journals are listed in parentheses):

Cost Engineering (AACEI)

Engineering Management Journal (ASEM)

Engineering Management Review (IEEE)

Industrial Management (IIE)

International Journal of Project Management (IPMA)

Journal of Engineering and Technology Management

Journal of Product Innovation management (PDMA)

Project Management Journal (PMI)

Transactions on Engineering Management (IEEE)

The following professional societies present annual conferences and specialty publications which include discussions on "project evaluation and selection":

American Association for Cost Engineering (AACEI International), 209 Prairie Avenue, Morgantown, WV 26501, (304) 296-8444; *www.aacei.org*

American Society for Engineering Management (ASEM), Rolla, MO 65401, (314) 341-2101, *www.asem.org*

Association of Proposal Management Professionals, 300 Smelter Avenue NE #1, Great Falls, MT 59404, Phone/Fax: (406) 454-0090; *www.apmp.org*

Institute of Electrical and Electronic Engineers (IEEE), East 47 St., New York, NY 10017-2394, (800) 678-4333, *www.ieee.org*

Institute for Industrial Engineering (IIE), *www.iie.org*

International Association for Management of Technology (IAMOT), *www.iamot.org*

International Project Management Association (IPMA), *www.ipma.org*

Product Development and Management Association (PDMA), 17000 Commerce Pkwy, Mount Laurel, NJ 08054, (800) 232-5241, *www.pdma.org*

Project Management Institute (PMI), Newtown Square, PA 19073, (610) 734-3330, *www.pmi.org*

Society for Concurrent Engineering and Product Development (SCEPD), Boston, Massachusetts.

Appendix

SUMMARY DESCRIPTION OF TERMS, VARIABLES, AND ABBREVIATIONS USED IN THIS CHAPTER

I. TERMS:

Cross-Functional. Actions which span organizational boundaries.

Phase Management. Projects are broken into natural implementation phases, such as development, production, and marketing, as a basis for project planning, integration, and control. Phase management also provides the framework for *concurrent engineering* and *Stage-Gate processes.*

Project Success. A comprehensive measure, defined in both quantitative and qualitative terms which includes economic, market, and strategic objectives.

Stage-Gate Process. Framework for executing projects within predefined stages (see also Phase Management) with measurable deliverables *(gates)* at the end of each stage. The gates provide the review metrics for ensuring successful transition and integration of the project into the next stage.

Weighing of Criteria. A multiplier associated with specific evaluation criteria.

II. VARIABLES AND ABBREVIATIONS:

A   *Annuity* is the present worth of a revenue or cost at the end of a period n.

CB   *Cost benefit* is the net present value of all ROIs in dollars.

i   Prevailing *interest rate.*

I   *Investment.*

IRR   *Internal rate of return*, the average return on investment realized by a firm on its investment capital.

MARR *Minimum attractive rate of return* on new investments acceptable to an organization.

NPV *Net present value* of a stream of future revenues or costs.

PBP *Payback period,* the time period needed to recover the original investment.

PR *Project rating factor,* a measure developed by Carlo Pacifico for predicting project success.

PW *Present worth* (also called annuity), the present value of a revenue or cost at the end of a period n.

ROI *Return on investment.*

z *Project rating factor,* a measure developed by Sobelman for predicting project success.

LEADING TECHNOLOGY TEAMS

Dieter Zetsche replaced James P. Holden as CEO of Chrysler in November 2000. Holden was fired with less then one year on the job because of severe financial problems that led to a $2 billion loss in 2001 and a no-profit projection until 2003. Since the merger with Daimler-Benz in 1998, Chrysler's fortune was shifting rapidly. The company owed much of its success to the ability to avoid head-on competition by creating such innovative, high-margin vehicles as the minivan. But, as more efficient rivals, such as Honda and Toyota moved swiftly into the U.S. market with similar vehicles, prices began to fall, and Chrysler started to suffer. Now Zetsche has to turn the company around. Part of his strategy is to overhaul the product development process, a highly interdisciplinary three-year team effort involving all functional areas across Chrysler, plus hundreds of suppliers, contractors, government agencies, and the Daimler-Benz parent company. Zetsche is determined to wring synergy out of the $36 billion merger by combining German engineering with American marketing. He is devoting a great deal of effort to achieving an agreement with Daimler to make diesel engines and other M-class SUV parts available for Chrysler's Cherokees and PT Cruisers. Going even further, Zetsche plans to install a wide array of Mercedes parts in a large number of Chrysler cars, ranging from the Grand Cherokee to the Dodge Intrepid. Starting in 2004, the new

Crossfire, a two-seat roadster, will be built with a Mercedes transmission, axles, and engine. In addition Chrysler will work with Mitsubishi to develop small and midsized cars.

Zetsche admits that a "not-invented-here" syndrome kept Chrysler and Mercedes from sharing technology in the past. In fact, it took a group of senior executives from both companies several months to work out a cooperative technology agreement. While the policy draft of this executive agreement is still heavily debated by several newly formed committees in both companies, "a momentum of cooperation is developing," says Zetsche. Zetsche benefited from his boss's endorsement of this initiative when Jurgen Schrempp announced that "it's finally like someone is taking the foot off the brake!" With increasing competition, Zetsche needs to keep customers coming through the door without offering big price discounts. This will require combining new styles and features with thrift and agility in the market. In overhauling the new vehicle development process, Zetsche counts on W. Bernhard, his chief operating officer, to put more focus on the early stages of the car development. By pulling together teams from all areas of the company, including design, styling, engineering, manufacturing, marketing, purchasing, and field services, he hopes to increase resource efficiency and quality, without diminishing Chrysler's creativity and market responsiveness. A key ingredient in this new process is a system of "Quality Gates," borrowed from Mercedes. This will allow Zetsche and his team to monitor and review a vehicle via 11 checkpoints throughout its three-year development cycle. Zetsche is fully aware of the organizational complexities and team challenges involved in bringing Daimler-Chrysler's numbers into the black: "To my knowledge, no one has ever turned such a grim situation around without a near-death experience." However, Zetsche has no choice. If successful, he can pretty well name his price. Otherwise, his name will likely join the list of casualties of the $36 billion merger experiment.

Source: Web site *www.DaimlerChrysler.com* and "Can This Man Save Chrysler?" *Business Week,* September 17, 2001.

9.1   CHALLENGES OF TECHNICAL TEAMWORK

The complexities and challenges faced by Daimler-Chrysler are quite common in today's world of business (Armstrong 2000, Dillon 2001, Gray and Larson, 2000, Thamhain 2002). While any organization that wants to compete today must pay attention to teamwork, those most challenged seem to be managers in complex and technology-intensive situations, characterized by high speed, great change, and high uncertainty (Shim and Lee 2001, Zhang et al. 2003). As indicated in the Chrysler scenario, these technical team efforts often span many organizational lines, including an intricate functional spectrum of assigned personnel, support groups, subcontractors, vendors, partners, government agencies, and customer organizations. Uncertainties and risks introduced by technological, economic, political, social, and regulatory factors are always present and can be an enormous challenge to organizing and managing the project teams. Therefore, team leaders must be both technically and socially competent. In fact, it is interesting to note in findings from field research that performance problems involve mostly managerial, behavioral, and organizational issues, rather than technical difficulties (Belassi and Tukel 1996, Hartman and Ashrafi 2002, Whitten 1995). Critical success factors (CSF) for project teams span a wide spectrum of technological, organizational, and interpersonal issues, including gaining and maintaining cohesiveness, commitment, technology transfer, self-direction, dealing with innovation, changing requirements and resource limitations, and flexibility and speedy implementation. In these situations, traditional models of team management are often not effective, and can be even counterproductive.

9.1.1   Teamwork a Managerial Frontier

Most managers consider effective cross-functional teamwork a key determinant of project performance and success in today's competitive world of business. Virtually all managers recognize the critical importance of effective teamwork and strive for continuous improvement of team performance in their organizations. Yet, only 1 in 10 of these managers has a specific metric for actually measuring team performance. Obviously, this creates some tough challenges, especially in project-based environments where teamwork is crucial to business success. In these organizational environments, work teams must successfully integrate multidisciplinary activities, unify different business processes, and deal with cross-functional issues, such as innovation, quality, speed, producability, sourcing, and service. Managerial principles and practices have changed dramatically. Not too long ago, project management was considered to a large degree "management science." Project leaders *could* ensure successful integration for most of their projects by focusing on properly defining the work, timing, and resources, and by following established procedures for project tracking and control. Today, these factors are still crucial. However, they have become threshold competencies, critically important but unlikely to guarantee project success by themselves. Today's complex business world requires *project teams* that are fast and flexible, and can dynamically and creatively work toward established objectives in a changing environment (Bhatnager 1999, Jasswalla and Sashittal 1999, Thamhain 2001, Thamhain and Wilemon 1999). This requires effective networking and cooperation among people from different organizations, support groups, subcontractors, vendors, government agencies, and customer communities. It also requires the ability to deal with uncertainties and risks caused by technological, economic, political, social, and regulatory factors. In addition, project leaders have to organize and manage their teams across organizational lines. Dealing with resource sharing, multiple reporting relationships, and broadly based alliances is as common in today's business environment as e-mail, flex time, and home offices.

Because of these complexities and uncertainties, traditional forms of hierarchical team structure and leadership are seldom effective and are being replaced by self-directed, self-managed team concepts (Barner 1997, Thamhain and Wilemon, 1999). Often the project manager becomes a social architect who understands the *interaction of organizational and behavioral variables*, facilitates the work process, and provides overall project leadership for developing multidisciplinary task groups into unified teams and fostering a climate conducive to involvement, commitment, and conflict resolution. Typical managerial responsibilities and challenges of today's project team leaders are summarized in Table 9.1. This table can also be used as a tool for self-assessment, including establishing performance measures and training and organizational development needs.

Table 9.1   Responsibilities and Challenges of Project Team Leaders

- Bringing together the right mix of competent people who will develop into a team
- Building lines of communication among task teams, support organizations, upper management, and customer communities
- Building the specific skills and organizational support systems needed for the project team
- Coordinating and integrating multifunctional work teams and their activities to form a complete system
- Coping with changing technologies requirements and priorities, while maintaining project focus and team unity
- Dealing with anxieties, power struggles, and conflict
- Dealing with support departments; negotiating, coordinating, integrating
- Dealing with technical complexities
- Defining and negotiating the appropriate human resources for the project team
- Encouraging innovative risk taking without jeopardizing fundamental project goals
- Facilitating team decision making
- Fostering a professionally stimulating work environment where people are motivated to work effectively toward established project objectives
- Integrating individuals with diverse skills and attitudes into a unified workgroup with unified focus
- Keeping upper management involved, interested, and supportive
- Leading multifunctional task groups toward integrated results in spite of often intricate organizational structures and control systems
- Maintaining project direction and control without stifling innovation and creativity
- Providing an organizational framework for unifying the team
- Providing or influencing equitable and fair rewards to individual team members
- Sustaining high individual efforts and commitment to established objectives

9.2   WHAT WE KNOW ABOUT TECHNOLOGY-ORIENTED TEAMS

To be sure, teamwork is not a new idea. The basic concepts of organizing and managing teams go back to biblical times. However, it was not before the beginning of the twentieth century that work teams were formally recognized as an effective device for enhancing organizational performance. Specifically, the discovery of important social phenomena in the classic Hawthorne studies (Roethlingsberger and Dickinson 1939) led to new insight on group behavior and the benefits of workgroup identity and cohesion to performance (Dyer 1977). Indeed, much of the *human relations movement* that followed Hawthorne is based on the group concept. McGregor's (1960) theory Y, for example, spells out the criteria for an effective workgroup, and Likert (1967) called his highest form of management the participating group, or System 4. However, the process of team building becomes more complex and requires more specialized management skills as bureaucratic hierarchies decline and horizontally oriented teams and work units evolve.

9.2.1   Redefining the Process

In today's complex multinational and technologically sophisticated environment, the group has reemerged in importance as the *project team* (Fisher 1993, Nurick and Thamhain 1993, Thamhain and Wilemon 1999). Supported by modern information technologies, and consistent with the concepts of stakeholder management (Newell and Rogers 2002) and learning organizations (Senge and Carstedt 2001), the roles and boundaries of teams are expanding toward self-direction within more open and organizationally transparent processes. Work teams play an important role not only in traditional projects, such as new product development, systems design, and construction, but also in implementing organizational change, transferring technology concepts, and running election campaigns. Whether Yahoo! creates a new search engine, Sony develops a new laptop computer, or the World Health Organization rolls out a new information system, success depends to a large degree on effective interactions among the team members responsible for the new development. This includes support groups, subcontractors, vendors, partners, government agencies, customer organizations, and other project stakeholders (Armstrong 2000; Barkema, Baum, and Mannix 2002; Dillon 2001; Gray and Larson 2000; Karlsen and Gottschalk 2004; Thamhain 2003; Zanoni and Audy 2004). Globalization, privatization, digitization, and rapidly changing technologies have transformed our economies into a hypercompetitive enterprise system where virtually every organization is under pressure to do more things faster, better, and cheaper. Effective teamwork is seen as a key success factor in deriving competitive advantages from these developments. At the same time, the process of team building has become more complex and requires more sophisticated management skills as bureaucratic hierarchies and support systems decline.

In this transformation process, the goals and energies of individual contributors merge and focus on specific objectives. When describing an effective project team, managers stress consistently that high performance, although ultimately reflected by producing desired results, on time and within budget, is a derivative of many factors, which are graphically shown in Figure 9.1. Team building is an ongoing process that requires leadership skills and an understanding of the organization and its interfaces, authority, power structures, and motivational factors. This process is particularly crucial in environments where complex multidisciplinary or transnational activities require the skillful integration of many functional specialties and support groups with diverse organizational cultures, values, and intricacies (Oderwald 1996).

Typical examples of such multidisciplinary activities requiring unified teamwork for successful integration include:

Establishing a new program
Transferring technology
Improving project-client relationships
Organizing for a bid proposal
Integrating new project personnel
Resolving interfunctional problems
Working toward major milestones
Reorganizing mergers and acquisitions
Transitioning the project into a new activity phase
Revitalizing an organization.

Because of their potential for producing economic advantages, work teams and their development have been researched by many. Starting with the evolution of formal project organizations in the 1960s, managers in various organizational settings have expressed increasing concern with and interest in the concepts and practices of multidisciplinary team building. As a result, many field studies have been conducted, investigating workgroup dynamics and criteria for building effective, high-performing

Figure 9.1   Characteristics of high-performing project team.

project teams. These studies have contributed to the theoretical and practical understanding of team building and form the fundamental concepts discussed in this chapter. Prior to 1980, most of these studies focused just on the behavior of the team members, with limited attention given to the organizational environment and team leadership. While the qualities of the individuals and their interaction within the team are crucial elements in the teamwork process, they represent only part of the overall organization and management system that influences team performance, a fact that was recognized by Bennis and Shepard as early as 1956. Since 1980 an increasing number of studies have broadened our understanding of the team work process (Dumaine 1991, Tichy and Urlich 1984, Walton 1985). These more recent studies show the enormous breadth and depth of subsystems and variables involved in the organization, development, and management of a high-performing work team (Gupta and Wilemon 1996). These variables include planning, organizing, training, organizational structure, nature and complexity of task, senior management support, leadership, and socioeconomic variables, just to name the most popular ones. Even further, researchers such as Dumaine (1991), Drucker (1996), Peters and Waterman (1987, 1997) and Moss Kanter (1989), have emphasized the nonlinear, intricate, often chaotic, and random nature of teamwork, which involves all facets of the organization, its members, and the environment. These teams became the conduit for transferring information, technology, and work concepts across functional lines quickly, predictably, and within given resource restraints.

9.2.2   Team Life Cycle

The life cycle of a project team spans the complete project, not just a particular phase. For example, the Chrysler team responsible for creating a new model integrates activities ranging from assessing an opportunity to product research, feasibility analysis, development, and engineering, transferring technology to manufacturing, marketing, and field service. This work may also involve bid proposals, licensing subcontracting, acquisitions, and offshore manufacturing. The need for close integration of activities across the entire project life cycle requires that these multidisciplinary teams stay together as a unified, effective workgroup for most of the product life cycle, rather than just for a particular phase of core activities. For example, the primary mission of the product development team may focus on the engineering phase, but the team also supports activities ranging from the recognition of an opportunity to feasibility analysis, bid proposals, licensing, subcontracting, transferring technology to manufacturing, distribution, and field service. This creates managerial challenges in dealing effectively with resource leveling, priority conflicts, and long-range multifunctional commitment

9.3   TOWARD SELF-DIRECTION AND VIRTUAL TEAMS

Especially with the evolution of contemporary organizations, such as the matrix, traditional bureaucratic hierarchies have declined, and horizontally oriented teams and

Table 9.2   Self-Directed Teams

Definition: A group of people chartered with specific responsibilities for managing themselves and their work, with minimal reliance on group-external supervision, bureaucracy, and control. Team structure, task responsibilities, work plans, and team leadership often evolve based on needs and situational dynamics.

Benefits: Ability to handle complex assignments, requiring evolving and innovative solutions that cannot be easily directed via top-down supervision. Widely shared goals, values, information, and risks. Flexibility toward needed changes. Capacity for conflict resolution, team building, and self-development. Effective cross-functional communications and work integration. High degree of self-control, accountability, ownership, and commitment toward established objectives.

Challenges: A unified, mature team does not just happen, but must be carefully organized and developed by management. A high degree of self-motivation and sufficient job, administrative, and people skills must exist among the team members. Empowerment and self-control might lead to unintended results and consequences. *Self-directed* teams are *not* necessarily *self-managed*; they often require *more* sophisticated external guidance and leadership than conventionally structured teams.

work units have become increasingly important to effective project management (Fisher 1993, Marshall 1995, Shonk 1996). Increasingly, the team leader's role as supervisor has diminished in favor of more *empowerment and self-direction* of the team, as defined in Table 9.2. In addition, advances in information technology made it feasible and effective to link team members over the Internet or other media, creating a *virtual team* environment, as described in Table 9.3. *Virtual teams* and *virtual project organizations* are powerful managerial tools, especially for companies with geographically dispersed project operations and for linking contractors, customers, and regulators with the core of the project team.

These contemporary team concepts are being applied to different forms of project activities in areas of products, services, acquisition efforts, political election campaigns, and foreign assistance programs. They are also found in specialty task groups such as venture teams, skunk works, process action teams, and focus groups. For these kinds of highly multifunctional and nonlinear processes, researchers stress the need for strong integration and orchestration of cross-functional activities, linking the various workgroups into a unified project team that focuses energy and integrates all subtasks toward desired results. While these realities hold for most team efforts in today's work environment, they are especially pronounced for efforts that are associated with risk, uncertainty, creativity, and team diversity such as high-technology and/or multinational projects. These are also the work environments that first departed from traditional hierarchical team structures and tried more self-directed and network-based virtual concepts (Fisher 1993).

Table 9.3   Virtual Teams

Definition: A group of project team members, linked via the Internet or media channels to each other and various project partners, such as contractors, customers, and regulators. Although they are physically separated, technology links these individuals so they can share information and operate as a unified project team. The number of elements in a virtual team and their permanency can vary, depending on need and feasibility. An example of a virtual team is a project review conducted among the team members, contractors, and a customer through an Internet Web site.

Benefits: Ability to share information and communicate among team members and organizational entities on geographically dispersed projects. Ability to share and communicate information in a synchronous and asynchronous mode (application: communication across time zones, holidays, and shared workspaces). Creating unified visibility of project status and performance. Virtual teams, to some degree, bridge and neutralize the cultural and value differences that exist among different task teams of a project organization.

Challenges: The effectiveness of the virtual team depends on the team members' ability to work with the given technology. Information flow and access are not necessarily equal for all team members. Information may not be processed uniformly throughout the team. The virtual team concept does not fit the culture and value system of all members and organizations. Project tracking, performance assessment, and managerial control of project activities are often very difficult. Risks, contingencies, and problems are difficult to detect and assess. Virtual organizations often do not provide effective methods for dealing with conflict, power, candor, feedback, and resource issues. Because of the many limitations, more traditional team processes and communications are often needed to augment virtual teams.

9.4   MEASURING PROJECT TEAM PERFORMANCE

"A castle is only as strong as the people who defend it." This Japanese proverb also applies to organizations. They are only as effective as their unified team efforts. However, team performance is difficult to measure because it involves highly complex, interrelated sets of variables, including attitudes, personal preferences, and perceptions that are difficult to quantify. Yet, in spite of the existing cultural and philosophical differences among organizations, research shows[1] that a general agreement exists among managers on certain performance measures for project teams and their results.

9.4.1   Project Performance Measures

Starting with the "bottom line" of *team results* and *project success*, a considerable agreement exists among managers on the following metrics of nine measures:

---

[1]In fact over 90% of the project managers interviewed during a survey by Thamhain and Wilemon (1987) and Thamhain (2004) mentioned three factors, (1) project success according to agreed-on results, (2) on-time performance and (3) on-budget performance, as being among the most important criteria of team performance.

1. Project success according to agreed-on results
2. On-time performance
3. On-budget performance

An estimated 90 percent of project managers include these factors among the three most important measures of project success. The majority of managers rank these factors in the order shown. In addition, other factors are often mentioned as important to project success. They include:

4. Overall customer or sponsor satisfaction
5. Responsiveness and flexibility to customer requirements and changes
6. Dealing effectively with risk and uncertainty
7. Positioning the project for future business
8. Stretching beyond planned goals
9. Organizational learning to benefit future projects

9.4.2 Team Effectiveness Measures

Team characteristics drive project performance. However, this relationship is not "linear." Moreover, project performance is influenced by many "external" factors, such as technology, socio-economic factors, and market behavior, making it difficult to determine exactly how much the team characteristics influence project perform-ance. Yet, lessons from field research[2] strongly suggest specific factors, such as those graphically shown in Figure 9.1, characterizing high performing teams. More specif-ically, Table 9.4. breaks the characteristics of a high-performing teams into four cat-egories: (1) work and team structure, (2) communications and control, (3) team leadership, and (4) attitude and values. These broad measures can provide a frame-work for benchmarking. Teams that score high on these characteristics are also seen by upper management as most favorable in dealing with cost, quality, creativity, schedules, and customer satisfaction. They also receive favorable ratings on the more subtle measures of team performance, such as flexibility, change orientation, inno-vative performance, high morale, and team spirit.

The significance of determining the association between team characteristics and project performance lies in two areas. First, it offers some insight as to what an effective team environment looks like, providing the basic framework for team assessment, benchmarking, and development. Second, a better understanding of how team characteristics affect project performance provides building blocks for further research on organization development, such as defining drivers and barriers to team performance. It also provides a framework for leadership style development.

---

[2]For more detailed discussions of the field research see Thamhain (2003, 2004).

Table 9.4   Benchmarking Your Team Performance

Work and Team Structure

- Team participates in project definition, work plans evolve dynamically
- Team structure and responsibilities evolve and change as needed
- Broad information sharing
- Team leadership evolves based on expertise, trust, respect
- Minimal dependence on bureaucracy, procedures, politics

Communication and Control

- Effective cross-functional channels, linkages
- Ability to seek out and process information
- Effective group decision making and consensus
- Clear sense of purpose and direction
- Self-control, accountability, and ownership
- Control is stimulated by visibility, recognition, accomplishments, autonomy

Team Leadership

- Minimal hierarchy in member status and position
- Internal team leadership based on situational expertise, trust, and need
- Clear management goals, direction, and support
- Inspires and encourages

Attitudes and Values

- Members are committed to established objectives and plans
- Shared goals, values, and project ownership
- High involvement, energy, work interest, need for achievement, pride, self-motivated
- Capacity for conflict resolution and resource sharing
- Team building and self-development
- Risk sharing, mutual trust and support
- Innovative behavior
- Flexibility and willingness to change
- High morale and team spirit
- High commitment to established project goals
- Continuous improvement of work process, efficiency, quality
- Ability to stretch beyond agreed-on objectives

9.5   A MODEL FOR TEAM BUILDING

Figure 9.2 provides a simple model for organizing and analyzing the variables that influence the team's characteristics and its ultimate performance, as baselined in Figure 9.1 and Table 9.1. The influences shown in Figure 9.2 are divided into four sets: (1) drivers of and barriers to high team performance; (2) managerial leadership style, including components of authority, motivation, autonomy, trust, respect, credibility, and friendship; (3) organizational environment, such as working conditions, job content, resources, and organizational support factors; and (4) the social, political, and economic factors of the firm's external business environment. All four sets of variables are intricately interrelated. However, using the systems approach allows researchers and management practitioners to break down the complexity of the process, thus helping in analyzing team performance and in developing leadership effectiveness.

9.5.1   Drivers of and Barriers to High Team Performance

Management tools such as benchmarking and root-cause analysis can be helpful in identifying the drivers of and barriers to effective teamwork. *Drivers* are factors that influence the project environment favorably, such as interesting work and good project leadership. These factors are perceived as enhancing team effectiveness, and therefore correlate positively with team performance. *Barriers* are factors that have an unfavorable influence, such as unclear objectives and insufficient resources, therefore impeding team performance. Based on field research (Thamhain 2003, 2004), the 10 strongest drivers and 15 strongest barriers are listed in Table 9.5. All of these factors have been listed alphabetically to avoid too narrowly drawn conclusions. While the actual statistics yielded different performance correlation levels

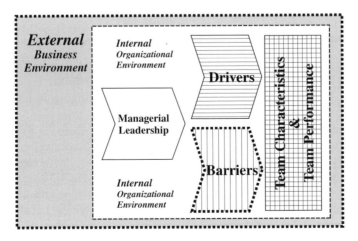

Figure 9.2   Model for analyzing team performance.

Table 9.5   Strongest Drivers of and Barriers to Project Team Performance

Drivers	Barriers
1. Clear project plans and objectives	1. Communication problems
2. Good interpersonal relations and shared values	2. Conflict among team members or between team and support organizations
3. Good project leadership and credibility	3. Different outlooks, objectives, and priorities perceived by team members
4. Professional growth potential	
5. Professionally interesting and stimulating work	4. Poor qualification of team/project leader
	5. Poor trust, respect, and credibility of team leader
6. Project visibility, high priority	
7. Proper technical direction and team leadership	6. Insufficient resources
	7. Insufficient rewards
8. Qualified, competent team personnel	8. Lack of project challenge and interest
9. Recognition of sense of accomplishment	9. Lack of senior management support, interest, and involvement
10. Management involvement and support	10. Lack of team definition, role conflict, and confusion
	11. Lack of team member commitment
	12. Poor project team/personnel selection
	13. Shifting goals and priorities
	14. Unclear team leadership, power struggle
	15. Unstable project environment, poor job security, anxieties

for different drivers and barriers, they represent the strongest association observed in the field studies. All correlations are statistically significant at $p = .05$ or better. Collectively, they explain over 85 percent of the variance in project team performance.

It is further interesting to note that many of the factors in Table 9.5 are, to a large degree, based on the perception of team members. That is, team members *perceive* "good personal relations" or "communication problems." Since this perception is the reality that influences the team behavior, management must deal with the conditions as seen by the people and foster a project environment conducive to the needs of the team. Such a favorable work environment not only enhances the drivers of and minimizes the barriers to project performance, it is also associated with the *15 measures that characterize a high-performing project team*, as discussed earlier and shown in Figure 9.1.

These field-research-based observations provide some focus and support to the broad range of field observations and criteria for effective team management discussed throughout this chapter. An important lesson follows from the analysis of these observations. Managers must foster a work environment supportive of their team members. Creating such a climate and culture conducive to quality teamwork involves multifaceted management challenges, which increase with the complexities of the project and its organizational environment. No longer will technical expertise or good leadership alone be sufficient, but excellence across a broad range of skills and sophisticated organizational support are required to manage these project teams effectively. Hence, it is critically important for project leaders to understand, identify, and minimize the various barriers to team development.

9.6 BUILDING HIGH-PERFORMING TEAMS

What does all this mean to managers in today's work environment with high demands on efficiency, speed, and quality? Project teams are becoming increasingly more important. However, exploiting the team potential can rarely be done "top-down." Given the realities of today's business environment, its technical complexities, cross-functional dependencies, and the need for innovative performance, more and more project leaders have to rely on information from and judgments made by their team members for developing solutions to complex problems. Especially with decision processes distributed throughout the team and solutions often evolving incrementally and iteratively, power and responsibility are shifting from managers to the project team members, who take higher levels of responsibility, authority, and control over project results. That is, these teams become *self-directed*, gradually replacing the more traditional, hierarchically structured project team. These emerging team processes are seen as a significant development for orchestrating the multifunctional activities that come into play during the execution of today's complex projects. These processes rely strongly on group interaction, resource and power sharing, group decision making, accountability, self-direction, and control. Leading such self-directed teams also requires a great deal of team management skills and overall guidance by senior management. In addition, managers must recognize the organizational dynamics involved during the various phases of the team development process. A four-stage model, originally developed by Hersey and Blanchard, and graphically shown in Figure 9.3, is often used by management researchers and practitioners as a framework to analyze the team development process. The four stages are labeled (1) team formation, (2) team startup, (3) partial integration, and (4) full integration. These stages are also known as *forming*, *storming*, *norming*, and *performing*, giving an indication of team behavior at each one of the stages.

No workgroup comes fully integrated and unified in its members' values and skill sets, but rather needs to be skillfully nurtured and developed. Leaders must recognize the professional interests, anxieties, communication needs, and challenges of

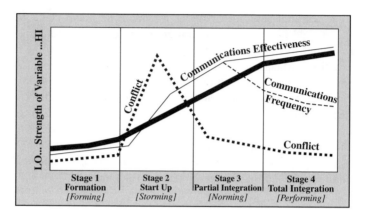

Figure 9.3 The four stages of team development.

their team members and anticipate them as the team goes through the various stages of integration. Moreover, team leaders must adapt their managerial style to the specific situation of each stage.[3] That is, team leaders must recognize what works best at each stage, and what is most conducive to the team development process. Many of the problems that occur during the formation of the new project team or during its life cycle are normal and often predictable. However, they present barriers to effective team performance. The problems must be quickly identified and dealt with. Early stages, such as the *team formation* and *startup*, usually require a predominately directive style of team leadership. Providing clear guidelines on the project mission and its objectives and requirements and creating the necessary infrastructure and logistics support for the project team are critically important in helping the team to pass through the first two stages of their development quickly. During the third stage, *partial integration*, or *norming*, the team still needs a considerable amount of guidance and administrative support, as well as support in dealing with the inevitable human issues of conflict, power and politics, credibility, trust, respect, and the whole spectrum of professional career and development issues. This is the stage where a combination of *directive* and *participative leadership* will produce the most favorable results. Finally, a team that has reached the fully integrated stage, by definition, becomes "self-directed." That is, such a fully integrated, unified team can work effectively with a minimum degree of external supervision and administrative support, as described earlier in Table 9.2. While at this stage the team often appears to have very little need for external managerial intervention, it requires highly sophisticated external leadership to maintain this delicate state of team effectiveness and focus.

[3]The discussions on leadership style effectiveness are based on field research by H. J. Thamhain and D. L. Wilemon, published in "A High-Performing Engineering Project Team," *The Human Side of Managing Innovation* (R. Katz, editor), New York: Oxford Press, 1997.

In summarizing the criteria for effective team management, *three propositions* can be stated to highlight the organizational environment and managerial style conducive to high project team performance:

- *Drivers and Barriers.* The degree of project success seems to be primarily determined by the strength of specific driving forces and barriers that are related to *leadership*, *job content*, *personal needs*, and the *general work environment*, as shown in Table 9.4..
- *Team Environment.* The strongest single driver of team performance and project success is a *professionally stimulating team environment*, characterized by interesting and challenging work, visibility and recognition of achievements, growth potential, and good project leadership.
- *Work Challenge Dividend.* A professionally stimulating team environment is also favorably associated with *low conflict*, *high commitment*, *highly involved personnel*, *good communications*, *willingness to change*, *innovation*, and *on-time* and *on-budget performance*.

To be effective in organizing and directing a project team, the leader must not only recognize the potential drivers and barriers, but also know when in the life cycle of the project they are most likely to occur. The effective project leader takes preventive actions early in the project life cycle and fosters a work environment that is conducive to team building as an ongoing process.

The effective team leader is a social architect who understands the interaction of organizational and behavioral variables and can foster a climate of active participation and minimal dysfunctional conflict. This requires carefully developed skills in leadership, administration, and organization and technical expertise. It further requires the project leader to have the ability to involve top management to ensure organizational visibility, resource availability, and overall support for the new project throughout its life cycle.

9.7 RECOMMENDATIONS FOR EFFECTIVE TEAM MANAGEMENT

Managing technology-based project teams is not for the weak and faint of heart. However, the large number of successful implementation of complex technology projects from all segments of business and society, new product concepts provides clear evidence that these challenges can be met with the right organizational ambience. Observations from best practices show that specific working conditions and managerial processes appear most favorably associated with teamwork, in spite of the complexities, organizational dynamics, and cultural differences among companies.[4]

---

[4]Although each organization is unique with regard to its business, operation, culture, and management style, field studies show a general agreement on the types of factors that are critical to effectively organizing and managing projects in concurrent multiphase environments (Denker 2001, Harkins 1998, Nellore and Balachandra 2001, Pillai et al. 2002, Prasad 1998, Thamhain and Wilemon 1999).

These conditions serve as bridging mechanisms that enhance team performance. A number of specific recommendations may help managers in facilitating working conditions conducive to multidisciplinary task integration and to building high-performing project teams. The sequence and focus of these recommendations follow to some degree the chronology of a typical product development life cycle.

Early-Project-Life-Cycle Team Involvement. As for any project, effective project planning and early team involvement are crucial to successful project team performance. This is especially important for product developments where parallel task execution depends on continuous cross-functional cooperation for dealing with incremental work flow and partial result transfers. Team involvement, early in the project life cycle, will also have a favorable impact on the team environment, building enthusiasm toward the assignment, team morale, and ultimately team effectiveness. Because project leaders have to integrate various tasks across many functional lines, proper planning requires the participation of all stakeholders, including support departments, subcontractors, and management. Modern project management techniques, such as phased project planning and Stage-Gate concepts, plus established standards, such as *PMI's Project Management Body of Knowledge (PMBOK)*, provide the conceptional framework and tools for effective cross-functional planning and for organizing the work for effective execution.

Define Work Process and Team Structure. Successful project team management requires an infrastructure conducive to cross-functional teamwork and technology transfer. This includes properly defined interfaces, task responsibilities, reporting relations, communication channels, and work transfer protocols. The tools for systematically describing the work process and team structure come from the conventional project management system; they include (1) *project charter*, defining the mission and overall responsibilities of the project organization, including performance measures and key interfaces, (2) *project organization chart*, defining the major reporting and authority relationships, (3) *responsibility matrix* or *task roster*, (4) *project interface chart, such as the N-Squared Chart discussed earlier*, and (5) *job descriptions*.

Develop Organizational Interfaces. Overall success of a project team depends on effective cross-functional integration. Each task team should clearly understand its task inputs and outputs, interface personnel, and work transfer mechanism. Team-based reward systems can help to facilitate cooperation with cross-functional partners. Team members should be encouraged to check out early feasibility and system integration. QFD concepts, *N*-square charting, and well-defined phase-gate criteria can be useful tools for developing cross-functional linkages and promoting interdisciplinary cooperation and alliances. It is critically important to include in these interfaces all of the support organizations, such as purchasing, product assurance, and legal services, as well as outside contractors and suppliers.

Staff and Organize the Project Team. Project staffing is a major activity, usually conducted during the project formation phase. Because of time pressures, staffing is often done hastily and prior to defining the basic work to be performed. The result is team personnel that are suboptimally matched to the job requirements, resulting in conflict, low morale, suboptimum decision making, and ultimately poor project performance. While this deficiency will cause problems for any project organization, it is especially unfavorable in a product development environment that relies on strong cross-functional teamwork and shared decision making, built on mutual trust, respect, and credibility. Team personnel with skill sets poorly matched to the job requirements are seen as incompetent. This affects the trust, respect, and credibility among team members, and ultimately team performance. For best results, project leaders should *negotiate the work assignment* with their team members one to one, at the outset of the project. These negotiations should include the overall task, its scope, objectives, and performance measures. A thorough understanding of the task requirements often develops as the result of personal involvement in the front-end activities, such as requirements analysis, bid proposals, project planning, interface definition, and overall product planning. This early involvement also has positive effects on the buy-in to project objectives, plan acceptance, and the unification of the task team.

Communicate Organizational Goals and Objectives. Management must communicate and update the organizational goals and project objectives. The relationship and contribution of individual work to the overall product development, the business plans, and the importance to the organizational mission must be clear to all team personnel. Senior management can help in unifying the team behind the project objectives by developing a "priority image" through their personal involvement in, visible support of, and emphasis on project goals and mission objectives.

Build a High-Performance Image. Building a favorable image for an ongoing project in terms of high priority, interesting work, importance to the organization, high visibility, and potential for professional rewards is crucial for attracting and holding high-quality people. Senior management can help develop a "priority image" and communicate the key parameters and management guidelines for specific projects. Moreover, establishing and communicating clear and stable top-down objectives helps in building an image of high visibility, importance, priority, and interesting work. Such a pervasive process fosters a climate of active participation at all levels, helps attract and hold high-quality people, unifies the team, and minimizes dysfunctional conflict.

Build Enthusiasm and Excitement. Whenever possible, managers should try to accommodate the professional interests and desires of their personnel. Interesting and challenging work is a perception that can be enhanced by maintaining the visibility of the work, management attention and support, a priority image, and the overlapping of personnel values and perceived benefits with organizational objectives. Making work more interesting leads to increased involvement, better communication, lower conflict, higher commitment, a stronger work effort, and higher levels of creativity.

Define Effective Communication Channels. Poor communication is a major barrier to teamwork and effective project performance, especially in concurrent engineering environments, which depend to a large degree on information sharing for their concurrent execution and decision making. Management can facilitate the free flow of information, both horizontally and vertically, by proper workspace design, regular meetings, reviews, and information sessions. In addition, modern technology such as voice mail, e-mail, electronic bulletin boards, and conferencing can greatly enhance communication, especially in complex organizational settings.

Create Proper Reward Systems. Personnel evaluation and reward systems should be designed to reflect the desired power equilibrium and authority/responsibility sharing needed for the concurrent engineering organization to function effectively. Creating a system and its metrics for reliably assessing performance in a product development environment is a great challenge. However, several models, such as the Integrated Performance Index (Pillai, Joshi, and Rao 2002), have been proposed and provide a potential starting point for customization. A *quality function deployment (QFD)* philosophy, where everyone recognizes the immediate "customer" for whom a task is performed, helps to focus efforts toward desired results and customer satisfaction. This customer orientation should exist, both downstream and upstream, for both company-internal and -external customers. These "customers" should score the performance of the deliverables they receive and therefore have a major influence on the individual and team rewards.

Ensure Senior Management Support. It is critically important that senior management provide the proper environment for a project team to function effectively (Prasad 1998). At the onset of a new project, the responsible manager needs to negotiate the needed resources with the sponsor organization, and obtain commitment from management that these resources will be available. An effective working relationship among resource managers, project leaders, and senior management critically affects the credibility, visibility, and priority of the engineering team and their work.

Build Commitment. Managers should ensure team member commitment to their project plans, specific objectives, and results. If such commitments appear weak, managers should determine the reason for such lack of commitment of a team member and attempt to modify possible negative views. Anxieties and fear of the unknown are often major reasons for low commitment (Stum 2001). Managers should investigate the potential for insecurities, determine the cause, and then work with the team members to reduce these negative perceptions. Conflict with other team members and lack of interest in the project may be other reasons for such lack of commitment.

Manage Conflict and Problems. Product development activities are highly disruptive to any organization. Conflict is inevitable. Project managers should focus their efforts on problem avoidance. That is, managers and team leaders, through experience, should recognize potential problems and conflicts at their onset and deal with them before they become big and their resolutions consume a large amount of time and effort (Haque et al. 2003).

Conduct Team-Building Sessions. A mixture of *focus-team* sessions, *brainstorming*, *experience exchanges*, and *social gatherings* can be a powerful tool for developing the workgroup into an effective, fully integrated, and unified project team (Thamhain and Wilemon 1999). Such organized team-building efforts should be conducted throughout the project life cycle. Intensive team-building efforts may be especially needed during the formation stage of a new project team. Although formally organized, these team-building sessions are often conducted in a very informal and relaxed atmosphere to discuss critical questions such as (1) how are we working as a team? (2) what is our strength? (3) how can we improve? (4) what support do you need? (5) what challenges and problems are we likely to face? (6) what actions should we take? and (7) what process or procedural changes would be beneficial?

Provide Proper Direction and Leadership. Project managers and team leaders can influence the attitude and commitment of their people toward the project objectives by their own actions. Concern for the project team members and enthusiasm for the project can foster a climate of high motivation, involvement with the project and its management, open communication, and willingness to cooperate with the new requirements and use them effectively.

Foster a Culture of Continuous Support and Improvement. Successful project management focuses on people's behavior and their roles within the project itself. Companies that effectively manage product development have cultures and support systems that demand broad participation in their organizational development efforts. Encouraging management throughout the enterprise to be proactive and aggressive toward change is not an easy task. Yet, our continuously changing business environment requires that provisions be made for updating and fine-tuning the established project management process and for developing its teams. Such organizational developments must be undertaken on an ongoing basis to ensure relevancy to today's project management challenges. It is important to establish support systems—such as discussion groups, action teams, and suggestion systems—to capture and leverage the lessons learned, and to identify problems as part of a continuous improvement process. Tools such as the *Project Maturity Model* and the Six Sigma Project Management Process can provide the framework and toolset for analyzing and fine-tuning the team development and its management process.

9.8   SUMMARY OF KEY POINTS AND CONCLUSIONS

The major points that have been made in this chapter include:

- The increasing complexities of today's project environment, both internally and externally, create enormous managerial challenges for directing, coordinating, and controlling teamwork. With the expansion of self-directed team concepts in particular, additional managerial tools and skills are required to handle the burgeoning group dynamics and infrastructure.
- Effective teamwork is a critical determinant of project success and of the organization's ability to learn from its experiences and position itself for future growth.
- To be effective in organizing and directing a project team, the leader must not only recognize the potential drivers of and barriers to high-performance teamwork, but also know when in the life cycle of the project they are most likely to occur.
- Four major conditions must be present for building effective project teams: (1) a professionally stimulating work environment, (2) good project leadership, (3) qualified personnel, and (4) a stable work environment.
- Effective project leaders take preventive actions early in the project life cycle and foster a work environment that is conducive to team building as an ongoing process.
- The new business realities experienced today force managers to focus increasingly on cross-boundary relations, delegation, and commitment, in addition to establishing the more traditional formal command and control systems.
- Project leaders must involve top management in their project to ensure organizational visibility, resource availability, and overall support for the new project throughout its life cycle.
- Building effective project teams requires the whole spectrum of management skills and company resources, and is the shared responsibility of functional managers and the project leader.
- The effective team leader is a social architect who understands the interaction of organizational and behavioral variables and can foster a climate of active participation, minimal dysfunctional conflict, effective communication, and conduciveness to change, commitment, and self-direction.
- By understanding the criteria and organizational dynamics that drive people toward effective team performance, managers can examine and fine-tune their leadership style, actions, and resource allocations toward continuous organizational improvement.

In summary, managerial leadership, at both the project level and the senior management level, has a significant impact on the team environment, which ultimately affects team and project performance. Effective team leaders understand the interaction of organizational and behavioral variables. They can foster a climate of active participation and minimal dysfunctional conflict. This requires carefully developed skills in leadership, administration, organization, and technical expertise. It further requires the ability to involve top management and ensure organizational visibility, resource availability, and overall support for the project throughout its life cycle and across the enterprise and its support functions, suppliers, sponsors, and partners.

9.9 CRITICAL THINKING: QUESTIONS FOR DISCUSSION

1. Using the lead-in case scenario for this chapter, identify Chrysler's organizational challenges and the issues that require effective project teamwork for resolution.

2. Identify and profile the type of leadership style that is needed for effectively managing a new product team at Chrysler.

3. Write a charter for the manager in charge of a new car development project at Chrysler.

4. What are some of the characteristics of a fully integrated team? Develop a list of performance measures or criteria for a specific team you know, and evaluate this team against your benchmark list.

5. Identify specific project situations where team building is critical. Why is team building such an important and challenging task for today's managers?

6. What are some of the organizational variables that influence project team characteristics? How can you as project leader influence these variables?

7. List and discuss some of the task- and people-related qualities of successful project teams. How could you measure these qualities? How can you influence these qualities?

8. Discuss the importance of a project manager's understanding of the drivers of and barriers to team performance.

9. Do you accept the three propositions related to project success given in this chapter? Explain.

10. What are the characteristics of a self-directed, self-managed team? As a project manager, how can you promote such team behavior?

11. How do you develop team leadership skills?

12. How do you identify the "best" team members for a newly formed project?

13. How do you unify a newly formed team and move it toward the "performing" stage fast?

14. How do you develop a well-performing project team further?

15. Identify a workgroup that you know. Define the metrics for measuring team performance.

16. What means and methods, other than money, would you consider to motivate your team toward high project performance.

17. Is it more challenging to lead project teams today than 25 years ago? Why or why not?

18. How does the team leadership style change over a typical project life cycle?

19. Why are team commitment and ownership important to team performance?

20. As a team leader, how can you build team commitment to the project objectives?

21. As a project team leader, how can you integrate senior management into your team?

22. Two team leaders have identical projects in the same organizational environment. One team perceives the project as highly interesting and important with a lot of opportunities for career advancement, while the other team perceives the project as boring, routine work with little opportunity for career advancement. What do these two project leaders do differently?

23. How can the proper application of conventional project management tools and techniques help in building high-performing teams?

24. How can senior management help in building a high-performing project team?

25. What kind of changes do you anticipate over the next 10 years regarding project team challenges, characteristics, and leadership style?

9.10   REFERENCES AND ADDITIONAL READINGS

Armstrong, David (2000) "Building teams across borders," *Executive Excellence*, Vol. 17, No. 3 (March), p. 10.

Barkema H., Baum J., and Mannix, E. (2002) "Management challenges in a new time," *Academy of Management Journal*, Vol. 45, No. 5, pp. 916–930.

Barner, R. (1997) "The new millennium workplace," *Engineering Management Review (IEEE)*, Vol. 25, No. 3, Fall 1997, pp. 114–119.

Belassi, W. and Tukel, O. (1996) "A new framework for determining critical success/failure factors in projects," *International Journal of Project Management*, Vol. 14, No. 3, pp. 141–151.

Bennis, W. and Shepard, H. (1956) "A theory of group development," *Human Relations*, No. 9, pp. 415–437.

Bhatnager, Anil (1999) "Great teams," *Academy of Management Executive*, Vol. 13, No. 3 (August), pp. 50–63.

Bond, E., Walker, B., Hutt, M., and Reigen, P. (2004) "Reputational effectiveness in cross-functional working relationships," *Journal of Product Innovation Management*, Vol. 21, No. 1 (January), pp. 44–60.

Chell, E. and Allman, K. (2003) "Mapping the motivations and intentions of technology orientated entrepreneurs," *R & D Management*, Vol. 33, No. 2 (March), pp. 117–134.

DiBella, Anthony J. (1995) "Developing learning organizations: a matter of perspective." *Academy of Management Journal* (Best Papers Proceedings), pp. 287–290.

Dillon, P. (2001) "A global challenge," *Forbes Magazine*, Vol. 168 (September 10), pp. 73+.

Drucker, P. (1996) *The Executive in Action: Managing for Results, Innovation and Entrepreneurship*, New York: Harper.

Dumaine, B. (1991) "The bureaucracy buster," *Fortune*, June 17, pp. 30-35.

Dyer, W. G. (1977) *Team Building: Issues and Alternatives*, Reading, MA: Addison-Wesley.

Eisenhardt, K. M. (1989) "Building theories from case study research," *Academy of Management Review*, Vol. 14, No. 4, pp. 532–550.

English, K. (2004) "The changing landscape of leadership," *Research Technology Management*, Vol. 45, No. 4 (July/August), p. 9.

Fisher, K. (1993). *Leading Self-Directed Work Teams*, New York: McGraw-Hill.

Glaser, B. G. and Strauss, A. L. (1967) *The Discovery of Grounded Theory: Strategies for Qualitative Research*, Chicago: Aldine.

Gray, Clifford and Larson, Erik (2000) *Project Management*, New York: Irwin McGraw-Hill, 2000.

Gupta, A. and Wilemon, D. (1996) "Changing patterns in industrial R&D management," *Journal of Product Innovation Management,* Vol. 13, No. 6 (November), pp. 497-511.

Haque B., Pawar K., and Barson, R. (2003), "The application of business process modeling to organizational analysis of concurrent engineering environments," *Technovation*, Vol. 23, No. 2 (February), pp. 147-162.

Hartman, F. and Ashrafi, R. (2002) "Project management in the information systems and technologies industries," *Project Management Journal*, Vol. 33, No. 3, pp. 5–15.

Jasswalla, A. R. and Sashittal, H. C. (1998) "An examination of collaboration in high-technology new product development processes," *Journal of New Product Innovation Management*, Vol. 15, No. 3, pp. 237–254.

Jassawalla, Avan R. and Sashittal, Hemant C. (1999) "Building collaborate cross-functional new product teams," *Academy of Management Executive*, Vol. 13, No. 3 (August), pp. 50–63.

Karlsen, J. and Gottschalk, P. (2004) "Factors affecting knowledge transfer in IT projects," *Engineering Management Journal*, Vol. 16, No. 1. pp. 3-11.

Katz, Nancy (2001) "Sports teams as model for workplace teams," *Academy of Management Executive*, Vol. 15, No. 3 (August), pp. 70–72.

Keller, R. (2001) "Cross-functional project groups in research and new product development," *Academy of Management Journal*, Vol. 44, No. 3, pp. 547–556.

Kostner, Jaclyn (2001) *Bionic eTeamwork: How to Build Collaborative Virtual Teams at Hyperspeed*, Dearborn, MI: A Kaplan Professional Company.

Kruglianskas, Isak and Thamhain, Hans (2000) "Managing technology-based projects in multinational environments," *IEEE Transactions on Engineering Management*, Vol. 47, No. 1 (February), pp. 55–64.

Likert, R. (1967) *New Patterns of Management*, New York: McGraw-Hill.

MacCormack, A., Verganti, R., and Iansiti, M. (2001) "Developing products on Internet time," *Management Science*, Vol. 47, No. 1, pp. 22–35; *Engineering Management Review*, Vol. 29, No. 2, pp. 90–104.

McGregor, D. (1960) *The Human Side of the Enterprise*, New York: McGraw-Hill.

Marshall, E. (1995) *Transforming the Way We Work,* New York: AMACOM.

Moss Kanter, R. (1989) "The new managerial work," Harvard Business Review, Vol. 65 No. 6 (November-December), pp. 36–47.

Nellore, R. and Balachandra, R. (2001) "Factors influencing success in integrated product development (IPD) projects," *IEEE Transactions on Engineering Management*, Vol. 48, No. 2, pp. 164–173.

Newell, F. and Rogers, M. (2002), *loyalty.com: Relationship Management in the Era of Internet Marketing*, New York: McGraw-Hill.

Nurick, A. J. and Thamhain, H. J. (1993) "Project team development in multinational environments," Chapter 38 in *Global Project Management Handbook* (D. Cleland, ed.), New York: McGraw-Hill.

Oderwald, S. (1996) "Global work teams," *Training and Development*, Vol. 5, No. 2 (February), pp. 64-76.

Parker, Glenn, McAdams, Jerry, and Zielinski, David (2000) *Rewarding Teams: Lessons from the Trenches*, New York: Jossey-Bass/Wiley and Sons.

Peters, T. and Waterman, R. (1987, 1997) *In Search of Excellence,* New York: Harper & Row.

Pillai, A., Joshi, A., & Raoi, K. (2002). "Performance measurement of R&D projects in a multi-project, concurrent engineering environment." *International Journal of Project Management,* Vol. 20, No. 2, 165–172.

Prasad, B. (1998). "Decentralized cooperation: a distributed approach to team design in a concurrent engineering organization," *Team Performance Management,* Vol. 4, No. 4, 138–146.

Roethlingsberger, F. and Dickerson W. (1939) *Management and the Worker,* Cambridge, MA: Harvard University Press.

Senge, Peter (1994), *The Fifth Discipline: The Art and Practice of the Learning Organization,* New York: Doubleday/Currency.

Senge, P. and Carstedt, G. (2001) "Innovating our way to the next industrial revolution," *Sloan Management Review,* Vol. 42, No. 2, pp. 24–38.

Sethi, R. and Nicholson, C. (2001) "Structural and contextual correlates of charged behavior in product development teams," *Journal of Product Innovation Management,* Vol. 18, No. 3, pp. 154–168.

Shim, D. and Lee, M. (2001) "Upward influence styles of R&D project leaders," *IEEE Transactions on Engineering Management,* Vol. 48, No. 4, pp. 394–413.

Shonk, J. (1996) *Team-Based Organizations,* Homewood, IL: Irwin.

Stum, D. (2001) "Maslow revisited: building the employee commitment pyramid," *Strategy and Leadership,* Vol. 29, No. 4, 4–9.

Thamhain, H. (1983) "Managing engineers effectively," *IEEE Transactions on Engineering Management,* Vol. 30, No. 4, pp 231–237.

Thamhain, H. (1990) "Managing technologically innovative team efforts toward new product success," *Journal of Product Innovation Management,* Vol. 7, No. 1 (March), pp. 5–18.

Thamhain, H. (1990) "Managing technology: the people factor," *Technical & Skill Training Journal,* (August/September), pp. 24–31.

Thamhain, H. (1996) "Managing self-directed teams toward innovative results," *Engineering Management Journal,* Vol. 8 No. 3, pp. 31–39.

Thamhain, H. (2001) "Team management," Chapter 19 in *Project Management Handbook* (J. Knutson, ed.), New York: Wiley & Sons.

Thamhain, H. (2002) "Criteria for effective leadership in technology-oriented project teams," Chapter 16 in *The Frontiers of Project Management Research* (Slevin, Cleland, and Pinto, eds.), Newton Square, PA: Project Management Institute, pp. 259–270.

Thamhain, H. (2003) "Managing innovative R&D teams," *R&D Management,* Vol. 33, No. 3 (June), pp. 297–312.

Thamhain, H. (2004). "Leading technology-based project teams," *Engineering Management Journal,* Vol. 16, No. 2, pp. 42–51.

Thamhain, Hans and Wilemon, David (1987) "Building high-performance engineering project teams," *IEEE Transactions on Engineering Management,* Vol. 34, No. 1 (February), pp. 130–142.

Thamhain, Hans and Wilemon, David (1987) "Leadership, conflict, and project management effectiveness," *Executive Bookshelf on Generating Technological Innovations, Sloan Management Review,* Fall 1987, pp. 68–87.

Thamhain, H. and Wilemon, D. (1999) "Building effective teams for complex project environments," *Technology Management,* Vol. 5, No. 2 (May), pp. 203–212.

Tichy, N., and Ulrich, D. (1984) "The leadership challenge—call for the transformational leader," *Sloan Management Review,* Vol. 26, No. 1 (Fall), pp. 59-69.

Verma, Vijay K. (1997) *Managing the Project Team,* Newton Square, PA: Project Management Institute.

Walton, R. (1985) "From control to commitment in the workplace," *Harvard Business Review,* Vol. 61, No. 2 (March/April), pp. 65–79.

Whitten, N. (1995) *Managing Software Development Projects*, 2nd Edition, New York: John Wiley & Sons.

Zanoni, R. and Audy, J. (2004) "Project management model for physically distributed software development environment," *Engineering Management Journal*, Vol. 16, No. 2, pp. 28–34.

Zhang, P., Keil, M., Rai, A., and Mann, J. (2003) "Predicting information technology project escalation," *Journal of Operations Research*, Vol. 146, No. 1, pp. 115–129.

Appendix

FIELD RESEARCH SUMMARY ON TEAM LEADERSHIP

The Study. This thumbnail sketch summarizes the results of a field study of the critical success factors of technology-based team performance. Action research methodology was used to collect data from 76 project teams with a total sample population of 895 professionals such as engineers, scientists, and technicians, plus their managers, covering over 180 projects in 27 companies of the "FORTUNE-500" category.[5] The projects involved mostly high-technology product/service developments, such as information system, computer, and pharmaceutical products, and financial services, with average budgets of $1,200,000. In addition to the statistical correlation, content analysis has been used for evaluating the qualitative part of the interviews, questionnaires, observations, and action research.[6]

Results. Table 9.A1 summarizes the associations among factors of the organizational environment and project team performance. As indicated by the strongest correlations, factors that fulfill professional esteem needs, clarify objectives, and provide needed skill sets seem to have a particularly favorable influence on project team performance. The five most significant associations are (1) professionally

---

[5] 65% of the companies in the sample fall in to the Fortune-500 classification, 23% are Fortune-1000 companies, while the remainder are smaller firms. None of the companies in the sample can be classified as "small or medium size."

[6] For a more detailed discussion of the background, method, and results of this field study, see H. Thamhain (2004), "Leading technology-based project teams," *Engineering Management Journal*, Vol. 16, No. 1, pp. 22–31.

stimulating and challenging work environments [τ=.45], (2) opportunity for accomplishments and recognition [τ=.38], (3) the ability to resolve conflict and problems [τ=.37], (4) clearly defined organizational objectives relevant to the project [τ=.36], and (5) job skills and expertise of the team members appropriate for the project work [τ =.36]. These influences appear to deal effectively with the integration of goals and needs between the team member and the organization. In this context, the more subtle factors seem to become catalysts for cross-functional communication, information sharing, and ultimate integration of the project team with a focus on desired results. Other favorable factors relate to overall directions and team leadership [τ=.35], trust, respect, and credibility among team members and their leaders [τ=.30], and business process, as reflected by cross-functional cooperation and support [τ=.27], communications [τ=.27], clear project plans [τ=.25], clearly defined authority relations, and sufficient autonomy and freedom of actions in line with the managerial expectations and accountabilities [τ=.23]. To a lesser degree, opportunities for career development and advancement [τ=.12], as well as job security [τ=.12], seem to have a positive influence. All associations are significant at $p =.1$ or better, with the most significant correlations of $p = .01$ or stronger shown in bold italics. It is interesting to note that the same conditions, which are conducive to overall team performance, also lead to (1) a higher ability to deal with risks and uncertainties and (2) a stronger personal effort and commitment to established objectives and their team members, as shown in the correlation table. The field data analysis moreover supports the expectation that project teams that are perceived as effective by their management are also seen as (3) creative problem solvers who can (4) effectively utilize time and resources. In fact, a high degree of cross-correlation exists among the set of four of variables, as measured via Kruskal-Wallis analysis of variance by rank.[7] The test shows that managers agree on the ranking of team performance factors at a confidence level of 98 percent. That is, managers who rate the team performance high in one category are likely to give high ratings also to the other three categories.

In addition to the 13 most significant factors reported in Table 9.A1, it is interesting to note that many other characteristics of the work environment that were perceived by managers as important to effective team performance *did not correlate significantly*, as measured by a p-level threshold of .10. Among the *factors of lesser influence* to project team performance are (1) salary, (2) time off, (3) project visibility and popularity, (4) maturity of the project team, measured in terms of time worked together as a team, (5) project duration, (6) stable project requirements with minimum changes, (7) stable organizational structures and business processes that

---

[7]The Kruskal-Wallis One-Way Analysis of Variance by Rank is a test for deciding whether k independent samples are from different populations. In this field study, the test verifed that managers perceive in essence the same parameters in judging high team performance.

Table 9.A1 *Strongest* Drivers Toward Project Team Performance (Kendall's Tau Rank-Order Correlation)

Team Environment	Project Team Performance		
	Ability to Deal with Risk	Effort & Commitment to Results	Overall Team Perfor-mance
1 Interesting, Stimulating Work	*.39*	*.43*	*.45*
2 Accomplishment & Recognition	.27	.35	*.38*
3 Conflict & Problem Resolution	.33	.30	*.37*
4 Clear Organizational Objectives	.21	.28	*.36*
5 Job Skills & Expertise	.32	.15	*.36*
6 Direction & Leadership	.27	.22	.35
7 Trust, Respect, Credibility	.08	*.40*	.30
8 Cross-Funct'l Coop & Support	*.37*	.28	.27
9 Effective Communications	.34	.27	.27
10 Clear Project Plan & Support	*.36*	*.36*	.25
11 Autonomy & Freedom	.34	*.36*	.23
12 Career Developmt/Advancement	.10	.07	.12
13 Job Security	.30	.12	.12

All variables were measured with descriptive statements on a 5-point Likert scale: (1) strongly disagree, (2) disagree, (3) neutral, (4) agree, (5) strongly agree.

Statements were judged by team members [*] and senior management [#], as indicated. *Statistical Significance:* $p=.10$ ($\tau \geq 20$), $p=.05$ ($\tau \geq 31$), $p=.01$ ($\tau \geq 36$); correlations of $p=.01$ or stronger are marked in bold italics.

result in minimal organizational changes, such as those caused by mergers, acquisitions, and reorganization, (8) minimum technological interdependencies, such as those caused by the dependency on multiple technologies, technological disciplines, and processes, and (9) project size and project complexity. This argues that project scope, size, and implementation challenges, by themselves, do not necessarily translate into lower team or project performance. It is further interesting to see that several of the weaker influences actually seem to have opposite effects to those popularly held to be true by managers. For example, it appears that the more stable the project requirements, the less overall team performance is to be expected. While these correlations are clearly insignificant from a statistical point of view, they shed some additional light on the subtle and intricate nature of project team performance in technology-intensive environments. From a different perspective, it is interesting to observe that influences that support *intrinsic professional needs* show a strong favorable performance correlation, while the findings give only weak support to the benefit of "extrinsic influences/motivators" (such as salary increases, bonuses, and time off) and metrics-related factors (such as team tenure, project duration, and changes) as well as complexity and technology factors. This is so in spite the fact that all influences discussed here were perceived by most managers as critically important to team performance. This finding suggests that managers are more accurate in their perception of team members' intrinsic rather than extrinsic needs. It also seems to be more difficult to assess the impact of project parameters, such as size, duration, or complexity, than the impact of human needs on project work performance.

Discussion and Implications. The empirical results presented in this paper show that specific conditions in the team environment appear most favorable to project team work. These conditions serve as bridging mechanisms, helpful in enhancing project performance in technology-based organizations. Considering the exploratory nature of this study, an attempt is being made to go beyond the obvious results of the statistical data and to integrate some of the lessons learned from the broader context of the field research. The interviews and observations conducted for proper formulation, introduction, and follow-up of the questionnaires were especially useful in gaining additional perspective on and insight into the processes and challenges of teamwork; they also helped in gleaning lessons for effective technical project management.

Succeeding in today's ultracompetitive word of business is not an easy feat. No single set of broad guidelines exists that guarantees success. However, project success is not random! A better understanding of the criteria and organizational dynamics that drive project team performance can assist managers in developing a better, more meaningful insight into the organizational process and critical success factors that drive project team performance.

One of the most striking findings is that many of the factors that drive project team performance are derived from the human side. Organizational components that satisfy personal and professional needs seem to have the strongest effect on commitment, the ability to deal with risk and contingencies, and overall team performance. Most significant are those influences derived from the work itself. People who find their assignments professionally challenging, leading to accomplishments, recognition, and professional growth, also seem to function more effectively in a technology-intensive team environment. Such a professionally stimulating ambience also lowers communication barriers, increases the tolerance for conflict and risk taking, and enhances the desire to succeed.

Other influences on project team performance are derived from the organizational process, and have their locus outside the project organization and are controlled by senior management. Organizational stability, availability of resources, management involvement and support, personal rewards, and stability of organizational goals, objectives, and priorities are all derived from organizational systems that are controlled by general management. Project team leaders must work with senior management to ensure an organizational ambience conducive to effective teamwork. Leaders of successful project teams create a sense of community across the whole enterprise. That is, they understand the factors that drive team performance and create a work environment conducive to such behavior. Effective project leaders can inspire their people, making everyone feel proud to be part of the project organization and its mission. Both clarity of purpose and alignment of personal and organizational goals are necessary for a unified team culture to emerge. Encouragement, personal recognition, and visibility of the contributions to customer and company values helps to refuel and sustain commitment and unites the team behind its mission.

Taken together, managerial leadership, at both the project level and senior management, has a significant impact on the team environment, which ultimately affects team and project performance. The effective team leader is a social architect who understands the interaction of organizational and behavioral variables and can foster a climate of active participation and minimal dysfunctional conflict. This requires carefully developed skills in leadership, administration, and organization and technical expertise. It further requires the ability to involve top management in the project, to ensure organizational visibility, resource availability, and overall support for the project throughout its life cycle, and managing across the entire work process, including support functions, suppliers, sponsors, and partners.

10

MANAGING R&D AND INNOVATION

REINVENTING CORPORATE R&D

Every month at Merck & Co., groups of scientists in different areas of disease research gather to evaluate the latest breakthroughs. But the innovations they haggle over don't come from Merck's own laboratories. They're generated someplace, anyplace, else: journals, conference reports, patent literature, and visits to other labs. "We scour the world," says Merv Turner, Merck's senior vice president for external research.

Scouting for innovative ideas: At the end of the meetings, each group typically flags a couple of innovations for a small team of scientists to investigate. Last year, one of the brain trusts tagged Amrad Corp., a tiny Australian biotech company that's working on a promising drug to treat respiratory diseases. By June, Merck had ponied up $5 million to seal an exclusive license and a multiyear research collaboration with Amrad that one day could be worth $112 million. It's just one payoff from Merck's "very aggressive antenna function to survey what's going on in the world, Turner says.

Welcome to the future of corporate research. In recent years, there has been much hand wringing over the slowdown in industrial spending by America's big corporate labs. Legendary science and engineering bastions such as Xerox Corp.'s Palo Alto Research Center and the old AT&T Bell Labs have gone through painful pruning. And for the first time in more than a decade, there is an actual decline in overall corporate research and development funding—$192 billion for 2003, or 0.7 percent less than 2002, notes F. M. Ross Armbrecht Jr., president of Industrial Research Institute Inc.

Rather than drying up, however, industrial research is bubbling up all over the place, with more big companies following Merck. The downsizing by industrial giants makes headlines. But at the same time, a new R&D model is emerging, dubbed *open innovation*. Companies of all sizes are rounding up more partners, big and small, than ever before, and they're casting wide research nets, snapping up work at diverse corporate, government, and academic labs. So despite the gloomy dollar-spending numbers, U.S. R&D may be heading for a healthy makeover. While big companies have been a source of major breakthroughs, small and midsized companies have always been the main font of new products. Today, they're more innovative than ever—not least because the rapid growth of computer power has put unprecedented research resources in the hands of thousands of teams around the globe.

Furthermore, scientists everywhere are increasingly motivated by commercial payoffs. As a result, the time between new discoveries and product rollouts is collapsing. "Fundamental science breakthroughs now have fairly rapid commercial applications," says Walter W. 'Woody' Powell, a guru in organizational behavior at Stanford University. Meanwhile, industry-university consortiums are spinning off engineering startups at a quicker pace than ever before, says J. David Roessner, associate director of the Science & Technology Policy Program at SRI International.

At the same time, big companies are becoming more inventive in tapping researchers employed elsewhere to solve vexing problems. Giant chemical companies such as Dow Chemical Co. and BASF are posting research problems at a Web site run by Eli Lilly & Co.'s InnoCentive Inc., which tries to match scientists with research problems. Even Xerox, which set the standard for do-it-yourself research, has turned to outsiders to help develop optical-network technology. "We were definitely not doing that 20 years ago," says Herve Gallaire, Xerox's chief technology officer. Xerox has contributed its expertise in imaging technology to a joint project with Intel Corp. that produced Intel's latest microprocessor, tailored for applications in document imaging.

Chalk it up mainly to corporate heavyweights learning that they can't develop all the breakthroughs in-house. "It's very hard for any company, even one that spends as much as we do on R&D, to do everything," says Paul Horn, senior vice president and director of research at IBM, which last year spent $4.8 billion on research.

In high tech, belt tightening also is pushing companies to find cheaper alternatives to conventional research. A study last year by MIT's *Technology*

Review magazine found that most leading companies in struggling industries—aerospace, computers, semiconductors, and telecommunications—had trimmed R&D outlays. But they haven't turned their backs on innovation; they've just farmed it out. "It's not necessary for companies to invest in basic research to make money from it," says Henry W. Chesbrough, a Harvard Business School professor and author of *Open Innovation*, widely cited by research execs such as Xerox's Gallaire and IBM's Horn. Is there a danger in corporations relying too much on outside help? Merrilea J. Mayo, director of the Government-University-Industry Research Roundtable at the National Academy of Sciences, is worried because so much research is moving offshore. American innovativeness, she warns, could be hobbled more by loss of higher-skill jobs than by those lost in NAFTA. "That 'giant sucking sound' that Ross Perot heard is now happening in R&D," Mayo says.

The challenge is to find the optimum balance of research and development. Procter & Gamble Co. thrived for years relying on the 7,500 employees in its R&D group to gin up new products. But as the pace of innovation elsewhere increased, P&G faltered. To fix the problem, CEO Alan G. Lafley has decreed that half of the company's ideas must come from outside, up from about 10 percent when he took over in 2000. "We are probably as good as the next guy in inventing," Lafley says. "But we are not absolutely and positively better than everybody else." Today, P&G has 53 "technology scouts" who search beyond company walls for promising innovations. Other companies are scouting Web-based research services for talent. A year ago, Dow Chemical's Dow AgroSciences unit turned to Lilly's InnoCentive Web site to solve a dozen research riddles. InnoCentive has signed up 30,000 scientists worldwide and pays bounties of up to $100,000 for solutions to problems posted on the site by Dow and other seekers. Dow just made its first payout—$50,000. It was a bargain, says Mark W. Zettler, manager of new-product development for Dow AgroSciences. Besides, he adds, "I would rather find the person who has the answer today than somebody that needs to work for three to six months in our lab." Since it was launched in mid-2001, InnoCentive has paid out roughly $500,000 to its problem solvers.

Another way to optimize research: Be more innovative in cozying up to universities. Intel is pioneering a new approach. It has set up satellite research facilities adjacent to Carnegie Mellon University, University of California at Berkeley, University of Washington, and Britain's Cambridge University. The idea is to move the brightest minds in academia temporarily into these "lablets" and focus on long-term projects. "We're only going to create a small fraction of the good ideas, and only a small fraction are going to pass through our labs unless we take extra-strong action," explains David L. Tennenhouse, Intel's director of research.

Ultimately, when the corporate R&D makeover is finished, U.S. industry may be in its best fighting trim ever. America was the innovation champ long before companies started splurging on basic research in the postwar era. Recent cutbacks in corporate funding of research are disturbing to many scientists. But the vibrant, open model now taking shape will make the United States "tremendously productive," says IBM's Horn. As long as companies don't forsake the quest for the next big thing.

Source: Business Week, 9/22/2003, Issue 3850, p.74, 2p. reprinted with permission.

For additional information on corporate R&D of companies cited in the above scenario, please visit AT&T Labs *www.att.com/attlabs*, Dow Chemical *www.com/automotive/capab/prod_rd.htm*, IBM *www.research.ibm.com*, IRI *www.iriinc.org/webiri/index.cfm*, Lilly's InnoCentive *www.innocentive.com*, Merck *www.merck.com/mrl*, P&G *www.pg.com/about_pg/science_tech/research_development/category_main .jhtml*, and Xerox *www.parc.xerox.com*.

10.1 THE NEED FOR INNOVATIVION IN BUSINESS: CHANGES AND CHALLENGES

Innovation is the last remaining frontier in today's world of business, helping companies to achieve lower cost, superior performance, and new products and services[1] (Pospisil 1996; Reed, Lemak, and Montgomery 1996). However, deriving competitive advantages from innovation is a highly intricate process, as shown by various examples in the lead-in synopsis of this chapter. It involves technical complexities, functional interdependencies, evolving solutions, high levels of uncertainty, and highly complex forms of work integration (Sherma 1999, Stringer 2000, Drucker 1998, Brown and Eisenhardt 1995, Pospisil 1996). Investment in R&D activities alone does not guarantee success, as evidenced by the high failure rate of products and services that do not make it through the commercialization process.[2] The reality is that few companies can meet all of their research goals alone. Therefore, it is not surprising that companies like Merck are adopting an "*open innovation*" model. These companies try to supplement their own R&D efforts with "ready-to-use" product concepts and scientific breakthroughs from other companies, rather than trying to invent everything in-house.

In fact, scouting for new product and service ideas is not new. It has been around for a long time. Sources such as federal Labs, trade shows, Web sites, and competitive products have been explored for innovative content for a long time, in hopes of finding useful information for generating new products for the ongoing business. For

---

[1]U.S. companies spend over $75 billion annually on R&D to gain competitive advantages and economic benefits through innovation transformed into the marketplace (Holman, et al., 2003 and McKinsey).

[2]Less than 15 percent of R&D generated concepts actually succeed in the market (McKinsey, Holman, et al., 2003).

similar reasons of leveraging corporate R&D budgets, alliances, consortia, and partnerships have been formed with universities, the government, and other firms.

For technology companies, the key challenge is not so much the generation of innovative ideas at the R&D stage, but the effective transfer of technology from the discovery stage to the market. This brings focus to the management side of R&D. One of the primary requirements for such technology transfer is effective interdisciplinary teamwork across all functions of the enterprise, including customers and suppliers, as a critical factor of success (Sawhney and Pradelli 2000). Management must facilitate the establishment of a team environment conducive to market-oriented innovation, consistent with the existing business dynamics and complexities (Tomkovich and Miller 2000, Debruyne et al. 2002). The large number of successful commercializations of new product concepts provides clear evidence that these challenges can be met with the right organizational ambience. While most studies examine innovative organizational performance as a function of company policy or organizational parameters, some researchers have focused on the human side of innovative performance and variables that drive or inhibit innovative team results in an R&D environment. In fact, a growing body of data suggests that managerial leadership style and work environment significantly influence innovative performance, especially within self-directed work teams (DiBella 1995, Shim and Lee 2001). Other studies by Thamhain (1990, 1996, 2002), supported by others (McDonough 1993, Thamhain and Wilemon 1996, 1999), found that innovative performance in technical project teams is not only strongly associated with project success, but also favorably influenced by leadership and professional attitude.

This chapter discusses the principal factors that influence innovative performance in high-technology, team-based work environments. We will explore the innovative characteristics and overall performance of R&D and high-tech product development teams in technology-oriented environments.

10.2 WHAT WE KNOW ABOUT MANAGEMENT OF TECHNICAL INNOVATION

The important role of technological innovation for a company's business success has been long recognized (Crawford 1983, Johne and Snelson 1988). Innovation can generate a competitive advantage for one firm, while eroding the market position for another (Bennis and Nanus 1985; Conway and McGuiness 1986; Drucker 1985, 1996; Bugelman et al. 1988). However, deriving such a competitive advantage from innovation is an intricate and risky process. The companies that will survive and prosper in the decades ahead will be those that can *manage* innovation and derive business benefits from it. They must do this in spite of the complex organizational processes, rapidly changing technology, increasing risks, uncertainties, cost, demands for better market responsive, and relatively low barriers to entry into almost every business.

Research on innovation has traditionally focused on the qualities of the individuals who play crucial roles in the innovation process. In recent years, an increasing number of studies have broadened the investigations into other areas. In a research review article, Johne and Snelson (1988) show the enormous breadth and depth of subsystems and variables involved in managing successful technology-oriented innovation. Managers recognize today that successful R&D must not only be innovative but also be transferable into the marketplace (Sharma 2003, Zhao 1995). Moreover, as the sources of innovation become more diverse and distributed throughout industries and geographic regions, firms shift the focus beyond their own organizational boundaries, often engaging in collaboration to create joint intellectual property. The processes involved in generating innovative ideas, and ultimately transferring them into the market, are highly sophisticated and complex. For innovation to succeed in such a dynamic environment, some managers argue, requires both discipline and flexibility (Drucker 1998). In fact, strong arguments have been made for the critical importance of traditional managerial controls in support of the innovation transfer process. These control systems include planning, entrepreneurship (Brown and Eisenhard 1995), product champions, top management involvement (Bartlett and Ghoshal 1995), marketing (Gupta and Wilemon 1987, Souder 1988, Brown and Eisenhard 1995), and more recently business strategy (Zhao 1995, Tushman and O'Reilly 1997, Mintzberg and Lampel 1999, Kim and Maubourgne 2004). Moreover, researchers such as Edward Roberts (1988) have attempted to model the macroprocess of innovation, including its interfaces, people, organizational processes, and strategy. In spite of all the research, scholars and practitioners concede that the management of innovative processes is highly intricate, messy, and difficult to control. At best, innovation is the result of interrelated multifunctional efforts that require strong cross-functional integration and orchestration (Abernathy and Clark 1985). On the more challenging side, researchers question whether innovation can be managed by conventional methods. Writers such as Warren Bennis, Peter Drucker, Tom Peters, and J. Brian Quinn quite commonly emphasize the nonlinear, intricate, and often chaotic nature of innovation, especially for technology-based environments, and characterize it as a process that involves all facets of the organization, its members, and its environment, suggesting the need to look beyond established practices for the managing toward innovative results.

10.2.1 An Increased Focus on Teamwork

More than any other process, teamwork affects innovation and organizational performance.[3] Because of their potential for producing an economic advantage, work teams have been studied by many (Katz, 1993, Nurick and Thamhain 1993,

---

[3]In response to this challenge many researchers have investigated teamwork and its relationship to the innovation process (Abbey and Dickson 1983; Gupta, Raj, and Wilemon 1987; Kozar 1987; Larson 1988; Thamhain 1998; Thamhain and Wilemon 1999). Often, such research is especially related to technology-oriented developments because these multidisciplinary team efforts rely on interaction among various organizational, managerial, and environmental subsystems. Team members come from different organizations with different needs, backgrounds, interests, and expertise. To be effective, they must be transformed into an integrated work group that is unified so that they work toward the project's objectives.

Spilen 1991; Dumaine 1994; DiBella 1995), producing a considerable body of knowledge on the characteristics and behavior of teams in various work settings.[4]

It is interesting to note that, in spite of changing leadership styles and continuously emerging new management practices, this established body of knowledge has formed an important and solid basis for guiding managers in our contemporary, demanding work environment (Dumaine 1994, Katzenbach and Smith 1994). It also forms the basis for new management research, theory development, and tools and techniques applied to the management of projects, technology, and innovation.[5]

However, building a workgroup into a unified team requires strong social skills, organizational ability, and leadership. Team members come from different organizations with different needs, backgrounds, interests, and expertise. These differences create barriers to unified team behavior. While these barriers are predictable and natural, as summarized in Table 10.1, they must be removed to transform the collection of people into an integrated team that is unified toward the project objectives. Because of the complexities involved, most formal studies of innovative team performance focus on one particular aspect of innovation or its applications. Typical subareas of studies include a focus on product development, service, technology, or process. However, some researchers, most noticeably Roberts (1988), Crawford (1983), Martin (1994), and Peters (1987), have strongly emphasized the need for linking several microsystems and integrating the findings into a more macro-oriented conceptual view. More recently, research has gained additional insight into the *distributed nature of knowledge and idea generation,* shifting the focus of innovation beyond the boundaries of the firm, leading to new concepts such as *outsourcing of innovation* (Quinn 1999) and *communities of creation* that promote collaboration among individuals and firms to create *joint intellectual property* (Sawhney and Prandelli 2000), as is typical for the scenarios discussed in the lead-in vignette of this chapter.

For management, the challenge is to facilitate a team environment conducive to innovation, in spite of the existing organizational dynamics and complexities. In fact, a growing body of data suggests that both managerial leadership style and work environment significantly influence innovative performance (DiBella 1995, Thamhain and Wilemon 1996, 1999).

---

[4]The characteristics of a high-performing technical project team have been studied extensively by Thamhain and Wilemon (1996, 1999). The studies found a strong association among project success, innovative performance, and certain leadership criteria that include the ability to (1) provide clear directions, (2) unify the team so that it works toward a common project goal, (3) foster clear communication channels and interfaces with other work groups, (4) provide stimulating work, (5) provide professional growth potential, (6) facilitate mutual trust and good interpersonal relations, and (7) involve management.

[5]Work teams have long been considered an effective device to enhance organizational effectiveness. Since the discovery of the importance of workgroups as a social phenomenon in the classic Hawthorne studies, management theorists and practitioners have tried to enhance group identity and cohesion in the workplace. Indeed, much of the "human relations movement" that occurred in the decades following Hawthorne is based on a group concept. McGregor's Theory Y, for example, spells out the criteria for an effective workgroup, and Likert called his highest form of management the *participative group,* or System 4. For further discussion of these issues, see Chapter 9.

Table 10.1   Barriers to Effective Team Performance

Workgroups, such as R&D teams, are subject to all of the phenomena known as group dynamics. They are highly visible and focused, and often take on a special significance and status commensurate with expectations of performance. Although these groups bring significant energy and perspective to a task, the possibilities of malfunctions are great. A myth is that the assembly of talented and committed individuals automatically results in synergy and renders such a team impervious to many of the barriers commonly found in a project team environment. These barriers are quite natural and predictable, but they must be managed. Understanding these barriers, their potential causes, and influencing factors is an important prerequisite for managing teams effectively to achieve desired innovative results. The most common barriers to innovative team performance are summarized below with a focus on R&D-oriented project environments.

1. Different Points of View. To generate innovative results, a project team must harness divergent skills and talents. Coming from different parts of the organization, there is the strong likelihood that team members will naturally see the world from their own unique point of view. There is a tendency to stereotype and devalue "other" views. Such tendencies are heightened when the project involves members from the broader organization, such as manufacturing, marketing, and legal, with different work cultures, norms, values, needs, and interests. Further, these barriers are often particularly strong in highly technical project situations where members speak in their own codes and languages.

2. Role Conflict. Project teams create ambiguity. Team members must often act in multiple roles and report to different leaders, with dual accountabilities and possibly conflicting loyalties. Especially in self-directed team environments, the "home" group or department has a set of expectations that might be at variance with the project team's. For example, a department may be run in a very mechanistic, hierarchical fashion, while the project team may be more democratic, participatory, and self-managed. Team members might also experience time conflicts due to multiple task assignments that overlie and compete with "functional" job responsibilities. The pull from these conflicting forces can either be exhilarating or a source of considerable tension for individual team members.

3. Power Struggles. Conflict can also occur vertically as different authority levels are often represented on the team. Individuals who occupy powerful positions elsewhere in the organization might exercise that influence in the group. Often such attempts to impose ideas or to exert leadership over the group are met with resistance, especially in self-directed groups that operate with a minimum of hierarchical structure, command, and control. While some struggle for power is inevitable in a diverse group, it must be managed to minimize potentially destructive consequences.

4. Group Think. This common group phenomenon refers to the tendency for a highly cohesive team to develop a sense of detachment and elitism. It can particularly afflict groups that work on special highly visible projects. In an effort to maintain cohesion, the group creates the illusion of invulnerability and unanimity. This affects particularly decision making and creativity. There is a reluctance to examine different points of view because these are seen as dangerous to the group's existence. As a result, group members compromise their ability to deal with changes and creativity.

10.3   MEASURING INNOVATIVE PERFORMANCE

Increasingly, a key issue facing managers is how to judge organizational effectiveness. Part of this challenge is to determine and award innovative performance fairly and equitably. According to many managers, innovativeness is difficult to measure, even at the organizational level; it seems nearly impossible to measure meaningfully for task teams or individual contributors. These managerial observations are also confirmed by field research, as summarized in Table 10.2. Challenges exist especially in flatter, less hierarchically structured organizations, where the entire workforce is engaged in developing or improving products and services. It is also a challenge in organizations where R&D contributes only a small portion to the new product development and success or failure is shared among many groups and individuals. Yet, in spite of its intricate nature, some framework can be established for measurability of innovative performance.

Table 10.2   The Difficulties of Measuring Innovative R&D Performance–Some Management Research Perspective

How should one measure innovative R&D performance? This is difficult. All components—project team, innovation, and R&D—involve highly complex sets of intricately related variables. Researchers have consistently pointed at the nonlinear, often random nature of these processes, which involve many facets of the organization, its members, and its environment (Cooper and Kleinschmidt 1988, Danneels and Kleinschmidt 2001, MacCormack et al. 2001, Nellore and Balachandra 2001, Sethi and Nicholson 2001). Investigating these organizational processes simultaneously is not a simple task. The enormous breadth and depth of subsystems and variables involved in managing innovation is laid out in a research review article by Johne and Snelson (1988). The complexities are further emphasized in technology-based environments, where endogenous and exogenous forces of innovation cannot be isolated easily. All of these factors compound the nonlinear nature of the innovation process, making it even less likely to find simple models for researching these environments. Moreover, as the sources of innovation become more diverse and distributed throughout industries and geographic regions, firms shift the focus beyond their own organizational boundaries, often engaging in collaboration to create joint intellectual property. The processes involved in generating innovative ideas, and ultimately transferring them into the market, are highly sophisticated and complex. Successful innovation in such a dynamic environment is a function of many organizational subsystems, including planning, entrepreneurship (Brown and Eisenhard 1995), product champions, top management involvement (Bartlett and Ghoshal 1995), marketing (Gupta et al. 1987, Souder 1988, Brown and Eisenhard 1995), and more recently business strategy (Zhao 1995, Tushman and O'Reilly 1998, Mintzberg and Lampel 1999, Chan and Maubotgne 1999). Therefore, it is not surprising that research in the area of innovation and R&D has traditionally focused on selected dimensions, such as *characteristics of the individual, management style*, or *organizational environment* (Keller 2001, Jasswalla and Sashittal 1998). However, in recent years, a number of studies have

Table 10.2 (Continued)

broadened the investigations, recognizing that successful innovation and team performance cannot be measured as a function of team results alone, but must also take into account the transferability of the results into the marketplace (Li and Atuahene-Gima 2001; Fulmer, Gibbs, and Goldsmith 2000; Sherma 1999; Zhao, 1995). Moreover, researchers such as Edward Roberts (1988, 2004) have attempted for some time to model the macroprocess of innovation, including its interfaces, people, organizational processes, and strategies. Taken together, all of this creates great challenges for both practitioners and management researchers of R&D and innovation.

10.3.1 Innovative Performance Measures

Because of its mix and intricate relationships with teamwork, time lines, multidisciplinary contributions, and value perception, meaningful *measures of innovative performance* are difficult to establish and to apply. An additional challenge is ensuring consistency and fairness across the organization. Perception of the degree and value of innovativeness fluctuates considerably with the cultural and philosophical differences among managers, departments, and companies. Yet, most managers agree that certain metrics, such as those shown in Table 10.3, are *commonly used as indicators of innovative performance for a profit center or a company as a whole.* The following measures are most frequently cited by managers: (1) number of new products/services introduced to market, (2) time to market, (3) cost and performance improvements, and (4) patent disclosures. *For task teams and individual contributors,* meaningful performance measures are even more difficult to define. However, from interviews with R&D and new product development managers we know that the most common metrics for assessing innovative performance for teams or individuals are[6] (1) judgment of innovative performance, (2) number of innovative ideas, (3) patent disclosures and papers, and (4) effort and commitment to established objectives. Yet, the assessment of how R&D and the individuals involved in it are contributing to a new product or service and its competitive advantage is very difficult and often impossible to obtain with any degree of confidence. Even more difficult are the measurements of innovative performance outside of R&D. Functional areas such as manufacturing, marketing, and product assurance depend a great deal on innovation and creativity for meeting customer expectations or delivering results according to plan. Traditional measures of innovative performance seldom apply. In many cases, the end product or deliverable is the result of collaborative efforts among many departments and individuals. Often, no meaningful metrics exist for measuring the contributions such as effort, multifunctional cooperation, agility, risk taking, change orientation, and customer satisfaction. Therefore, it is not surprising that most managers, especially those in technology-based companies, use *overall judgment* as the

[6]For source see Thamhain (2002).

Table 10.3   Typical Metrics of Innovative Company Performance

Variable	Typical Measures
Response Time	• Time to market (T2M) • Time to fill order • R&D response time to in-house request
New Product or Service	• Number of new product/service concepts identified • Number of new product/service concepts introduced to market • Price-performance of new product/service • Revenue ratios: new vs. old products
Product Features	• Feature catalogue • Market price • Judgment by critics
Cost Reduction	• Cost reduction on existing product or service • Cost reduction of internal business process
Technology Transfer	• Cost of bringing a new product to market • Cost and time to transfer a new product to manufacturing, marketing, or field operations
Customer Satisfaction	• Satisfaction measured by survey or other feedback • Repeat business • Referrals • Critics' report
Quality	• Reliability measures • Customer returns • Customer satisfaction measures • Warranty measures • QFD models
Continuous Improvement	• Number of suggestions • Hours of training • Learning curve, e.g., upc/yr • Cost reduction • Quality improvements
Patents and Publications	• Number of patents and publications • Number of citations • Royalties from patents
Revenues from Intellectual Property	• Licensing, royalties • User fees
Business Success	• Market share • Sales volume • Profitability measures • Payback period and ROI • Corporate score card • World leadership image • Press, media coverage

Table 10.4 Factors Influencing Innovative Performance

External Factors:	Internal Factors:
• Competition	• Leadership
• Economy	• Organizational Environment
• Market, customers	• People
• Regulations	• Processes
• Social/political ambience	• Strategy
• Suppliers	• Tasks
• Technology	• Technology
• Timing	• Tools and Techniques

principal measure of innovative performance. However, in support of such an overall judgment, specific subsets of parameters are often developed based on the business objective to be achieved. For this purpose, the metrics defined in Table 10.3 can provide some guidelines. Although it is difficult to quantify the metrics for individuals, they can be used to (1) articulate desired innovative behavior and characteristics to members of the work team, (2) benchmark innovative performance, especially on the department or team level, (3) engage in focus-group discussions aimed toward organizational improvement of innovative performance, and finally (4) support managerial judgment of overall innovative performance and salary reviews.

10.3.1 Influences on Innovative Performance

Based on field studies,[7] I see two major sources of *influences affecting innovative performance, influences from the "external" and "internal" environment*, as shown in Table 10.4. While all of these variables are being seen as complex and interrelated, the "internal" environment is controllable by management. Through their actions, policies, and leadership, managers can influence many of the internal variables. As a framework for discussion, benchmarking, and management policy development, these variables can be grouped into three categories: (1) people, (2) tasks, and (3) process, tools, and technology, as shown in Figure 10.1. Many of these variables can be measured via perception of the various players, depending on whose reality one tries to determine. The graphical presentation in Figure 10.1 is useful for benchmarking and managerial decision making; in addition, Figure 10.1 provides a simple model for research, analysis, and discussion.

[7]For specific scope, methods, and results please see (Thamhain 2000, 2003).

Figure 10.1   Major influences on innovative performance.

10.4   CHARACTERISTICS OF AN INNOVATIVE WORK ENVIRONMENT

One of the major changes in the process of innovation is the involvement of the entire workforce. Especially in technology companies, innovation is no longer the result of individual geniuses. Rather, innovation is a multidisciplinary effort, involving teams of people and support organizations interacting in a highly complex, intricate, and sometimes even chaotic way. The process requires experiential learning, trial and error, risk taking, as well as cross-functional coordination and the integration of technical knowledge, information, and components. Most managers in today's organizations see innovation as a fuzzy process that cannot be described objectively, and whose results cannot be predicted with certainty. Yet, research shows that the work environment influences innovative performance. Specific characteristics of the team environment seem to be conducive to technological innovation, while others seem to hinder the process. Appendix A, "Voices from the Field," summarizes some field research findings, shedding some light on the various drivers of and barriers to innovation that ultimately affect the mission objectives and enterprise performance.

10.5   MANAGING FOR INNOVATIVE PERFORMANCE

In spite of the complexities of the innovation process and the cultural differences among companies, specific working conditions appear most favorable to innovation. These conditions serve as bridging mechanisms, helpful in enhancing innovative performance in technology-based organizations. Field research, such as that summarized in "Voices from the Field," helps to gain additional insight into the processes and challenges of innovative teamwork and effective R&D management. To provide focus, the following discussion is divided into three parts: (1) *people*, (2) *organizational process, tools, and technology,* and (3) *R&D work/tasks.* This breakdown is consistent with the earlier discussion on *influences on innovative performance,* summarized in Figure 10.1:

People-Oriented Influences. As shown in Table 10.5, factors that satisfy personal and professional needs seem to have the strongest effect on the innovative performance of an organization. Statistically, the most significant drivers are *derived from the work itself*, including *personal interest, pride and satisfaction with the work, professional work challenge, accomplishments*, and *recognition.* Other important influences include effective communications among team members and with support units across organizational lines, good team spirit, mutual trust and respect, and low interpersonal conflict, plus opportunities for career development and advancement and, to some degree, job security. All of these factors help in building a unified project team that can leverage the organizational strengths and competencies effectively and produce integrated results that support the organization's mission objective. These factors seem to foster a work environment conducive to innovation and an ambience that ultimately helps to transforms collective team efforts into business results.

Table 10.5   Rank-Order Correlations of Work Environment and Innovative Team Performance (Strong Performance Associations[+])

CHARACTERISTICS OF WORK ENVIRONMENT*	INNOVATIVE R&D TEAM PERFORMANCE[#]				
	Predicting Market & Technology Change τ_1	Market Focus, Mkt Responsiveness and Speed τ_2	Cross-Functional Communication Effectiveness τ_3	Resource Effectiveness τ_4	Overall Innovative Performance τ_5
Interesting, Stimulating Work	.32	*.38*	*.43*	*.36*	*.41*
Accomplishment & Recognition	*.39*	*.42*	*.40*	*.41*	*.39*
Low Conflict & Prompt Problem Resolution	.30	.27	.35	*.39*	*.37*
Clear Organizational Objectives	*.46*	*.37*	.28	.34	*.36*
Job Skills and Expertise	*.42*	.22	.21	.28	*.36*
Direction and Leadership	*.40*	*.37*	*.37*	.32	.35
Trust, Respect, Credibility	.33	.33	.31	.28	.33
Cross-Functional Cooperation & Support	*.42*	*.40*	*.63*	*.37*	.32
Effective Communications	*.40*	.33	.36	.26	.29
Clear Project Plan & Support System	.33	.31	.34	*.38*	.25
Autonomy and Freedom	.31	.28	.26	.30	.24
Career Development & Advancement	.19	.17	.23	.21	.22
Stable Long-Range Goals and Priorities	.29	.34	.26	.33	.20
Job Security	.10	.12	.34	.15	.17

[+] Associations of variables were measures by Kendall's Tau Rank-Order Correlation, with statistical significance as indicated below

*As perceived by project team members on a five-point scale: (1) strongly disagree, (2) disagree, (3) neutral, (4) agree, and (5) strongly agree.

[#] As perceived by senior management on a four-point scale: (1) poor, (2) marginal, (3) good, and (4) excellent.

Statistical Significance: $p = .10$ $(\tau \geq .20)$, $p = .05$ $(\tau \geq .31)$, $p = .01$ $(\tau \geq .36)$; correlations of $p = .01$ or stronger are shown in bold italics.

Organizational Process, Tools, and Technology. These influences include the organizational structure and the technology transfer process, which relies by-and-large on modern project management techniques. While the research did not favor specific project structures and processes, it specifically pointed at *cross-functional cooperation, effective communication, effective project planning and support systems, stable long-range organizational goals and priorities,* and overall managerial leadership as important conditions for effective R&D team performance and innovative results. An effective project management system also includes effective functional support, joint reviews and performance appraisals, and the availability of the necessary resources, skills, and facilities. Other crucial components that affect the work/business process are team structure, managerial power, command and control and its sharing among the team members and organizational units, autonomy and freedom, and most importantly technical direction and leadership. The findings also provide food for thought for top management. Many of the variables in the organizational process, management tools, and technology have their locus outside the R&D team organization. These variables are often a derivative of the company's business strategy, developed and controlled by senior management. It is important for management to recognize that these variables can directly affect the quality of the work environment, as shown in Table 10.5, including the perception of organizational stability, availability of resources, management involvement and support, personal rewards, and stability of organizational goals, objectives, and priorities. Because all of these influences are images of personal perception, it is important for management to understand the personal and professional needs of their team members, and to foster an *organizational environment* conducive to these needs. *Proper communication* of organizational vision and perspective is especially important. For example, a company merger might be perceived as an opportunity or threat, a stabilizer or a destabilizer, depending on how it is communicated. In the relationship of managers to staff and people in their organizations, mutual trust, respect, and credibility are all critical factors in building an effective partnership between the R&D team and its sponsor organization.

Work- and Task-Related Influences. Innovative performance also has its locus in the work itself. This is highlighted with the correlations in Table 10.5. It shows that in particular those variables associated with the personal aspects of work—such as interest, ability to solve problems, job skills, and experience—are statistically significant in driving innovative performance. Many other work-related variables from the structural side, such as project size, work complexity, and work process, had little statistical significance in influencing innovative performance. This finding is important in two areas. First, managers must be able to attract and hold people with the right skill sets, appropriate for the work to be performed. They must also invest in maintaining and upgrading job skills and support systems. Second, managers must effectively assign the work, leveraging their control over work partitioning and results. That is, while the total task structure and the development process are fixed and difficult to change, the way that managers distribute, assign, and present the work is flexible. This is an important finding with managerial implications: Promoting a climate of high interest, involvement, and support might be *easier* to achieve than redefining the R&D project or reengineering the work process, yet it might have an *equal or higher* impact on innovative performance.

10.6 CONCLUSIONS

This chapter examined the complexities and challenges involved in managing technology-based workgroups toward innovative results. At the core of these creative workgroups are the "knowledge workers," such as engineers, scientists, designers, and architects, typically associated with R&D and new product development functions. Yet, business performance is rarely enhanced just by a lone genius, but by collaborate and creative efforts of many people throughout the organization. This includes professionals from design, production, marketing, field services, finance, law, human resources, and other support groups. Collectively, these people "transfer" new ideas toward a business application, creating new technology, products, services, or content. The challenges of managing innovation, on an individiual or organizational level, seem to originate, in part, from the multidisciplinary nature and complexity of the work, its association with risks, uncertainty, and nonlinear processes. These challenges are also part of the complex interfaces between R&D and the total enterprise which must be carefully developed and managed to enable the transfer and integration of innovative ideas. In addition, projects of different nature, size, duration, and complexity require different organizational support and team environments conducive to innovative results. A new telephone product by Nokia and the movie production *Lord of the Rings* both require a great deal of technology-based innovation but, obviously, different organizational support, management style, and leadership. This points toward the situational effectiveness of managerial leadership.

Conclusions in this chapter concur with the discussions on people, organizations, and teams in earlier chapters. Innovative performance too is derived from the human side. Organizational components that satisfy personal and professional needs seem to have the strongest effect on the innovative performance of R&D teams. The most statistically significant drivers are social influences rooted in the work itself. People who find their assignments professionally challenging, leading to accomplishments, recognition, and professional growth, also seem to perform more creatively and function more effectively in solving complex problems. A professionally stimulating environment seems to lower communication barriers and conflict, and enhance the desire to succeed. It strengthens the collective awareness of changes and trends in the business environment, and the desire to respond to resulting threats and opportunities effectively.

Other important factors that affect innovative R&D performance are related to the social setting of the team (cf. discussion in previous chapter). Effective communication among team members and support units across organizational lines, good team spirit, mutual trust and respect, and low interpersonal conflict, plus opportunities for career development and advancement and, to some degree, job security, seem to *unify the team*, a necessary condition for integrating creative efforts toward desired results. It also increases the individual team members' tolerance for risk and uncertainties, and stimulates commitment toward the team mission.

In addition, field research points out the critical importance of effective project management systems. Effective project planning, functional support systems, project tracking, reviews, and cross-functional support were among the factors that were associated very favorably with desired R&D results.

Many of the influences on R&D performance have their locus outside of the R&D organization. They are controlled by other management groups, such as marketing, manufacturing, field operations, legal, and strategic planning. Organizational stability, availability of resources, management involvement and support, personal rewards, overall organizational goals, objectives, and priorities are all derived from enterprise systems that are orchestrated and controlled by general management. Both R&D leaders and senior management must recognize these influences and work together to ensure an organizational ambience conducive to innovative R&D performance.

Finally, leaders of successful, continually innovative organizations create a sense of community across the whole enterprise. That is, management understands the factors that drive innovative performance and creates a work environment conducive to such behavior. Innovative organizations have leaders who can inspire their people and make everyone feel proud of being part of the company and the innovative process. Both clarity of purpose and alignment of personal and organizational goals are necessary for an innovative culture to emerge and to prosper. Encouragement, personal recognition, and the visibility of one's contributions to customers and company values help to unite the R&D team as part of the whole enterprise and its mission, and to refuel the commitment to desired results. Leadership also means personal involvement. Effective management is action oriented, providing the needed resources, proper plans, and directions for program implementation and helping in identifying and solving problems in their early stages. Effective R&D team leaders continuously monitor feedback from their project environment and focus their efforts on early problem detection and resolution. While managing all potential problem scenarios may be difficult or impossible and undesirable, the effective manager can keep an eye on situations that may cause problems, proactively intercepting and minimizing them wherever possible.

Taken together, success in today's continually changing word of business is not an easy feat. No single set of broad guidelines exists that guarantees R&D success. However, the innovation process is not random! A better understanding of the interaction of organizational and behavioral variables, and the criteria that drive innovative performance, can assist managers in developing a better, more meaningful insight into the enterprise dynamics that affect R&D performance, thus helping to fine-tune leadership styles, managerial actions, and resource allocation.

10.7   **SUMMARY OF KEY POINTS AND CONCLUSIONS**

The key points that have been made in this chapter include:

- Innovation is the last remaining frontier, helping companies to achieve lower cost, superior performance, and new products and services.
- Deriving competitive advantages from innovation is a highly intricate process, involving technical complexities, functional interdependencies, evolving solutions, high levels of uncertainty, and highly complex forms of work integration.
- Investment in R&D activities alone does not guarantee success, as evidenced by the high failure rate of new products and services.
- For technology companies, the key challenge is not so much the generation of innovative ideas, but the effective transfer of technology from the discovery stage to the market.
- While researchers question the manageability of innovation by conventional methods, R&D success is not random and can be managed to some extent.
- More recently, the focus of research and innovation has shifted beyond the boundaries of the firm. New insight into the distributed nature of knowledge and idea generation has led to new concepts such as *outsourcing of innovation*, *communities of creation*, and *open R&D* to create *joint intellectual property*.
- Innovativeness is difficult to measure, even at the organizational level, and nearly impossible to measure for task teams or individual contributors.
- Most managers agree on *innovative performance measures for a profit center or a company as a whole*: (1) number of new products/services, (2) time to market, (3) cost and performance improvements, and (4) patent disclosures.
- *For task teams and individuals* the most common metrics for assessing innovative performance are (1) judgment of innovative performance, (2) number of innovative ideas, (3) patent disclosures and papers, and (4) effort toward and commitment to established objectives. Yet, the assessment of how R&D and its individuals are contributing to a new product or service and its competitive advantage is very difficult, and often impossible, to obtain. Even more difficult is the measurement of innovative performance outside of R&D.
- Managers can influence many of the variables that drive innovation through their actions, policies, and leadership. These variables can be grouped into three categories: (1) people, (2) tasks, and (3) process, tools, and technology.
- The most significant drivers to innovative performance are *derived from the work itself*, including *personal interest, pride and satisfaction with the work, professional work challenge, accomplishments, and recognition*. Other important influences include effective communication among team members and support units across organizational lines, good team spirit, mutual trust and respect, low interpersonal conflict, and opportunities for career development.

- Successful R&D requires effective project management, including project planning, functional support, project tracking, and reviews.
- Effective R&D managers understand the interaction of organizational and behavioral variables and can foster a climate of active participation, minimal dysfunctional conflict, and effective communication.

10.8   CRITICAL THINKING: QUESTIONS FOR DISCUSSION

1. Define innovation in a business context.
2. What is the role of innovation in company competitiveness and business performance?
3. Is R&D a necessary function for all high-tech companies? Why or why not?
4. What are some of the risk factors that make R&D performance so unpredictable?
5. What is the role of teamwork in innovation and R&D work?
6. What are the critical success factors (CSF) for corporate R&D? Are there some common factors among companies? Select a few successful high-tech enterprises. Can you find common factors (not just generalities)?
7. Discuss the advantages and challenges of "joint R&D ventures" and "collaborative R&D communities."
8. Use the metrics and components related to innovative performance, given in this chapter, to develop your criteria for measuring innovativeness in your work environment.
9. Is managerial perception a good indicator of innovative performance? Why or why not?
10. Select a high-tech company that you judge to be "innovative." Examine the organizational subsystems, such as structure, work process, management style, and support systems. Then discuss the factors which you feel are critical to innovative performance of this company. Could the "innovative characteristics" be transferred to another company (e.g., your employer)?
11. What kind of personal traits would you look for in a "highly innovative" person? How can you evaluate a job applicant for these desirable traits?
12. Describe a management style conducive to the innovative performance of (1) a lab technician, (2) a product development engineer, and (3) a biochemical researcher. Discuss any differences in managerial style and their rationale.
13. Discuss how an organization can foster a work environment that supports innovation/innovators. Write some managerial guidelines for stimulating/enhancing innovative behavior within your workgroup at your company.
14. Long-range R&D projects often don't produce significant results for a long time. How do you motivate the research team in the absence of specific recognizable results?
15. Define a mission statement or charter for an R&D department of a high-tech company (you select and define the type of company and its business).
16. How would you evaluate the performance of an R&D director?

10.9 REFERENCES AND ADDITIONAL READINGS

Abbey, A. and Dickson, J. W. (1983) "R&D Work Climate and Innovation in Semiconductors." *Academy of Management Journal*, Vol. 26, No. 4, pp. 362–368.

Abernathy, W. and Clark, K. B. (1985) "Innovation—Mapping the Winds of Creative Destruction." *Research Policy*, Vol. 14, No. 1, pp. 3–22.

Bartlett, C. and Ghoshal, S. (1995) "Rebuilding behavioral context: turn process reengineering into people rejuvenation," *Sloan Management Review*, Vol. 37, No. 1, pp. 11–23.

Bennis, W. and Nanus, B. (1985) *Leaders: The Strategies for Taking Charge*, New York: Harper & Row.

Bloch, E. (2003) "Securing U.S. research strength," *Issues in Science and Technology,* Vol. 19, No. 4 (Summer), p. 20.

Bonner, J., Ruekert, R., and Walker, O. (2002). "Upper management control of new product development projects and project performance," *Journal of New Product Innovation Management*, Vol. 19, No. 3 (May), pp. 233–245.

Bonner, J., Ruekert, R., and Walker, O. (2004) "Selecting influential business-to-business customers in new product development," *Journal of New Product Innovation Management*, Vol. 21, No. 3 (May), pp.155–169.

Brown, S. and Eisenhardt, K. M. (April 1995) "Product development: past research, present findings, and future directions," *Academy of Management Review*, Vol. 20, No. 2, pp. 343–378.

Bugelman, R., Modesto, M. A., Maidique, A., and Wheelwright, S. C. (1988) *Strategic Management of Technology and Innovation*, Chicago, IL: Irwin.

Castro, M., Foster, R., and Roberts, B. (2003). "Managing R&D alliances within government: the virtual agency concept," *IEEE Transactions on Engineering Management*, Vol. 50, No. 3 (August), pp. 297–306.

Conway, A. and McGuinness, N. W. (1986) Idea generation in technology-based firms," *Journal of Product Innovation Management*, Vol. 3, No. 4, pp. 276–291.

Cooper, R. and Kleinschmidt, E. J. (1995) "Winning business in product development: the critical success factors," *Research-Technology Management,* Vol. 39, No. 4, pp. 18–29.

Crawford, C. (1983) *New Products Management*, Homewood, IL: Richard D. Irwin

Danneels, E. and Kleinschmidt, E. J. (2001) "Product innovativeness from the firm's perspective," *Journal of Product Innovation Management*, Vol. 18, No. 6, pp. 357–374.

Debruyne, M., Moenaert, R., Griffin, A., Hart, S., Hultink, E. J., and Robben, H. (2002) "The impact of new product launch strategies on competitive reaction in industrial markets," *Journal of Product Innovation Management*, Vol. 19, No. 2, pp. 159–170.

DiBella, Anthony J. (1995) "Developing learning organizations: a matter of perspective," *Academy of Management Journal* (Best Papers Proceedings), pp. 287–290.

Donlon, J. P. (1996) "Are you empowering innovation." *Chief Executive*, No. 115, pp. 66–82.

Drucker, P. (1985) *Innovation and Entrepreneurship: Practice and Principles*, New York: Harper & Row.

Drucker, P. (1998) "The discipline of innovation," *Harvard Business Review*, Vol. 76, No. 4, pp. 73–84.

Dumaine, B. (1994) "The trouble with teams," *Fortune,* Vol. 130, No. 5, pp. 86–92.

Eisenhardt, K. M. (1989) "Building theories from case study research," *Academy of Management Review*, Vol. 14, No. 4, pp. 532–550.

Fulmer, R., Gibbs, P., and Goldsmith, M. (2001) "Developing leaders: how winning companies keep on winning," *Sloan Management Review*, Vol. 42, No. 1, pp. 49–59.

Garcia, R. and Calatone, R. (2002). "A critical look at technological innovation typology and innovativeness terminology: a literature review," *Journal of New Product Innovation Management*, Vol. 19, No. 2 (March), pp.110–132.

Glaser, B. G. and Strauss, A. L. (1967) *The Discovery of Grounded Theory: Strategies for Qualitative Research*, Chicago: Aldine.

Gupta, A., Raj, S. P., and. Wilemon, D. L. (1987) "Managing the R&D-marketing interface," *Research Management*, Vol. 30, No. 1, pp. 38–43.

Holmann, R., Kaas, H. and Keeling (2003) "The future of product development," *McKinsey Quarterly*, 2003, No. 3, pp. 33–42.

Jasswalla, A. R. and Sashittal, H. C. (1998) "An examination of collaboration in high-technology new product development processes," *Journal of New Product Innovation Management*, Vol. 15, No. 3, pp. 237–254.

Johne, F. and Snelson, P. (1988) "Success factors in product innovation: a selected review of the literature," *Journal of Product Innovation Management,* Vol. 5, No. 2 (June), pp. 114-128.

Karlsen, J. and Gottschalk, P. (2004) "Factors affecting knowledge transfer in IT projects," *Engineering Management Journal*, Vol. 16 No. 1, pp. 30–38.

Katzenbach, J. R. and Smith, D. K. (1994) "Teams at the top," *McKinsey Quarterly*, Vol. 4, No. 2, pp. 71–79.

Keller, R. (2001) "Cross-functional project groups in research and new product development," *Academy of Management*, Vol. 44, No. 3, pp. 547–556.

Kim, C. W. and Mauborgne, R. (1999) "Strategy, value innovation, and the knowledge economy," *Sloan Management Review*, Vol. 40, No. 3, pp. 41–53.

Kim, C. W. and Mauborgne, R. (2004) "Blue ocean strategy," *Harvard Business Management Review*, Vol. 82, No. 10 (October), pp. 76-85.

Leenders, A. and Wierenga, B. (2002) "The effectiveness of different mechanisms for integrating marketing and R&D," *Journal of New Product Innovation Management*, Vol. 19, No. 4 (July), pp. 305–317.

Li, H. and Atuahene-Gima, K. (2001) "Product innovation strategy and the performance of technology ventures in China," *Academy of Management Journal*, Vol. 44, No. 6, pp. 1123–1134.

Lorch, C. and Tapper, U. (2002) "Implementing a strategy-driven performance measurement system for an applied research group," *Journal of New Product Innovation Management*, Vol. 19, No. 3 (May), pp. 185–198.

Maschitelli, R. (2002) "Building a project-driven enterprise," *Journal of New Product Innovation Management*, Vol. 19, No. 3 (May), pp. 199-232.

MacCormack, A., Verganti, R., and Iansiti, M. (2001) "Developing products on Internet time," *Management Science*, Vol. 47, No. 1, pp. 22–35; *Engineering Management Review*, Vol. 29, No. 2, pp. 90–104.

Martin, M. (1994) *Managing Innovation and Entrepreneurship in Technology-Based Firms*. New York: Wiley & Sons.

McDonough, E. (1993) "Faster new product development: investigating the effects of technology and characteristics of the project leader and team," *Journal of Product Innovation Management*, Vol. 10, No. 3, pp. 241–250.

Mintzberg, H. and Lampel, J. (1999) "Reflecting on the strategy process," *Sloan Management Review*, Vol. 40, No. 3, pp. 12–20.

Moss Kanter, R. (1989) "Swimming in newstreams: mastering innovation dilemmas," *California Management Review*, Vol. 31, No. 4, pp. 45–69.

Nellore, R. and Balachandra, R. (2001) "Factors influencing success in integrated product development (IPD) projects," *IEEE Transactions on Engineering Management*, Vol. 48, No. 2, pp. 164–173

Nemeth, C. (1997), "Managing innovation: when less is more," *California Management Review*, Vol. 40, No. 1 pp. 37–48.

Nunes, S. (2004) "IBM research: ultimate source for new businesses," *Research Technology Management,* Vol. 47, No. 2 (March/April), pp. 20–23.

Nurick, A. and Thamhain, H. (1993) "Project team development in multinational environments," Chapter 38 in *Global Project Management Handbook* (David I. Cleland, ed.), New York: McGraw-Hill.

Olson, E. M., Walker, O. C., Ruekert, R. W, and Bonner, J. M. (2001) "Patterns of cooperation during new product development among marketing, operations and R&D," *Journal of New Product Development*, Vol. 18, No. 4, pp. 258–71.

Peters, T. (1987) *Thriving on Chaos: Handbook for a Management Revolution*, New York: Knopf.

Pospisil, V. (1996) "New constellations." *Industry Week*, Vol. 245, No. 14, p. 26.

Quinn, J. B. (1985) "Managing innovation: controlled chaos." *Harvard Business Review*, Vol. 61, No. 3, pp. 73–84.

Quinn, J. B. (1999) "Outsourcing innovation: the new engine of growth," *Sloan Management Review*, Vol. 40, No. 3, pp. 63–72.

Reed, R., Lemak, D. L., and Montgomery, J. C. (1996) "Beyond process: TQM content and firm performance." *Academy of Management Review*, Vol. 21, No. 1, pp. 173–202.

Reid, S. and Brentani, U. (2004) "The fuzzy front end of new product development for discontinuous innovations: a theoretical model," *Journal of New Product Innovation Management*, Vol. 21, No. 3 (May), pp. 155–169.

Richey, J. and Grinnell, M. (2004) "Evolution of roadmapping at Motorola," *Research Technology Management*, Vol. 47, No. 2 (March/April), pp. 37–41.

Roberts, E. (2004). "Linkage, leverage and leadership drive successful technological innovation," *Research Technology Management*, Vol. 47, No.3 (May/June), pp. 9–11.

Roberts, E. (1988) "Managing inventions and innovation." *Technology Management*, Vol. 31, No. 1, pp. 11–29.

Sawhney, M. and Prandelli, E. (2000) "Communities of creation: managing distributed innovation in turbulent markets," *California Management Review*, Vol. 42, No. 4, pp. 45–69.

Seidel, E. and Thamhain, H. J. (2001) "Managing environmental quality at the enterprise," *Environmental Engineering and Policy*, Vol. 3, 1, pp. 19–32.

Sen, F. and Engelhoff, W. G. (2000) "Innovative capabilities of a firm and the use of technical alliances," *IEEE Transactions on Engineering Management*, Vol. 47, No. 2, pp. 174–183.

Senge, P. and Carstedt, G. (2001) "Innovating our way to the next Industrial Revolution," *Sloan Management Review*, Vol. 42, No. 2, pp. 24–38.

Sethi, R. and Nicholson, C. (2001) "Structural and contextual correlates of charged behavior in product development teams," *Journal of Product Innovation Management*, Vol. 18, No. 3, pp. 154–168.

Shapiro, S. M. (2002) *24/7 Innovation: A Blueprint for Surviving and Thriving in an Age of Change*, New York: McGraw-Hill.

Sharma, B. (2003) "R&D strategy and Australian manufacturing industry: an empirical investigation of emphasis and effectiveness," *Technovation*, Vol. 23, No. 12 (December), pp. 929–937.

Sherma, A. (1999) "Central dilemmas of managing innovation in large firms," *California Management Review*, Vol. 41, No. 3, pp. 65–85.

Shim, D. and Lee, M. (2001), "Upward influence styles of R&D project leaders," *IEEE Transactions on Engineering Management*, Vol. 48, No. 4, pp. 394–413.

Shuman, J. and Twombley, J. (2001) *Collaborative Communities*, Chicago: Dearborn Trade Publishing.

Smith, P. and Blanck, E. (2002) "Leading dispersed teams," *Journal of New Product Innovation Management*, Vol. 19, No. 4 (July), pp. 294–304.

Souder, W. (1988) "Managing relations between R&D and marketing in new product development projects," *Journal of Product Innovation Management*, Vol. 5, No. 1, pp. 6–19.

Stringer, R. (2000) "How to manage radical innovation," *California Management Review*, Vol. 42, No. 4 (Summer), pp. 55-68.

Thamhain, H. J. (1990) "Innovative performance in research, development, and engineering," *Engineering Management Journal*, Vol. 2, No. 1, pp. 3–11.

Thamhain, H. J. (1990) "Managing technologically innovative team efforts toward new product success," *Journal of Product Innovation Management*, Vol. 7, No. 1, pp. 5–18.

Thamhain, H. J. (1996), "Managing self-directed teams toward innovative results," *Engineering Management Journal*, Vol. 8, No. 3, pp. 31–39.

Thamhain, H. J. (1996) "Managing technology-based innovation." Chapter 9 in *Handbook of Technology Management* (G. Gaynor, ed.), New York: McGraw-Hill.

Thamhain, H. J. (2002) "Criteria for effective leadership in technology-oriented project teams," Chapter 16 in *The Frontiers of Project Management Research* (Slevin, Cleland, and Pinto, eds.), Newton Square, PA: Project Management Institute, pp. 259–270.

Thamhain, H. (2003) "Managing innovative R&D teams," *R&D Management*, Volume 33, Number 3 (June), pp. 297–312.

Thamhain, H. (2004). "Linkages of project environment to performance: lessons for team leadership," *International Journal of Project Management,* Vol. 23, No. 8 (November), pp. 480–488.

Thamhain, H. J. and Wilemon, D.L. (1996) "Building high performing engineering project teams," in *The Human Side of Managing Technological Innovation* (R. Katz, ed.), New York: Oxford University Press.

Thamhain, H. J. and Wilemon, D. L. (1999) "Building effective teams for complex project environments." *Technology Management*, Vol. 5, No. 2, pp. 203–212.

Tomkovich, C. and Miller, C. (2000) "Riding the wind: managing new product development in the age of change," *Product Innovation Management*, Vol. 17, No. 6, pp. 413–423.

Tushman, M. and O'Reilly, C. A. (1997) *Winning Through Innovation*, Boston: HBS Press.

Van den Ende, J. and Wijinberg, N. (2003). "The organization of innovation and market dynamics: managing increasing returns in software firms," *IEEE Transactions on Engineering Management*, Vol. 50, No. 3 (August), pp. 374–382.

von Hippel, E. (2001) "User toolkit for innovation," *Journal of Product Innovation Management*, Vol. 18, No. 4, pp. 247–257.

Zhao, L. (1995) "Integrating technology management with business strategy," *Advances in Applied Business Strategy*, Vol. 4, No. 1, pp. 11–30.

Appendix

VOICES FROM THE FIELD: SOME APPLIED RESEARCH

Using Kendall's Tau Rank-Order Correlation of field data,[*] Tables 10.5 and 10.6 summarize the association among factors of the organizational environment and innovative R&D team performance, listed in order of importance to overall innovative performance. The presence and strength of these organizational variables were measured on a five-point scale as a perception of project team members, while innovative performance was measured as a perception of senior management on a four-point scale. As indicated by the strong positive correlation, factors that fulfill professional esteem needs seem to have a particularly strong influence on innovative R&D performance. The three most significant associations are (1) professionally stimulating and challenging work environments [τ=.41], (2) opportunity for accomplishments and recognition [τ=.39], and (3) the ability to resolve conflict and problems [τ=.37]. All of these favorable correlations to innovative performance appear to deal effectively with the integration of goals and needs between the team member and the organization. In this context, the more subtle factors seem to become catalysts for cross-functional communication, information sharing, and ultimate integration of the project team with focus on desired results. The other factors in Table 10.5 with favorable correlation relate to job knowledge, skills, management, and business process. All associations are significant at $p = .1$ or better, with the most significant correlations of $p = .01$ or stronger shown in bold italics. It is interesting to note that

[*]An exploratory field study into technology-oriented R&D environments, conducted between 1997 and 2001, collected data from 27 R&D organizations, most of them part of large corporations of the Fortune 500 category. The field study yielded data from 74 project teams with a total sample population of 935 professionals such as engineers, scientists, and technicians, plus their managers, including 18 supervisors, 74 project team leaders, 18 product managers, 10 directors of R&D, 8 directors of marketing, and 10 general management executives at the vice presidential level. Together, the data covered over 180 projects in 27 companies. The R&D projects involved mostly high-technology product/service developments with budgets averaging $1,200,000 each. For specific method and results see Thamhain (1990, 1996, 2003).

the same conditions that are conducive to innovative performance also lead to (1) high predictability of changes in markets and technology, (2) high market focus, responsiveness, and favorable time-to-market performance, (3) effective cross-functional communication and networking, and (4) effective resource utilization. In fact, a high degree of cross-correlation exists among the four sets of variables, τ_1 through τ_4, as measured via Kruskal-Wallis analysis of variance by rank.* The test shows that managers agree on the ranking of team performance factors in Table 10.5 at a confidence level of 98 percent. That is, managers who rate the team performance high in one category are likely to give high ratings also to the other three categories. Further, people who are seen as innovative are also seen as having the ability to predict market and technology changes, respond to market needs quickly, and communicate and use resources effectively.

In addition to the 14 most significant characteristics reported in Table 10.5, it is interesting to note that many other characteristics of the work environmental that were perceived by managers as important to innovative performance *did not correlate significantly* as measured by a *p*-level threshold of .10. As summarized in Table 10.6, among the *factors of lesser influence* to innovation are (1) salary, (2) time off, (3) project visibility and popularity, (4) maturity of the project team, measured in terms of time worked together as a team, (5) project duration, (6) stable project requirements with minimum changes, (7) stable organizational structures and processes with minimal organizational changes such as those caused by mergers, acquisitions, and reorganization, (8) minimum technological interdependencies, such as those caused by the dependency on multiple technologies, technological disciplines, and processes, and (9) workable project size and project complexity. This argues that projects that are perceived by the team to exceed a comfort level of complexity or size will lead to lower innovative R&D performance. It is further interesting to see that several of the weaker influences shown in Table 10.4 actually have a negative association with R&D performance. For example, it appears that the more stable the project requirements or the organizational environment, the less overall innovative R&D performance seems to occur. While these correlations are clearly insignificant from a statistical point of view, they shed some additional light on the subtle and intricate nature of R&D performance and the difficulties of defining specific functions that drive such performance. From a different perspective, it is interesting that most influences supporting *intrinsic professional needs* show a strong favorable performance correlation, while the findings give only weak support to extrinsic influences and factors derived from the project metrics, as shown in Table 10.6. This is so in spite the fact that both sets of influences (Table 10.5 and Table 10.6) were perceived by R&D managers as critically important to team performance. The findings suggest that managers are more accurate in the perception of their people's intrinsic needs than their extrinsic ones. It also seems to be more difficult to predict the impact of project parameters, such as size, duration, or complexity, than the impact of human needs on R&D work performance.

---

*The Kruskal-Wallis One-Way Analysis of Variance by Rank is a test for deciding whether k independent samples are from different populations. In this field study, the test verified that managers perceive in essence the same parameters in judging high team performance.

Table 10.6 Rank-Order Correlations of Work Environment and Innovative Team Performance (Weak Performance Associations[+])

CHARACTERISTICS OF WORK ENVIRONMENT*	INNOVATIVE R&D TEAM PERFORMANCE[#]				
	Predicting Market & Technology Change τ_1	Market Focus, Mkt Responsive-ness and Speed τ_2	Cross-Functional Communication Effectiveness τ_3	Resource Effectiveness τ_4	Overall Innovative Performance τ_5
Salary Increases and Bonuses	.11	.07	.18	.25	.15
Compensatory Time Off	.02	.04	.00	.07	.15
Project Visibility and Popularity	.18	.25	.26	.13	.12
Team Maturity & Tenure	.07	.10	.20	.10	.10
Project Duration	.18	.25	.02	.05	-.08
Stable Project Requirements (Min Changes)	.07	-.12	-.22	.08	-.10
Stable Organizational Structures & Processes	-.09	.05	-.17	.10	-.12
Minimum Technological Interdependencies	.08.	-.11	.02.	.10.	-.15
Workable Project Size & Complexity	05	.26	06	35	-.18

[+] Associations of variables were measured by Kendall's Tau Rank-Order Correlation, with statistical significance as indicated below

*As perceived by project team members on a five-point scale: (1) strongly disagree, (2) disagree, (3) neutral, (4) agree, and (5) strongly agree.

As perceived by senior management on a four-point scale: (1) poor, (2) marginal, (3) good, and (4) excellent.

Statistical Significance: $p = .10$ ($\tau \geq |.20|$), $p > .10$ ($\tau < |.20|$) ... *is considered not significant.*

11

MANAGING ENVIRONMENTAL QUALITY

HUDSON RIVER PCB SUPERFUND SITE.[a]

In early 2003, New York State Department of Environmental Conservation (DEC) Commissioner Erin M. Crotty announced DEC's proposed cleanup plans for the General Electric Hudson Falls state Superfund site. "Restoration of the Hudson River to its full potential as an environmental resource is one of Governor Pataki's top priorities," Commissioner Crotty said. "The cleanup of the PCB contamination at the GE Hudson Falls site and eliminating a continuing source of PCBs to the Hudson River represents significant progress in our restoration efforts." Remedial Investigations and Feasibility Studies (RI/FS) conducted by GE under the state Superfund program with state oversight showed soil containing high levels of polychlorinated biphenyls (PCB) contamination in the vicinity of the former manufacturing buildings at the site; highly PCB-contaminated groundwater and PCB oil in the overburden soils; highly contaminated groundwater containing PCB oil in the bedrock beneath the site; and seeps of PCB oil in the bottom of the Hudson River adjacent to the site.

DEC's PRAP calls for the treatment and/or disposal of the PCB-contaminated soils, the installation of tunnel and drain groundwater and PCB oil control systems, and completion of any ongoing interim remedial measures (IRMs) as the preferred cleanup strategy for the site. Elements of DEC's preferred cleanup strategy include (1) appropriate decommission and demolition of the former manufacturing buildings at the site; (2) active remediation of contaminated soils at the site by treatment and/or off-site disposal; (3) installation of a vegetated soil cover over the entire site; (4) enhancement of the existing groundwater and PCB oil management system through the installation and operation of a tunnel and drain system constructed along the western portion of the site, adjacent to and beneath the Hudson River, to capture, treat, and prevent the migration of PCB-contaminated groundwater and PCB oil from the site to the Hudson River and other off-site areas; (5) expansion of the wastewater treatment plant at the site to manage the additional wastewater generated at the site; (6) a Declaration of Limited Use and Restrictions would be implemented, which restricts the site to only industrial and commercial uses and prohibits the extraction of groundwater for uses other than cleanup purposes; and (7) institution of a long-term monitoring program.

Numerous IRMs have already been undertaken by General Electric at the Hudson Falls site, including the removal and off-site disposal of PCB-contaminated sediment found within the mill and raceway structures adjacent to the site, and adjacent to the former 002 outfall location; collection of PCB oil seeps within the mill and raceway structures and on the bottom of the Hudson River adjacent to the site; construction and operation of a state-of-the-art wastewater treatment plant at the site; and installation and operation of an overburden and bedrock groundwater and PCB oil recovery system.

These are some of the environmental challenges faced by General Electric today, which developed over 30 years, and came to light in 1976 with charges by the State Department of Environmental Conservation and the placement of the Hudson River PCB Site on the Superfund National Priorities List (NPL) in 1981. How did this happen so suddenly? For approximately 30 years, beginning in 1947 at Fort Edward and in 1952 at Hudson Falls, New York, polychlorinated biphenyls (PCBs) were *legally discharged* to the upper Hudson River from capacitor manufacturing plants operated by the General Electric Company, which had obtained federal and state permits for the dumping.

Potentially, the environmental damages caused by GE's discharge, even though legal at the time, could cost the company billions of dollars, in addition to enormous damage to the company's image. However, in spite of numerous interim remedial measures (IRMs) already undertaken by General Electric—including the removal of PCB-contaminated sediment, the collection of PCB oil seeps, construction and operation of a state-of-the-art wastewater treatment plant at the site, and installation and operation of a PCB oil recovery system—the public's opinion and judgment of GE's environmental performance are quite dismal. Over the years, the debate has gone from PCBs to a more fundamental crusade. "Extremists have latched on to issues like PCBs to challenge the basic role of the corporation," says Jack Welch, former CEO of General Electric.[b] "These people often see companies as inanimate objects, incapable of values and feelings. GE isn't made up of bricks and buildings. It's nothing more than the flesh and blood of the people who make it come alive. Corporations are human. When they're big, they're an easy target. Facts are, GE has the best environmental record of any company in the world. . . . We're not perfect, nobody is, but we're always striving to do the best."

[a]Background and site information for the Hudson River PCB Superfund Site can be found on the Web by visiting *www.dasp.noaa.gov/neregion/hudsonr.htm* and *www.epa.gov/hudson/background.htm.*

[b]Quotes from Jack Welch (2001), *Jack: Straight from the Gut,* New York: Warner Business Books, 283–294.

11.1 NEW STANDARDS AND POLICIES TOWARD ENVIRONMENTAL QUALITY

The operational complexities, organizational liabilities, and issues of ethics and politics discussed in the preceding scenario are typical of today's business climate. The managerial challenges of dealing responsibly with environmental concerns are enormous. Most challenged are, of course, those companies that have an impact on the natural environment directly through their operations, causing discharge, temperature changes, depletion of natural resources, or disturbances of the natural environment. However, in comparison to a few decades ago, today these challenges are being worked on more cooperatively and conjointly between industry and society. The emergence of better scientifically based knowledge, intensified public pressure, and tougher regulations have led to higher levels of awareness among senior managers and overall more environmentally friendly and cooperative business behavior.[1]

[1]This chapter summarizes some of the most prevalent concepts and practices of environmental cost accounting, with a primary focus on North American and European industries. The chapter further integrates the results of an ongoing field study on environmental quality management, which examines the way business leaders use their business processes, including environmental cost accounting systems, tools, and techniques, to manage natural environmental quality in today's complex business world. The objective of reporting this field research is twofold: First, it provides insight into the established environmental cost accounting methods, beyond their structures and procedures, hence exploring their effectiveness and limitations. Second, the field study examines modern management and leadership techniques toward environmental quality management. Hence, the field study broadens the framework of dealing with environmental quality, from the conventional, predominately quantitative and control-oriented construct, to one that includes a balance of quantitative *and* qualitative measures such as managerial leadership, commitment, and other organization behavior components. Additional results of this ongoing study are reported more formally in Seidel and Thamhain (1999, 2002).

Not too long ago many companies had to choose between operating their businesses either in an environmentally responsible fashion or in an economically effective way. Today, few companies would consider such a dichotomy! Business performance has been redefined by consumers, competitors, regulators, and environmental activists (Dutton 1998; Wilson and Greeno 1993; Scott 1999; Hedstrom, Shopley, and DeDuc 2000). Government and public pressures have pushed companies toward treating environmental issues more realistically, focusing on longer-range cost-benefits, and integrating environmental challenges with the total business process (Seidel and Herbst 1999, Shrivastava 1995, Singh 2000, Smart 1992).

"Environmental protection is a long-term issue of survival for individuals, companies, and societies. Activities must be adapted to nature's own limitations in terms of resource use and pollution. Environmental concern must be the cornerstone to our operations." This statement by Michael Treschow, president and CEO of Electrolux, Sweden,[2] is typical of the new environmental stewardship being displayed by many companies today. It is an operating philosophy that clearly requires a look at the bigger picture, driven by both the continuously increasing cost of environmental protection (U.S. Department of Commerce 1994) and the opportunity to improve a company's competitive position (Hart 1997, Shirvastava 1995). Field studies have confirmed what many company executives have been saying for years: firms can reduce negative effects on the natural environment, while at the same time improving operational efficiency and their competitive position (Christmann 2000, Shirvastava 1995, Smart, 1992). Although European companies seem to be most advanced in effectively embracing environmental stewardship, most firms throughout the industrialized world today are taking an increasingly proactive and long-range business position in safeguarding the environment. As summarized in Table 11.1, Bayer AG, Bristol-Myer Squibb, Dow Chemicals, Eastman Kodak, Intel, McDonald's, Xerox, and 3M represent just a small sample of well-known companies that have demonstrated their commitment to the environment by going beyond simple compliance with regulatory pressures. Their actions range from establishing effective measurement systems to making alliances with environmental agencies, performing product life-cycle analysis (LCA), and undertaking managerial performance reviews tied to environmental goals.

11.1.1 Complex Problems Call for Innovative Solutions

In spite of the great progress made by industry toward environmental stewardship, most managers acknowledge the enormous magnitude and complexity of the issues involved, which go far beyond the quantifiable metrics, such as *environmental cost accounting systems*, *score cards*, and *regulatory systems*, which have been the cornerstone in our environmental protection efforts (Anderson and Bateman 2000, Christmann, 2000). Although these systems provide the necessary measurability critical for environmental quality control, they capture *only part* of the environmental picture, a picture that is often too intricate to be modeled strictly in quantitative

---

[2]For details see Electrolux's Web site, *www.electrolux.se.*

terms (Holmes and Ellis 1999). Therefore, it often becomes necessary to consider normative and judgmental factors, in addition to quantitative data, for environmentally responsive decision making (Ergi and Herman 2000). Starting in the early 1990s, a new vision of environmental stewardship emerged with the notion of *sustainable development* (Vig and Kraft 1997), which suggested a more comprehensive approach to environmental management, going far beyond regulatory compliance. This emerging managerial concept encourages firms to seek broad multidisciplinary solutions toward sustainable development, solutions that include the whole spectrum of business components, from operational process to organizational behavior, ethics, and strategy.

Table 11.1   Examples of Companies Demonstrating Environmental Stewardship

Bayer AG (USA) *Innovative Product Application*	Developed a new agricultural pesticide called *Gaucho*. The new product provides a precise coating of seeds before planting, thus eliminating the need for open air spraying of the pesticide, and virtually eliminating the harmful effects of the chemical on the environment, resulting in significant environmental and economic benefits.
Dow Chemical *External Advisory Committee*	Established an *external advisory committee*, consisting of environmental experts from government agencies, environmental interest groups, and academia, to ensure access to the most advanced thoughts and state-of-the-art environmental concepts for integration in their strategic business decisions and processes.
Eastman Kodak *Pollution Prevention* *Assessment Framework*	Devised a new method, *The Pollution Prevention Assessment Framework*, for predicting environmental effects for products under development. This enables the company to assess potentially harmful effects of new products and processes before product commercialization. Thus products and processes under development can be reformulated for improved environmental performance.
Intel *Process Improvements*	Designs all new manufacturing facilities, including the expansion into Costa Rica, for improved, more effective chemical, water, and energy usage, integrating the latest available technology. As a result, significant environmental *and* economic improvements have been gained, such as the 50 percent ultrapure water reduction for silicon wafer production at their facilities in Arizona, California, Ireland, and Israel.
McDonald's *EDF Alliance*	Formed an *alliance with the Environmental Defense Fund to fine-tune its pollution-prevention strategy*, resulting subsequently in the elimination of polystyrene for takeout food packaging.

Table 11.1   (Continued)

Myers Squibb *Product Life-Cycle Policy*	Initiated a company-wide policy, *Pollution Prevention, throughout the Product Life Cycle,* which became the cornerstone of the company's pollution prevention program.
Xerox *Environmental Cost Analysis*	Devised an *innovative environmental cost analysis process.* Starting with its 265 series of copiers, environmental issues, spanning the product life cycle, are considered and dealt with during the product design stage. As a result, total environmental stress is considerably reduced over the life of these products
3M *Product Responsibility* *Guidelines (Life-Cycle)*	Implemented a life-cycle analysis concept, the *Product Responsibility Guidelines Life-Cycle Model.* This policy/procedure helps management in analyzing operational scenarios for environmental, health, and safety performance proactively throughout the life of the product, from concept to disposal.

11.2 IMPORTANCE OF COST ACCOUNTABILITY TO ENVIRONMENTAL QUALITY

At the heart of any management system is the ability to identify and analyze specific costs and benefits associated with a company's business activities (Starik and Marcus 2000). The *environmental cost accounting system* provides management with a tool set to identify the cost of their operations, to the environment and their own business, in such terms as resource usage, waste, recycling, toxic releases, employee absenteeism, regulatory compliance, and lawsuits. Traditionally, all of these costs fell into an overhead classification, which made a detailed analysis of operational costs, benefits, *and savings* difficult, and an environmental impact assessment nearly impossible. However, today's more uniform standards and more sophisticated enterprise-wide information systems provide companies with a powerful framework for examining the environmental and economic impact of their business operations. This allows companies to look beyond such single factors as reducing waste or pollutants, or trying to minimize regulatory penalties, as was typical for earlier environmental protection efforts. Environmental cost accounting has made enormous progress, especially in areas that are quantifiable. Variables include a wide range of predominately end-of-process measures, such as (1) stakeholder satisfaction, (2) regulatory compliance, (3) incidents, accidents, and violations, (4) specific pollution statistics, (5) impact analysis, (6) energy use, and (7) energy cost.[3] However, in spite of the important contributions the cost accounting system has made to the management of environmental quality, the system has *major limitations,*

---

[3]These seven end-of-process measures had been suggested as a minimum set of measures for environmental accounting by Steven Potorzycki, "Measures for Environmental Effectiveness," *Prism,* 3rd Quarter (1993), pp. 33–41.

especially in dealing with organizational dynamics and the human side of environmental issues that must be recognized and dealt with effectively.[4]

11.2.1   Going Beyond Simple Cost: Focusing on the Total Life Cycle

Driven by the great potential for economic gains and losses, and pressures toward more effective environmental stewardship, many companies today take a "cradle-to-grave" management approach for their products and services. These more progressive companies consider the environmental impact of products over their complete life cycles, including disposal and beyond. This often requires new and innovative methods of producing, distributing, and maintaining goods and services. Examples of such innovative approaches include 3M's *Product Responsibility Guidelines Life Cycle Model* and Bristol Myers Squibb's initiative on *Pollution Prevention Throughout the Product Life Cycle*.[5] From the legislative side, published information and guidelines, such as the U.S. Environmental Protection Agency's *Life Cycle Design Guidance Manual*,[6] and CERES' data management system[7] encourage and direct companies to consider the environmental impact of their business activities in the broadest possible sense. Similarly, many industry associations, such as the Society for Environmental Toxicology and Chemistry (SETAC),[8] have long promoted the need for *environmental product life-cycle thinking*. Environmental standards and certification programs, such as the ISO 14000 family, ISO 14001, 14004, and most recently LCA Standard 14040,[9] provide a specific framework for an overall, long-range strategic approach to environmental policy, plans, and actions as part of a cross-functionally integrated environmental management system (EMS). In addition, think-tank organizations and environmental activist groups, such as The

---

[4]For a more detailed discussion of research findings and the important role of managerial leadership for environmental quality see Anderson and Bateman 2000, Ergi and Herman 2000, and Starik and Markus 2000.

[5]For detailed discussions see the case study by the *World Business Council for Sustainable Development*, published at *www.wbcsd,ch/BMS.htm*.

[6]Keoleian, G. A. and Menerey, D. (1992), *Life Cycle Design Guidance Manual*, Washington, DC: U.S. Environmental Protection Agency, EPA 600/R-92.

[7]The California Environmental Resource Evaluation System (CERES) is an information system developed by the California Resource Agency. Focusing on technology, data, and community, the goal of CERES is to improve environmental analysis and planning by integrating natural and cultural resource information, and encouraging cooperation among government, educational, and private groups. For more information visit *www.ceres.org/*.

[8]Many useful reports, papers, and newsletters are available on the Web site of the Society of Environmental Toxicology and Chemistry (SETAC) via its Life-Cycle Assessment (LCA) Advisory Group. The mission of this group is to advance the science, practice, and application of LCA, providing guidelines for the development and implementation of LCAs. For more information visit *www.setac.org/cla.html*.

[9]The ISO 14040–14043 standard provides a framework for compilation and evaluation of input and output data with the purpose of environmental impact assessment throughout a product life cycle. Currently the challenges are to build consensus on standards for measuring environmental impact and to integrate the new standards with other components of the environmental management system (EMS).

First Step,[10] have acted as a catalyst, helping to unify industry, government, and public interest groups, thus encouraging cooperation toward a better environment.

Today, many companies and communities around the world recognize that our common future depends on balancing our needs and finite resources via sustainable developments (Hart 1997, Schmidheiny 1992). As an integrated part of the environmental cost accounting system, *life-cycle analysis* provides yet another tool for assessing and minimizing the harmful side effects of a product throughout its various stages, from concept to manufacturing, distribution, usage, and disposal. Hence, the life-cycle approach is good for both environmental quality and business economics, reducing stress on our environment, while lowering the firm's operating costs.

11.3 THE ESTABLISHED ENVIRONMENTAL COST ACCOUNTING SYSTEM

Over the years, environmentally focused cost accounting has developed its own language, norms, and standards, which provide the backbone and integrity of the current system.

11.3.1 The Four Major Systems

The first three of the four *interrelated subsystems* are widely recognized, providing the principal methods and tools for environmental cost accounting in today's organizations:

- Conventional cost accounting and investment analysis
- Numerical indicators and comparative analysis
- Process-based input-output analysis

In addition, *more broadly integrated concepts* have been proposed under the category of

- Full-cost accounting

The characteristics of these subsystems are summarized in Table 11.2 and are briefly discussed below in terms of their values, strengths, and limitations in regard to environmental quality management.

11.3.1.1 *Conventional Cost Accounting and Investment Analysis*
This highly quantitative method provides the backbone for economically based environmental decision making. It uses cost-benefit analysis to rationalize operational

---

[10]*The Natural Step* is a frequently cited and discussed (Bradbury and Clair 2000) Swedish environmental education organization that is committed to creating operational strategies and processes with a favorable impact on both environmental quality and business economics. For more information on the company, its scope, and its mission, see *www.naturalstep.org*.

actions affecting environmental quality. The importance of quantifying economic benefits for justifying investments in environmental quality is illustrated below:

> *Siemens Corporation in Munich is reclaiming the energy from a water coolant before discharging it into the environment. The initial investment of $1,400,000 was easily justified by annual energy savings of $1,200,000, representing 86% ROI and 1.2 years pay-back.*[11]

This scenario is a typical example illustrating the close relationship between business economics and strategy. Any change in business operations most likely affects the cost of doing business and environmental quality. The environmental cost accounting system is *strong* in supporting *quantitative* aspects of environmental quality, but *weak* in dealing with *qualitative* issues and the human side of environmental quality.

Benefits.   The strength and benefits of conventional cost accounting are found in several areas. First, conventional cost accounting provides a tangible basis for cost-benefit analysis and a rationale for managerial decision making (Shaw, Fisher, and Randolph 1991; Singh, 2000). Second, the cost accounting methods are supported by well-established standards and principles, thus ensuring the reliability and integrity of the data and the decision processes. This allows for comparability among different situations and users. Third, because of the established standards and conventions, the environmental cost accounting system can be readily integrated with other business systems, such as banking, financial, tax, and shareholder reporting.

Table 11.2   Components and Subsystems of Environmental Cost Accounting

Subsystem & Characteristics	Strengths & Weaknesses	
Cost Accounting and Investment Analysis — Based on conventional cost accounting and investment analysis. Highly quantitative. Provides backbone to economically-based environmental decisions. Short-term orientation. *Application and Usefulness for Environmental Protection:* • Helps to identify, assess, and minimize environmentally harmful processes and practices • Identifies cost-benefits for environmental quality improvements • Provides benchmarks for comparison and improvement • Provides standards for regulatory guidelines and enforcement	*Strengths:* -Tangible cost-benefit analysis -Rational criteria for managerial decision making -Established standards allow uniform measurements and comparisons -Professional set of guidelines	*Weaknesses:* -Qualitative costs and benefits are difficult to assess -Potential (probabilistic) costs and benefits are difficult to assess -Business ethics are difficult to include in accounting -Narrow focus on quantitative factors leads to "game playing"

[11]Source: M. Gege's popular work *Cost Reduction through Environmental Quality Management*, 1997 (see References).

Table 11.2 (Continued)

Subsystem & Characteristics	Strengths & Weaknesses	

Numerical Indicators and Comparative Analysis

Standardized numeric indicators (e.g., ISO 14031) provide metrics for environmental comparisons, audits, and management. Highly quantitative. Measures are often normalized for "per-unit" comparison

Application and Usefulness for Environmental Protection:
• Helps to identify, assess, and minimize environmentally harmful processes and practices
• Provides benchmarks for comparison and improvement
• Provides standards for regulatory guidelines and enforcement

Strengths:
- Established standards and metrics for industry-wide comparison and quality management
- Rational criteria for managerial decision making
- Flexible for expansion and refinement

Weaknesses:
- Data collection methods and statistical models are not standardized.
- Impact of measures on environmental quality is not always clear. The "big picture" of environmental quality is difficult to see with focus on index numbers
- Narrow focus on index number metrics leads to "game playing"

Process-Based Input-Output Analysis

Process and activity-focused cost accounting. Focuses on cost savings gained from minimizing undesirable (hazardous) components. Includes both end-of-process and in-process measures.

Application and Usefulness for Environmental Protection:
• Establishes environmental costs as a derivative of material and energy used in the business process
• Environmental stress can be tracked within specified activity segments and process channels of the firm's operational system.
• Analysis integrates environmental and economic components by underscoring the economic benefits of environmental protection, achieved from minimizing undesirable factors.
• Provides benchmarks for comparison and improvement
• Provides standards for regulatory guidelines and enforcement

Strengths:
- The emphasis is on savings rather than increased operational costs
- Systematic treatment of cost variables
- Criteria for rational decision making
- Environmental cost-benefits include both end-of-process and in-process measures for specific direct and indirect costs

Weaknesses:
- Data collection methods and statistical models not standardized
- Impact of measures on environmental quality not always clear
- Weak in dealing with nonquantitative variables, including judgmental, ethical, behavioral, and political components

Limitations.   The limitations of conventional cost accounting for environmental quality management are related to the quantitative nature of cost accounting and its strong economic focus. In contrast to the benefits, the limitations are often subtle in nature, and in some cases can even lead to erroneous conclusions. An example is the limited ability to consider cross-functional effects and long-range and fully integrated business/environmental impacts. Specifically, conventional cost accounting does not deal well with qualitative benefits, such as good will, public opinion, quality of work, or quality of life in general. Second, probabilistic costs or benefits are difficult to quantify under conventional methods. Although the cost or benefit may become real eventually, it is often difficult to justify an investment under traditional economic principles, or the probability of avoiding a lawsuit or environmental cleanup. Furthermore, the net benefit of a managerial action will depend on the ultimate disposition of the generated "savings." That is, the full environmental benefit will occur only if the savings (material, energy, etc.) are permanent, rather than result in a transfer toward other applications. Examples include (1) less packaging material leads to cheaper packaging, which leads to a higher usage of packaging; (2) the diversion of material savings from one product to another; (3) environmentally unfriendly side effects of environmentally beneficial actions; and (4) increased environmental burden resulting from increased volume prompted by economies-of-scale operations. Third, because of the difficulties of assessing long-range costs and benefits, the system is more short range, with no consideration being given to full-life-cycle accountability. Finally, the quantitative nature of the system with its strong numerical focus invites the temptation to "manage the numbers rather than environmental quality." That is, under pressure to show favorable environmental statistics, management might focus its attention mostly on operations that have an impact on those statistics, ignoring other environmental factors, including long-range influences.[12]

11.3.1.2   *Numerical Indicators and Comparative Analysis*
This cost accounting subsystem relies on standardized operational measures, defined as numerical indicators, suitable to benchmark and manage environmental quality (Liebehenschel, Schanhorst, and Seidel 1998; Seidel 1999). Examples of numerical indicators are (a) the energy consumption of one business function as a percentage of total energy use for the enterprise, (b) the energy or material use per unit of product or service, (c) the water or material use per unit of product output, and (d) the waste per unit of product.

Benefits.   These mostly normalized measures of business activity, output, or consumption provide effective instruments for scenario comparisons, audits, and environmental quality control and management. For many years, efforts have increased to standardize environmental measures, resulting in managerial policies and guidelines such as the well-known set of 45 numerical indicators defined under the

---

[12]These aspects are discussed in *Industrial Environmental Performance Metrics: Challenges and Opportunities*, published by the National Academy of Engineering, National Research Council, Washington, DC: National Academic Press, 1999.

Environmental Performance Evaluation Standard, ISO 14031. All of these efforts have helped significantly to provide the ability to quantify, normalize, and compare industrial scenarios.

Limitations. The limitations of numerical indicators for environmental quality management are related to their quantitative and statistical nature. Data collection methods, measurements, and statistical models are often less uniform than assumed, and therefore comparisons of different situations or overall quality measures are often much less rigorous than their quantitative format suggests. However, while these limitations should be recognized and carefully considered in the formulation of any environmental conclusion, numerical indicators facilitate a relatively easy probabilistic analysis and ways of dealing with fuzzy logic uncertainties, cost, and benefits. Taken together, numerical indicators provide a useful framework for environmental quality accounting as an integrated part of the total enterprise.

11.3.1.3 *Process-Based Input-Output Analysis*

This environmental management subsystem is often referred to as the *Controlling-Rank Concept*. It focuses on cost savings gained from minimizing undesirable (i.e., hazardous) components (Arndt 1995, Fisher and Blasius 1995, Letmathe 1998). The conceptual framework is based on the fact that for many situations environmental stress is ultimately the derivative of the amount of material and energy used in the business process of generating, distributing, and maintaining goods and services. Consumption is converted into monetary units, which are tracked within specified activity segments and process channels of the firm's operational system. Hence, in this concept, environmental protection is not exclusively a function of "prevention cost" (overload), but is a cost component directly associated with business activities, thus integrating environmental and economic components into the cost accounting system. This analysis underscores the economic benefits of environmental protection achieved from minimizing undesired factors. The emphasis is on savings rather than increased operational costs (Bansal and Roth 2000). Environmental cost-benefits are treated in an even more systematic way by Lemathe (1998), who includes relevant in-process measures for specific direct and indirect costs, in addition to the more common end-of-process measures,[13] such as waste and energy use. This is an important refinement of the environmental cost accounting system that minimizes the limitations associated with the more common *end-of-process* measurements and analysis.[14]

Benefits. The benefits of Process-Based Input-Output Analysis to environmental quality management are derived, by and large, from its focus on cost savings. That

---

[13]For specific end-of-process measures and minimum set of variables suggested, see Steven Potorzycki (1993).

[14]Although conclusions can be drawn from analyzing end-of-process measures, the lessons learned often arrive too late to benefit ongoing operations. Managers need tools and techniques for economic and environmental analysis of their ongoing operations.

is, economic benefits are achieved from minimizing undesired outputs and remnants, such as waste. Hence, environmental protection measures lead to savings rather than increased operational costs. This is an important psychological benefit and driver toward environmentally friendly managerial decision making.

Limitations.   The limitations of Process-Based Input-Output Analysis for environmental quality management are, similarly to other parts of the cost accounting system, inherent in their weakness in dealing with nonquantitative variables, including judgmental, ethical, behavioral, and political components.

11.3.1.4   Full Environmental Cost Accounting

Full-cost accounting has gained popularity in recent years for treating the cost of environmental stress in a more comprehensive and integrated way. Full-cost accounting aims to reflect the true environmental cost of a product in its price. This would include its development, production, use, maintenance, and disposal. It offers a method for charging to both producers and users a "value-added cost" or "tax" for the environmental stress caused in each stage of converting raw material into a product, utilizing it, and disposing of it.

At the heart of the concept lies the ability to perform a true life-cycle analysis of a product, systematically identifying the environmental stress and cost for each step of the product's life, from raw material to disposal. Full-cost accounting holds many promises for more effective and appropriate environmental management. However, it also raises many issues. While it is relatively easy to capture the cost of materials, recycling, and cleanups for various stages of the product life cycle, it is significantly more difficult to assess the value of clean air, wetlands, endangered species, or noise pollution. Developing a useful system of true and full environmental cost accounting will require cooperation and integration among many interest groups from industry, government, and the environmental community, including a great deal of international teamwork. However, progress is being made. Efforts at the World Resource Institute, Environmental Defense Fund, United Nations, and many other industrial and government institutions continuously result in better-defined concepts, metrics, and guidelines. One of the biggest challenges is the integration and cooperation among all parties, gaining consensus for the concept and agreeing to an implementation plan (Ergi and Herman 2000, Jassawalla and Sashittal 1999, Ramus and Steger 2000, Thamhain and Wilemon 1998). There is a considerable risk that pressures toward a mandatory *full-cost accounting* system too early could result in polarization of the various interest groups and destroy the hope for *sustainable developments* in the near future.

Strength and Benefits of Full-Cost Accounting.   The concept promotes a more integrated and systemic approach to accounting for environmental cost that includes the total product life cycle from concept to disposal. Pricing reflects the true cost of product or service delivery, including all past and future environmental costs. Hence, the price provides at least a snapshot of the combined economic and environmental load of the product. Simply stated, under full-cost accounting, the lowest-cost product will be the most economical and most environmentally friendly choice.

Challenges and Limitations of Full-Cost Accounting. In spite of its excellent rationale and relative simplicity, full-cost accounting is highly complex and controversial in practice. One of the central challenges involves the distributed nature and interdependency of environmental factors among many stakeholders. To be effective, full-cost accounting must be expanded in its scope beyond the enterprise level. Some of the full cost accrues from other businesses and society, which are part of the product life cycle. Estimating and accounting for these costs is a difficult problem that will require new and innovative approaches to resolve. Cooperation and uniform environmental cost standards are required across each industry to maintain competitive parity in an open-market economy. Many new tools and support systems, such as product life-cycle analysis (LCA), standardized environmental cost metrics, and uniform pricing guidelines, need to be further developed before the concept is workable on an industry-wide scale. Many new developments, such as Streamlined Life Cycle Analysis (SLCA) and Life Cycle Analysis Scoping (LCAS), already provide methods that are simpler and more pragmatic than conventional LCA, and therefore more acceptable (Graedel 1998). In addition, a considerable amount of application software is available to support various forms of life cycle analysis.[15] Yet, in spite of the rapid progress, experts estimate that it will still take years to accomplish full-cost accounting. Ultimately, however, the system is expected to improve environmental performance more than any other program or regulation in use today.[16] Organizations currently engaged with full-cost accounting are laying the groundwork for building a critical toolset for managing companies in an environmentally responsible and sustainable manner.

11.3.2 Toward Sustainable Development

With increased environmental awareness, social pressure, and advances in technology, the collective values and beliefs of organizational leaders have changed toward higher levels of environmental responsibility (Collins and Starik 1995, President's Council 1997, Schmidheiny 1992). Over the last decade, environmental issues have become increasingly more important worldwide,[17] and a new breed of business

[15]A large number of software tools for environmental LCA are available. Many of these tools are being advertised and discussed on the Internet. A good summary of 24 LCA software products is currently shown on the Web site *www.life-cycle.org/LCA_soft.htm.*

[16]In field interviews and action research, we found that companies that are committed to Total Environmental Quality Management (TQEM) are more likely to succeed with their mission when they treat environmental issues like other business issues, considering cost, benefits, opportunities, and risks in the broadest possible sense. These business leaders question the time-honored assumption that environmental quality and economic gains are mutually exclusive. Full-cost accountability provides a framework and direction for companies to move their organizations toward environmental quality, while promoting economic self-interest, an important driver toward sustainable development.

[17]This increased awareness is visibly reflected in the number of new journals and books being published in the area of environmental quality, policy, and management. It is also reflected by the increased number of business and scientific journals carrying special articles or creating a special issue on environmental quality, such as the recent *Academy of Management Journal* issue on "The Management of Organizations in the Natural Environment," Volume 43, Number 4 (August 2000).

managers has emerged who are more willing to take proactive approaches toward environmental quality (Shirvastava 1995, Smart 1992). Starting in the early 1990s, the concept of *sustainable development* reemerged, providing a renewed framework for *environmental stewardship* (Hawken 1993; Shaw, Fisher, and Randolph 1991; Smart 1992). The basis for sustainable development is the efficient use of resources and an effective infrastructure that supports both environmental quality and economic development. The concept of sustainability is not a single plan, but a framework for dynamic and adaptable decision making. It is also a filter for viewing growth and development options. Pioneers such as Paul Hawken, Amory Lovins, and Hunter Loving (authors of *Natural Capitalism*, 1998) have broadened the concept of environmental quality from treating specific problems to dealing with the issues in an enterprise-wide context (Singh 2000, Vig and Kraft 1997). A major driver toward sustainable development is the increasing realization that clear competitive advantages accrue to companies that can integrate their environmental policies with their general business strategies. *Sustainable development* encourages a holistic, multi-disciplinary approach to environmental challenges with parallel solutions and broad cooperation among all stakeholders from action groups to government and industry (National Science and Technology Council 1995, Thamhain 2004). International standards for environmental performance, such as ISO 14001, and guidelines for institutional decision making, such as *CERES Principles* (see endnote 7) and *The Natural Step* (Bradbury and Clair 1999), provide the concepts and inspiration for managers to engage in environmentally responsive business practices. At the minimum, *sustainable development* includes the following elements:

- Business strategy based on *sustainable development* looks beyond the short-term economic advantage and competitive position. Management can balance current operations and priorities with long-term sustainable business goals to optimize its overall strategic position and economic sustainability.[18]
- Economic and strategic advantages derived from using renewable resources and environment-friendly processes.
- Environmental protection rules, regulations, and enforcement, structured so that cause, affects, and impact can be measured and treated specifically.
- Permitting trade-offs among facilities and their outputs with the objective of minimizing total environmental stress.
- Enterprise-wide environmental performance standards that could be extended to contractors, suppliers, and customers.

---

[18]An example of such sustainable development via strategic environmental stewardship is the *indirect economic benefit* gained by an oil company deciding to forgo exploration in a potentially deleterious environmental region. The company may utilize this information to showcase its environmental responsibility, making the company and its products more attractive to some of its customers. Another example is the joint venture formed by Dow Chemicals and Cargill to develop plastics based on carbohydrates from corn rather than hydrocarbons. Not only is corn a renewable resource, but it also requires less environmental management and control and has long-term economic benefits for both companies.

- Development of environmental standards based on environmental cost derived from the latest scientific knowledge.
- Encouraging of total product life-cycle approaches, including life-cycle analysis, design for the environment, and total product responsibility.
- Providing the public with clear and easy-to-understand information on the environmental impact of companies and their products.
- Encouraging cooperation among governmental, business, and other interest groups and allowing all environmental stakeholders to play a more significant role in the safeguarding of the public interest.
- Encouraging innovative solutions that integrate both operational and business processes.

Taken together, sustainable development suggests a comprehensive treatment of environmental concerns, similar to any other business scenario, representing opportunities, threats, costs, and benefits, as well as legal, moral, and ethnical issues. As expressed in the *United Nation's Brundtland Report* (World Commission on the Environment and Development 1986), this includes managerial commitment to and leadership in maintaining and enhancing natural capital, social capital, and the global capacity to generate and distribute wealth equitably, to meet the needs of the present without compromising the ability of future generations to meet their needs. Sustainable development builds on the idea of natural environmental cycles; waste from one organism is used to sustain another one. In industrial settings these cycles do not occur "naturally," but can be established and managed. To be successful, companies must deal with environmental challenges in a systemic way, examining each component of the business process with the objective of improving environmental performance for the entire organizational system. Finally, and critically important, successful implementation of sustainability depends on the development of ethical standards and behavior, creating an organizational environment that encourages people in environmental awareness, interest, and commitment to sustainable principles and policies.

11.4 NEW CONCEPTS OF ENVIRONMENTAL QUALITY MANAGEMENT IN TECHNOLOGY FIRMS

In response to these enormous complexities and challenges, companies have moved toward more sophisticated technology, management systems, and infrastructure, and a clearer environmental focus (Seidel and Thamhain 2002). Business leaders also realize that the management of environmental quality involves complex organizational processes and interfaces, intricate sets of deliverables, schedules, and budgets, as well as social, political, regulatory, and technological factors. In fact, the management of environmental quality involves all the issues that come into play during the execution of complex technology-based business operations and projects. Therefore, it is not surprising that many companies use their already established project management methods to implement their environmental quality programs. Specifically, research shows[19] that *modern project management* provides the tools, techniques, and managerial framework for developing and implementing environmental quality systems. By adopting standard project management systems, firms can benefit from established organizational processes, tools, and techniques, as well as from skill sets developed for the management of other multidisciplinary projects, including sophisticated administration and leadership skills needed for managing environmental quality. Further, project management provides a process for dealing with the complex operational issues typical for technology-based enterprises that rely to an increasing extent on innovation, cross-functional self-directed teamwork, distributed decision making,[20] and intricate multicompany alliances. In this highly complex form of work integration, environmental quality often depends to a considerable extent on buy-in, commitment, and team-generated performance norms, in addition to policies, procedures, and regulations.[21]

11.5   BENEFITS AND CHALLENGES OF THE NEW APPROACH

While this shift to more project-oriented tools and processes provides a framework for handling complex environmental challenges, it also requires radical departures from traditional management philosophy and practices for organization, motivation, leadership, and project control. As a result, traditional management tools and processes, designed largely for top-down control and centralized command and communications, are no longer sufficient for generating satisfactory results. The new management tools are often more integrated with the business process and offer more sophisticated capabilities for project tracking and control. These capabilities are crucial, especially for managing environmental quality projects, which are not only

[19]The results of a field study by Seidel and Thamhain (2002) show that effective environmental quality management depends to a considerable degree on the ability to deal with the multidisciplinary organizational, technical, social, and cultural issues involved in the integration of environmental cost accounting at the enterprise level. The study shows that the tools and techniques of modern project management provide an important framework for developing and implementing environmental quality control systems.

[20]Studies show the increasing importance of self-directed, self-managed teams in all areas of business and across all industries. These self-directed teams have gradually replaced the traditional, more hierarchically structured projects, and have become an important vehicle for orchestrating and managing projects. For specific discussions and research see Jassawalla, Avan R. and Sashittal, Hemant, C., "Building collaborative cross-functional new product teams," *Executive* (The Academy of Management), Vol. 13, No. 3 (August 1999), pp. 50–63; Shonk, J. H., *Team-Based Organizations*, Homewood, IL: Irwin, 1996; Thamhain, Hans J., and Wilemon, David L., "Building effective teams for complex project environments," *Technology Management*, Vol. 4, 1998, pp. 203–212.

[21]Barner, R. (1997) "The new millennium workplace," *Engineering Management Review* (IEEE), Vol. 25, No. 3 (Fall), pp. 114–119; Zenger, John H., Musselwhite, Ed, Hurson, Kathleen, and Perrin, Craig (1994) *Leading Teams*, Homewood, IL: Business One Irwin; Engel, Michael V. (1997) "The new non-manager manager," *Management Quarterly*, Vol. 38, No. 2 (Summer), pp. 22–29. Fisher, Kimball (1993), *Leading Self-Directed Work Teams*, New York: McGraw-Hill; Thamhain, Hans J. (1994) "Designing project management systems for a radically changing world," *Project Management Journal*, Vol. 25, No. 4 (December), pp. 6–7.

different in concept, but also have to deal with a broad spectrum of contemporary challenges, such as cross-functional dependencies, measurability, timing, innovation, resource limitations, technical complexities, environmental metrics, operational dynamics, risk, and uncertainty. All of this has a profound impact on the way project leaders must manage and lead. The methods of communication, decision making, soliciting commitment, and risk sharing are shifting constantly away from a centralized, autocratic management style to a team-centered, more self-directed form of project control. Equally important, project control must radically depart from its narrow focus on satisfying schedule and economic objectives to a much broader and more balanced managerial approach that focuses on the effective search for solutions to complex problems. This requires trade-offs among many parameters, such as creativity, change orientation, quality, and traditional business performance. Control also relies on team members' accountability and commitment to the project's objectives. Responding to the challenges of today's environmental projects, focus must shift from managing environmental cost to optimizing desired environmental quality in multifunctional cooperation with all members of the organization. Field studies consistently show that traditional management methods, focusing too narrowly on environmental metrics and time/cost tracking and control, are often useless and can be even counterproductive to improving overall environmental performance.[22] Success in achieving environmental quality objectives, and ultimately sustainability, will depend on creative solutions to complex problems, and cooperative, flexible, and innovative implementation of these solutions.

11.6 CRITERIA FOR MANAGING ENVIRONMENTAL QUALITY EFFECTIVELY

The basic concepts and methods for managing environmental quality have been known for a long time and have been used extensively. Yet, business managers are not always familiar with contemporary management tools and techniques, and are often not comfortable applying them effectively to their environmental quality programs (Cash and Fox 1992). Project management offers a framework and direction for implementing multidisciplinary programs. As for environmental missions, these programs are distributed across many organizations, and success depends not only on the effective use of established standards and managerial tools in *one* particular organizational environment, but equally important, on the effective use and coordination of these techniques across different organizational lines which often incorporate great differences in corporate cultures.

---

[22]Bahrami, Homa, (1992) "The emerging flexible organization: Perspectives from Silicon Valley," *California Management Review*, Vol. 34, No. 4 (Summer), pp. 33–52; Cash, Charles H., and Fox, Robert (1992) "Elements of successful project management," *Project Management Journal*, Vol. 23, No. 2 (June), pp. 43–47; Cespedes, Frank (1994) "Industrial marketing: Managing new requirements," *Sloan Management Review*, Vol. 35, No. 3 (Spring), pp 45–60; Scott, George M. (1999) "Top priority management concerns about new product development," *Executive* (The Academy of Management), Vol. 13, No. 3 (August), pp. 77–84.

Implementing environmental quality programs at the enterprise involves the issues of organizational change and development that were discussed in Chapter 7 for establishing project management systems and other managerial controls. Therefore, the recommendations advanced in Chapter 7 have been summarized in Table 11.3 below, with a focus on environmental quality programs. They should help in better understanding the complex interaction of organizational and behavioral variables involved in the management of environmental quality systems.

Table 11.3   Recommendations for Implementing Environmental Quality Programs

Align Environmental Programs with Company Goals. Organizational members are more likely to engage in and commit to an environmental program if they perceive it to be clearly related to the goals and objectives of the company. Clear linkages between the program and the company's mission, goals, and objectives create enthusiasm and the desire to participate, as well as lowering anxieties and helping to unify the people behind the program. Thus, communicating the significance of environmental programs and their relationship to the company's mission is critically important. Explicitly securing senior management's support and endorsement further enhances these benefits.

Clearly Define Environmental Policies and Implementation Plans. An effective environmental quality system requires clear managerial guidelines and implementation plans with specific milestones and performance metrics. Goals and objectives must be measurable and attainable. Visibility of accomplishments, recognition, and rewards are critically important to success.

Involve People Affected by the Intervention. Make people at all levels of the organization part of the company's environmental quality effort. Key personnel from all functions and levels of the organization should be involved in assessing environmental issues, searching for solutions, and evaluating new operations and procedures. Implementation teams should be encouraged to recommend changes to existing procedures, reporting relations, and work processes. It is crucial, however, that those team initiatives be integrated with the overall business process and supported by management. *Critical factor analysis, focus groups*, and *process action teams* are good vehicles for collective decision making.

Build on Existing Management Tools and Systems. Radically new methods are often greeted with anxiety and suspicion. If at all possible, *new* operational processes and management policies should be consistent with already established business processes and systems within the enterprise, or *be added incrementally*.

Unify Organizational Processes. Successful management of environmental programs requires the ability to build a unified managerial process across the enterprise.

Use Project Management Techniques. With organizational processes becoming more team-oriented, controls are based to a large extend on commitment and self-direction. Project management provides the operational framework for implementing environmental quality systems and should be used effectively. As an overlay to the functional organization, project management can also provide the network for integrating both local operations and corporate management, a condition that is essential for managing environmental quality.

Table 11.3 (Continued)

Involve Senior Management. Organizational change requires more than just another procedure. Enterprise-wide environmental quality programs require personal commitment from managers throughout the company. These partnerships are more readily achieved when managers in all organizations sense a strong degree of cross-organizational dependency. Senior managers need to work with their organizations to build strong linkages between the local management teams, their support systems, and the corporate leadership.

Anticipate Anxieties and Conflict, Manage Change. Especially for new or enhanced programs, personnel often perceive environmental quality systems as interfering with their work, with little or no benefits to their mission. These negative biases range from personal discomfort with skill requirements to anxieties over the impact of a new procedure on the work process, supervision, reduced personal freedom, overhead requirements, and personal performance evaluation. Management should anticipate these problems and deal with them in a straightforward manner as early as possible. For many companies, defining and implementing environmental programs amounts to *managing organizational change* with all of its dynamics, complexities, risks, and challenges.

Encourage Continuous Improvements. Any management system must be continuously fine-tuned and updated to stay current within our constantly changing world of business. Provisions should be made for reviewing and updating business operations, procedures, environmental measurements, and policies on an ongoing basis to ensure relevancy of the organizational operating environment to its environmental quality goals and processes. An important component of such continuous improvement is the ability to capture the lessons learned during ongoing business operations and programs, and to deal with surfacing problems as part of a continuous improvement process.

Manage and Lead. Organizational developments require top-down direction and support to succeed. People in the organization are more likely to support the implementation of a new management process if the criticality to business performance and the benefits to the organization and its members have been clearly articulated. Senior management involvement and encouragement helps to foster a climate of high motivation, open communication, and commitment toward the established mission objectives.

11.7 CONCLUSIONS

Managing environmental quality is complex and challenging. It involves a kaleidoscope of issues ranging from regulations, norms, and measurements to innovation, ethics, motivation, and leadership. These issues cannot easily be reduced to just regulatory compliance or cost management. For companies committed to sustainable development, management of these challenges requires a highly systemic, holistic approach, integrating all organizational components and subsystems that affect our natural environment. Modern project management offers a direction. It provides part of the tools and techniques and a managerial framework for dealing with these multifaceted challenges in a systematic way. Project management as an organizational system has the capability of integrating the large number of multidisciplinary activities and business processes that come into play during the implementation of today's environmental quality programs. It provides the tools and techniques for dealing effectively with both operational and behavioral challenges. The critical role of organizational support and managerial leadership cannot be overemphasized. Taken in the context of sustainable development, senior management support for and encouragement toward innovative, environmentally responsible behavior, competence building, goal setting, and measurability significantly influence employees' willingness to engage in ecological thinking and positive actions aimed at achieving effective environmental policies and sustainable development. Moreover, cooperation across the organization and commitment to environmental objectives are more likely to happen if the people in the organization perceive these objectives as personally desirable and consistent with their own values and associated with the potential for professional rewards. Further, these objectives should neither be threatening nor interfere with work assignments. Fostering such a work environment requires innovative managerial actions and sophisticated leadership.

11.7.1 Leveraging Technology and Organizational Science

By leveraging our environmental quality efforts with organizational science, we can optimize our resources and have a better chance to manage the complex challenges of our natural environment. Certainly, the development and implementation of next-generation environmental management systems cannot be accomplished in isolation. Modern project management concepts and cost accounting systems will provide some of the building blocks for future environmental quality systems, which will foster sustainable cooperation among various interest groups and motivate people to make a difference. Partnerships and the collaboration of industry, government, and environmental groups are critically important for solving the multifaceted challenges. Companies that will survive and prosper under pressures for a cleaner environment will be able to foster an organizational ambience high in cooperation. They also will be able to create buy-in and commitment across organizational lines aimed at proactive and environmentally beneficial actions. These will be organizations that have integrated their environmental programs and management systems with their business processes and are able to measure and demonstrate environmental responsibility and leverage their environmental performance strategically. These companies will lead the next generation of environmental management programs, controlling business risks and costs, while achieving both ecologically sustainable development and excellent business performance.

11.8 SUMMARY OF KEY POINTS AND CONCLUSIONS

This chapter summarizes some of the most prevalent concepts and practices of environmental cost accounting, with a primary focus on North American and European industries. It shows that effective environmental quality management depends to a considerable degree on the ability to deal with the multidisciplinary organizational, technical, social, and cultural issues involved in the integration of environmental cost accounting at the enterprise level. It further suggests that the tools and techniques of modern project management can provide an important framework for developing and implementing environmental quality control systems. The recommendations put considerable emphasis on the human side, focusing on the criteria under which environmental quality is best achievable. Strong senior management leadership, people skills, and organizational linkages and alliances all are critical for dealing with the complexities of managing responsibly to achieve sustained development in today's organizations. The significance of this chapter lies first in providing a model for assessing the situational effectiveness of various components of the environmental cost accounting system, thus providing some guidelines for managerial process development and fine-tuning in the areas of environmental cost accounting and business strategy. Second, the chapter provides insight into the managerial and leadership skills needed for applying and managing the cost accounting tools effectively as an integrated part of the overall environmental quality system across the total enterprise. Third, the chapter provides potential guidelines and a model for integrating operational management processes and tools from the well-established field of project management into the broader environmental quality management system. Finally, the chapter provides a framework for further training, organizational development, research, and comparative analysis and benchmarking of existing environmental management systems.

The following key points have been made in this chapter:

- Business performance has been redefined by consumers, competitors, regulators, and environmental activists.
- Government and public pressures have pushed companies toward treating environmental issues more realistically, focusing on longer-range cost-benefits.
- Firms can reduce negative effects on the natural environment, while at the same time improving operational efficiency.
- The environmental cost accounting system provides management with a toolset to identify the cost of its operations, and provides companies with a powerful framework for examining the environmental and economic impact of their business operations.
- Four interrelated subsystems provide the principal methods and tools for environmental cost accounting: (1) conventional cost accounting and investment analysis, (2) numerical indicators and comparative analysis, (3) process-based input-output analysis, and (4) full-cost accounting.
- *Conventional cost accounting* and investment analysis provide the backbone for economically based environmental decision making.
- While conventional cost accounting provides a tangible basis for cost-benefit analysis and a rationale for managerial decision making, it has a number of *limitations*, which are often subtle: limited ability to consider cross-functional, long-range, and fully integrated business/environmental effects and difficulty in

dealing with qualitative benefits, such as good will, public opinion, quality of work, or quality of life in general.

- *Numerical indicators and comparative analysis* offer methods for benchmarking and cost accounting, primarily based on resource usage, such as energy, material, and water consumption, or waste output.

- The *controlling-rank concept* is a *process-based input-output analysis* that focuses on cost *savings gained from minimizing undesirable components, such as waste.*

- *Full-cost accounting* reflects the true environmental cost of a product in its price. It offers a method for charging a "value-added cost" or "tax" for the environmental stress caused by a product during its life cycle.

- International standards for environmental performance, such as ISO 14001, and guidelines for institutional decision making, such as *CERES Principles* and *The Natural Step,* provide the concepts and inspiration for environmentally responsive business practices.

- Sustainable development suggests a comprehensive treatment of environmental concerns that addresses opportunities, threats, costs, and benefits, as well as legal, moral, and ethnical issues.

- *Modern project management* provides the tools and techniques and a managerial framework for developing and implementing environmental quality systems.

- Sustainable development requires a highly systemic, holistic approach that integrates all organizational components and environmental subsystems.

- Cooperation across the organization and commitment toward environmental objectives are more likely to happen if the people in the organization perceive these objectives as personally desirable and consistent with their own values and expectations for professional rewards.

- Partnerships and the collaboration of industry, government, and environmental groups are critically important for resolving the multifaceted environmental quality challenges.

- Companies that will survive and prosper under the new mandate for a cleaner environment will have integrated their environmental programs with their business processes, and will have leveraged their environmental performance strategically. They will also be able to measure and demonstrate environmental responsibility.

11.9 CRITICAL THINKING: QUESTIONS FOR DISCUSSION

1. What factors lead to a heightened awareness about our natural environment? Why was the environment much less of a public issue 20 or 30 years ago?

2. Discuss the relationship between a company's treatment of our natural environment and its competitiveness. Are there industries that have a stronger or weaker relationship between these two components?

3. Discuss the four major environmental cost accounting systems, including the benefits and limitations of each.

4. What are the major challenges with the *environmental full-cost accounting system*?

5. Discuss the concept of *sustainable development*.

6. How do you see project management as a tool for implementing new environmental quality programs? Discuss.

7. How can a CEO encourage environmental stewardship in his or her company? Select a specific industry and discuss it.

8. Write a management directive (less than 200 words) for initiating a new policy and procedure within your company for dealing with electrical dry-cell batteries (e.g., lithium, cadmium).

9. Discuss the guidelines (and method) for effectively implementing a new environmental policy that will require the redesign of an ongoing production process.

11.10 REFERENCES AND ADDITIONAL READINGS

Anderson, L. and Bateman, T. (2000) "Individual environmental initiative: championing natural environmental issues in U.S. business organizations," *Academy of Management Journal*, Vol. 43, No. 4, pp. 548–570.

Arndt, H. (1995) "Fußkostenrechnung—eine Umweltkostenkostenkonzeption für das Umweltmanagement," in *Die EG-Öko-Audit-Verordnung* (K. Fichter, ed.). Munich: Carl Hauser Verlag, pp. 249–259.

Bansal, P. and Roth, K. (2000) "Why companies go green: a model of ecological responsiveness," *Academy of Management Journal*. Vol. 43, No. 4, pp. 717–737.

Bradbury, H. and Clair, J. (1999) "Promoting sustainable organizations with Sweden's Natural Step," *Academy of Management Executive*. Vol. 13, No. 4, pp. 63–74.

Cash, C. and Fox, R. (1992) "Elements of successful project management," *Project Management Journal*. Vol. 23, No. 2, pp. 43–47.

Christmann, P. (2000) "Effects of best practices on environmental management cost advantage: the role of complementary assets," *Academy of Management Journal*. Vol. 43, No. 4, pp. 663–680.

Collins, D. and Starik, M. (Editors) (1995) *Sustaining the Natural Environment: Empirical Studies on the Interface between Nature and Organizations*, Greenwich, CT: JAI Press.

Dutton, G. (1998) "The green bottom line," *Management Review (AMA)*, Vol. 87, No. 8, pp. 59–63.

Ergi, C. and Herman, S. (2000) "Leadership in the North American environmental sector: values, leadership, styles and context of environmental leaders and their organizations," *Academy of Management Journal*. Vol. 43 No. 4, pp. 571–604.

Fischer H. and Blasius R. (1995) "Umweltkostenrechnung," in *Handbuch Umweltcontrolling* (Bundesumweltministerium, ed.), Munich: Carl Hauser Verlag, pp. 439-457.

Gege, M. (1997) *Cost Reduction through Environmental Quality Management*, Munich: Springer.

Graedel T. (1998) *Streamlining Life Cycle Assessment*, Englewood Cliffs, NJ: Prentice-Hall,

Hart, S. (1997) "Beyond greening: strategies for a sustainable world," *Harvard Business Review*, Vol. 75, No. 1, pp. 66–76.

Hawken, P. (1993) *The Ecology of Commerce*, New York: Harper.

Hawken, P, Amory, H., and Lovins, H. (1998) *Natural Capitalism: The Coming Efficiency Revolution*, New York: Hyperion Press.

Hedstrom, G., Shopley, J. and DeDuc, C. (2000) "Realizing the sustainable development premium," *Prism* (Arthur D. Little Publication), First Quarter, pp. 57-64.

Holmes, J. and Ellis, J. (1999) "An integrated assessment modeling framework for assessing primary and secondary impact from carbon dioxide stabilization scenarios," *Environmental Modeling and Assessment*, Vol. 4, No. 1, pp. 45–63.

Jassawalla A., and Sashittal H. (1999) "Building collaborative cross-functional new product teams," *Academy of Management Executive,* Vol. 13, No. 3 (August), pp. 50—63.

Lemathe, P. (1998) *Betriebliche Umweltkostenrechnung*, Munich, Germany: Springer.

Liebehenschel, T., Schanhorst, S., and Seidel, E. (1998) *Umweltzahlensysteme fuer den Handel*, Siegen, Germany: University of Siegen.

National Science and Technology Council (1995) *Bridge to a Sustainable Future: National Environmental Technology Strategy*, Washington, DC: National Science and Technology Council.

Potorzyeki, S. (1993) "Measurement for environmental effectiveness," *Prism (Arthur D. Little, Inc.)*, Third Quarter, pp. 33–42.

President's Council on Sustainable Development (1997) *Building on Consensus: A Progress Report on Sustainable America*. Washington, DC: President's Council on Sustainable Development.

Ramus, C. and Steger, U. (2000) "The role of supervisory support behaviors and environmental policy in employee 'ecoinitiatives' at leading-edge European companies," *Academy of Management Journal*. Vol. 43, No. 4, pp. 605–626.

Sarik, Mark and Marcus, A. (2000) "Special research forum on the management of organizations in the natural environment," *Academy of Management Journal*. Vol. 43, No. 4, pp. 539–547.

Schmidheiny, S. (1992) *Changing Course: A Global Business Perspective on Development and the Environment*. Cambridge, MA: MIT Press.

Scott, G. (1999) "Top priority management concerns about new product development," *Executive* (Academy of Management), Vol. 13, No. 4, pp. 77–84.

Seidel, E. (1999) *Betriebliches Umweltmanagement im 21. Jahrhundert*. Munich, Germany: Springer.

Seidel, E., Clausen, J., and Seifert, E. (1998) *Umweltkennzahlen: Plannungs, Steuerung- und Kontrollgroessen fuer ein Umweltorientiertes Management*. Munich, Germany: Environmental Conference.

Seidel, E. and Herbst, S. (1999) *Entwicklungen zum Kostenorientierten Umweltmanagement*. University of Siegen, Germany.

Seidel, E. and Thamhain, H. (1999) "Successful environmental cost accounting with a project management perspective." First Annual Conference of the Institute for Ecology Management, Siegen, Germany, September 29–October 1, 1999.

Seidel, E. and Thamhain, H. (2002) "Managing environmental quality at the enterprise: the role of project management," *Environmental Engineering and Policy*, Vol. 3, No. 1 (February), pp. 19–32.

Shaw, J., Fisher, C., and Randolph, A. (1991) "From maternalism to accountability," *Academy of Management Executive*, Vol. 5, No. 4, pp. 7–20.

Shirvastava, P. (1995) "Environmental technology and competitive advantage," *Strategic Management Journal*, Vol. 16, pp. 183–200.

Shrivastava, P. (1995) "The role of corporations in achieving ecological sustainability," *Academy of Management Review*, Vol. 20, No. 2, pp. 118–137.

Singh, J. (2000) "Making business sense of environmental compliance," *Sloan Management Review*, Vol. 41, No. 1, pp. 91–99.

Smart, B. (1992) *Beyond Compliance: A New Industry View of the Environment*, Washington, DC: World Resource Institute.

Starik, M. and Marcus, A. (Editors) (2000) "Introduction to the special research forum on the management of organizations in the natural environment," *Academy of Management Journal*, Vol. 43, No. 4, pp. 539–547.

Thamhaim, H. (2004) "Linkages of project environment to performance," *International Journal of Project Management*. Vol. 22, No. 7 (October), pp. 90–102.

Thamhain, H. and Wilemon, D. (1998) "Building effective teams in complex project environments," *Technology Management*, Vol. 5, No. 2 (May), pp. 203–212.

U.S. Department of Commerce (1994) *Pollution Abatement Costs and Expenditures*, Current Industrial Reports No. MA200 94:-1. Washington, DC: U.S. Government Printing Office.

Vig, N. and Kraft, M. (1997) "Environmental policy in the 1990's," *Congressional Quarterly*, Vol. 2, pp. 365–390.

Welch, J. (2001) *Jack: Straight from the Gut,* New York: Warner Business Books.

Wilson, J. and Grenno, J. (1993) "Business and the environment: The shape of things to come," *Prism (Arthur D. Little, Inc.)*, Third Quarter, pp. 15–18.

World Commission on the Environment and Development (1986) *The Brandtland Commission Report*, New York: United Nations.

12

MANAGING RISKS IN HIGH TECHNOLOGY

RISK TAKING IS PART OF STAYING COMPETITIVE AT INTEL

Three years ago, when Intel Corporation started pouring some $28 billion into new plant construction and R&D, CEO Craig R. Barrett had no idea when the semiconductor industry's longest slump would end. Weak demand could have meant doom as those new semiconductor plants began ramping up production. But Barrett plowed ahead. "The only question was when overall business would begin to pick up. A three-year recession in our industry is about twice as long as the worst I can remember," he says. While things looked pretty dicey for some time, at least for now it turns out that Barrett put his chips in the right place. Intel is working hard to satisfy demands for its Centrino notebook package, which includes the Pentium M processor and a WiFi chip. Barrett's push to build highly efficient factories means that Intel can churn out chips at costs way below those of competitors. With 62 percent gross margins, Intel is enjoying earnings of $5.4 billion in 2003, up 60 percent from last year.

Just 18 months before Barrett is expected to step down as CEO, he has become his industry's master and commander. After expanding into graphics chips, motherboards, and other PC-related fields, Intel plans to incorporate wireless routers directly into desktop computers. Intel is also pushing into consumer electronics, set-top boxes, and cell phones. High-risk strategies? Yes. But, important for Intel, even higher potential for gains.

For further information please visit Intel's Centrino Mobile Technology website at *www.intel.com/products/mobiletechnology/*.

12.1   THE HIGH STAKES OF PLAYING IN TECHNOLOGY

Uncertainty is a reality for most business ventures (Thieme et al 2003, Hillson 1999), and it shouldn't come as a surprise that many of these projects fail.[1] Yet, for tens of thousands of new product development efforts and complex projects, such as the Mars Explorer Rover, that succeed each year failure is not an option. What lessons can we learn? While high-tech projects and operating platform expansions, such as the Intel scenario, involve a complex and intricately linked system of variables (Nellore and Balachandra 2001, Thamhain 2002), a common denominator in business success appears to be the ability to deal effectively with the inevitable risks. *For the purpose of definition, risk occurs when uncertainties emerge that have the potential to adversely affect the development or business mission and ultimately impact one or more of the project objectives related to business performance.* Because of these challenges, many tools and techniques have been developed for dealing with risks (Barlett 2002, Dey 2002, Elliott 2001), covering a wide spectrum, ranging from the very simple to the highly complex, from the analytical to the behavioral, and from the quantitative to the qualitative. Methods include interviews, brainstorming, focus groups, online databases for categorizing and sorting risks, and sophisticated Monte Carlo analysis, all designed to determine specific outcomes at project life-cycle points more predictably.

There is little argument that all of these concepts provide an important toolset for enterprise risk management (ERM). Yet, there is a growing sense of disappointment and frustration, especially among technology-based managers, that not all techniques work equally well, nor are all techniques equally applicable to all situations (see Iansiti and MacCormack 1997, Piney 2002, Pappas 2002, Thamhain 1994, 2002). As a result, a "develop now and fix later" mentality often displaces scientifically based methods of risk management (Skelton 2002) in today's "faster, cheaper, better" business environment.

Why These Difficulties? Regardless of the specific tools available, seasoned managers have an intuitive sense of where uncertainties lurk. However, foreseeing specific events and neutralizing the factors that impede success are neither simple nor singular (Elliott 2001, Nellore and Balachandra 2001). Part of the reason for the difficulties in dealing with risk is the complexity of the work environment. Research studies have suggested that much of the root cause of product development–related risks can be traced to the organizational dynamics and multidisciplinary nature of today's business environment, especially for technology-based development projects. The involvement of many people, processes, and technologies, spanning different organizations, support groups, subcontractors, vendors, government agencies, and customer communities, compounds the level of uncertainty and distributes risk over a wide area of the enterprise and its partners (Thamhain and Wilemon 1999, 2000). As a result, many of the risk management tools that function well in traditional,

---

[1]Research by the Standish Group International (2004) and Gordon (1999) shows that on average 40% of new product development projects were cancelled prior to completion and of those that finished, 45% of the projects were late and/or over budget.

stable organizational environments fail in more dynamic, nonlinear settings that are typical for technology-based development projects.

12.1.1   Chapter Focus

In order to stay within the scope of this book and keep the topic manageable, we will explore the practices and business processes of an enterprise risk management (ERM) focus on *new product developments and other complex projects projects*. In addition, we will take a look at the *"fuzzy front end"* of *new product development projects*. We believe that a focus on the front end, or conceptual stage of new product developments, is especially useful, because decision making during this project phase has the greatest impact on the development outcome, with the least amount of resource commitment, as graphically shown in Figure 12.1.

12.2   HOW MANAGERS DEAL WITH RISKS—SOME FIELD RESEARCH FINDINGS

A recently concluded exploratory field study investigated 118 new high-technology product developments in 14 companies.[2] The results are summarized in this chapter to help both management practitioners and scholars to better understand the complex issues and organizational dynamics involved in enterprise risk management (ERM)

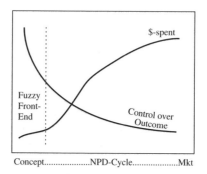

Figure 12.1   Determining project outcome during the conceptual phase.

---

[2]This exploratory field study was conducted between 2000 and 2004. Data were collected from 27 technology organizations as part of ongoing research in the area of technology-oriented product development and team-based project management. The field study yielded data from 74 project reams with a total sample population of 435 professionals such as engineers, scientists, and technicians, plus their managers, including 18 supervisors, 22 project team leaders, 18 product managers, 10 directors of R&D, 8 directors of marketing, and 10 general management executives at the vice presidential level. Together, the data covered over 118 projects in 14 companies. The projects involved mostly high-technology project/service developments with budgets averaging $1,2000,000 each. For detailed method and specific results see Skelton and Thamhain (1993).

of new product development and other project-oriented processes. The objective is to establish a framework for systematic risk management.

12.2.1 Risk Affects Product Success

Managers in our field study identified over 1,000 unique risk situations, which we grouped into 13 *generic risk categories of undesirable effects*, as shown in Figure 12.2, listed in order of frequency. These undesirable events contain *risk factors* which by definition can have an unfavorable impact on the project, its outcome, and its objectives.

Undesirable events are often *caused by a multitude of problems* that were not predictable or could not be dealt with earlier in the product development cycle. In a typical product development project, these problems often cascade, compound, and become intricately linked. Examples of how events such as the resignation of a key project team member can lead to design problems, confusion, organizational conflict, sinking team spirit, and fading commitment are summarized in Table 12.1. All of these factors contribute to schedule delays and system integration problems, causing time-to-market delays, missed sales opportunities, and ultimately an unsatisfactory profit performance of the new product in the market. Yet, by definition, risks are manageable, at least in principle, if the effects can be brought under control before they affect end objectives. This is the art and science of project and product management.

12.2.2 Managing Risks versus Dealing with Problems

The field observations clearly show that most project managers focus their efforts on fixing problems after they have had an impact on performance. That is, while project leaders understand the sources of risks well, they focus most of their attention on *monitoring* the "derivatives of the cause," such as schedules and budgets. They often try to manage problems and contingencies only after they have already affected project performance. We define these derivatives of the original problem that affect project objectives (e.g., schedule slips, budget overruns, and cash flow problems) *"Class-II risks,"* while we call the original problems that "cause" performance deterioration *"Class-I*

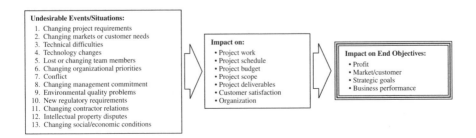

Figure 12.2   Impact of undesirable events on project work and objectives.

Table 12.1   How Uncertainties Affect New Product Success

- A *computer disk drive needed to be reworked* at a cost of $2M at the rollout stage to incorporate new technologies *not foreseeable at earlier development stages.*

- A *special instrument development*, although technically successful, *missed the NASA launch date due to technical difficulties* during the assembly and test stage.

- A new *ultraportable CD player failed* in the market because of higher-than-expected *unit production cost.*

- A *computer chip development* results in a marginally competitive product because of *unpredictable changes in IC support technologies.*

- A new *mutual fund product failed* in the market because of *changing investor needs and economic conditions.*

- A new *drug development was terminated* during the clinical trial stage because *FDA approval became very unlikely.*

- A *telecommunications satellite development* resulted in a large financial loss due to rework resulting from *regulatory changes.*

- An *automobile industry supplier lost a production contract to a competitor* after investing $5M in the product development effort.

- A *supersonic passenger jet development was canceled* after expenditures of $1B because of *changing conditions* in the airline industry.

- A *medical equipment development failed* to gain acceptance among MD users because of *operational complexities and costly maintenance procedures.*

- A *new chemical product development*, once announced as "critically important to the company's core business and long-range strategy," was *terminated because of changing corporate priorities.*

- A *new Web-based banking support system* received only a 12 percent satisfaction rating from the bank's customers, resulting in a major overhaul of the system, *doubling its original development budget.*

- An *application software development project failed* in the market because of *interconnectability problems and user-unfriendliness.*

- A *new computer operating system failed* at the system integration stage, requiring additional design work, resulting in *six months' schedule delay and a $2M budget overrun.*

- A *new oil refinery was delayed by two years* at the pilot operation because of newly discovered *environmental concerns.*

- A *new TV consumer product failed* in the market because of *reliability problems* that did not surface during the product development or rollout stage.

risks." On average, less than one-third of the project leaders interviewed could trace performance problems back to specific events or contingencies that occurred earlier in the work environment. Only one-quarter of these managers felt that they could have foreseen or influenced the events that eventually had an adverse impact on project performance. It is interesting to note that many of the organizational tools and techniques that support early risk detection and management, such as spiral processes, performance monitoring, early warning systems, and contingency planning, readily exist in many organizations, embedded in the planning, tracking, and reporting process of today's project management systems. However, they must be carefully cultivated. Understanding both (1) the potential risk factors and their organizational dynamics and (2) the project management process and its tools is an important prerequisite for *identifying and managing risk factors in their early stages, before they affect project performance irreversibly.*

It is further interesting to note that project leaders blame project performance problems and failures predominately on situations outside their sphere of control, such as scope changes, market shifts, and project support problems, as summarized in Table 12.2. Senior management, on the other hand, points more directly at project

Table 12.2 Top Reasons for Project Failure

Project Manager's Perspective:

- Too many changes and contingencies (scope, t, $, priorities, technology, market, sponsor, etc.)
- Lacking project support (inside and contractors)
- Suboptimal resources (competencies)
- Underestimating project complexity
- Cascading effects
- Interfering administrative processes and requirements

Senior Management Perspective:

- Inaccurate/insufficient planning
- Insufficient performance measurements
- Insufficient communication/escalation of problems
- Lacking change control
- Weak project leadership

Outsider Perspective (Field Research):

- *Difficulties in understanding/communicating complexities of project, its applications and support environment*
- Unrealistic expectations (scope, t, $)
- Underfunding
- Underestimating complexities
- Unclear requirements
- Weak sponsor commitment

managers, blaming them for insufficient planning, tracking, and control; poor communication; and weak leadership. Yet a third perspective is that of an outside observer. We found that the root causes of many of the project performance problems and failures could be traced to the broader issues and difficulties of understanding and communicating the complexities of the project, its applications, and its support environment, including unrealistic expectations for scope, schedule, and budget; underfunding; unclear requirements; and weak sponsor commitment. These findings have significance in several areas. First, the polarized perspective between project leaders and senior managers creates a potential for lots of organizational tension and conflict. It also provides an insight into the mutual expectations, especially for project leaders who are expected to "manage" the project to meet the agreed-on objectives. On the other side, it is clear that project leaders are often stretched too thin and placed in a no-win situation by unrealistic requirements, weak project support, and changing organizational conditions.

Many of the risk factors have their roots *outside the project organization,* controlled by senior management. Therefore, it is important for management to recognize these variables and their potential impact on the work environment. Organizational stability, availability of resources, management involvement and support, personal rewards, and stability of organizational goals, objectives, and priorities are all derived from enterprise systems that are controlled by general management. Project leaders must work with senior management, and vice versa, to ensure an organizational ambience conducive to innovative team performance, clearly an area that needs to be explored further in future research.

12.2.3   Enterprise Tools for Managing Risks

Driven by business pressures and advances in information technology, many companies have invested heavily in *risk management tools and techniques*, promising to improve the odds of project success by strengthening the ability to deal with contingencies more effectively. While analytical approaches are still a major and important component of the overall risk management toolset, they are predominately used for quantifying probabilities of risk and for translating these probabilities into specific schedule or budget parameters. Yet, the ultimate usefulness of many analytical methods depends on the assumption that the risk factors and their underlying parameters, such as economic, social, political, and market factors, can actually be quantified and reliably forecasted over the project life cycle. Because of these challenges, many of the more contemporary approaches to *risk management* go beyond the identification and quantification of risk factors, and try to deal with the broader issues of *eliminating the cause of risk.* In our field study, we observed many approaches that aimed effectively at the reduction or even elimination of risks, such as simplifying product designs and their processes, reducing development time, and testing product feasibility early in the development cycle. Other methods focus on the predictability of product success or failure at the very early stages of development, before substantial resources have been spent and organizational commitments have been made.

12.2.3.1 *Five Categories*

To establish a framework for discussion and future research, we can group risk management approaches for new product developments into five categories.

1. *Identifying and Managing Risk Factors.* An examples is the anticipation of changing requirements, market conditions, or technology. If the possibility of these changes is recognized, their probability and impact can be assessed, additional resources for mitigation can be set aside, and plans for dealing with the probable situation can be devised. This is similar to a fire drill or hurricane defense exercise, where specific risk scenarios are known and prevention measures, such as early warning systems, evacuation procedures, tool acquisition, and skill development, can be put in place as a measure of readiness to minimize the impact in case the risky situation actually occurs.

2. *Simplifying the Product and Product Design.* An example is the use of prefabricated components, subcontracting, snap-on assembly techniques, plastic as opposed to metal, a microprocessor instead of e-components, and compilers rather than machine language. Any innovation that reduces complexity, development time, resource requirements, testing, or production setup or assembly also reduces the chance of risky events occuring over the development cycle.

3. *Simplifying the Development Process.* Examples are innovations in work process or project management, such as the concurrent engineering, Stage-Gate, and integrated product development processes. All of these systems are designed to make the product development team work together as a unified group, minimizing organizational barriers and increasing communication effectiveness within the product development team and its host organization, while at the same time minimizing reliance on the team-external infrastructure.

4. *Reducing Product Development Time.* This can be accomplished through the combination of product and process simplification, but also via parallel development work, such as concurrent engineering or design-build processes. The logic for the resulting risk reduction is simple. The shorter the product development cycle, the fewer changes and contingencies that can occur in the product development and its social, economic, and technological subsystems.

5. *Testing Product Feasibility Early.* Traditionally, the crucial product viability tests, such as system integration, market acceptance, flight tests, and automobile crash tests, were performed toward the end of the development cycle. However, with the help of modern computer and information technology, it is possible to advance these tests to the very early stages of new product development. Examples are computer-aided design (CAD), computer-aided engineering (CDE), and computer-aided manufacturing (CAM), as well as simulations, emulations, and modeling of products in their final application environment. A simulated jet flight or automobile crash test is not only much less costly and time-consuming than the real thing, but also yields valuable

information for the improvement and optimization of the product design at its early stages, long before a lot of time and resources has been expanded. Technology also offers many other forms and methods of early testing and validation, ranging from stereolithography for model building to focus groups for early design usability testing. These technology-based methods also allow companies to test more new product ideas, and their underlying assumptions for success, in less time and with considerably fewer resources than traditional "end-of-the-development" test methods.

12.3 CRITERIA FOR EFFECTIVE RISK MANAGEMENT

In spite of the complexities of new product development processes and the inevitable uncertainties throughout the development cycle, the risk of product failure is not random and can be managed. Based on our field observations, the *major influences for dealing with risks* in product development projects are derived from three sources embedded in the subsystems of the enterprise: (1) work and its process, (2) analytical tools and methods, and (3) people. As shown in Figure 12.3 all three subsystems are interrelated as part of the organizational environment and its culture, which are driven by its managerial leadership. Each of the three subsystems offers a limited set of risk management capabilities. When integrated with each other and the enterprise as a whole, these subsystems can provide a powerful framework for managing risks as part of a new product development process. The three subsystems are discussed below with the focus on risk management.

Reducing Risks within Work and Organizational Processes. Risks and uncertainties originate from the work itself. This is highlighted by the examples listed in Table 12.1. Our field study shows that the complexity of the product and its design and work process contributes especially heavily to the uncertainties and

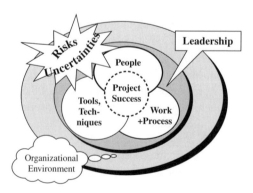

Figure 12.3 Organizational influences on risk factors and project success.

risks that affect product success. Whatever can be done to simplify the product, its design, and its work process will minimize the potential for problems, make the project more manageable, and increase its probability of success according to established plans. Many companies are focusing on the streamlining of their work processes by establishing new project management platforms, such as concurrent engineering or Stage-Gate processes. These organizational enhancements can not only simplify the work process but often also reduce development time, which further reduces the potential for changes and unexpected events. Yet another area of strong managerial focus is *early feasibility testing*. Modern computer and information technology makes it possible to test, at an early stage of product development, everything from functionality and interconnectability to environmental behavior and ultimate product performance in its target application. Whether a company undertakes an airplane development project, a computer design, or the rollout of a new insurance product, such rapid prototyping techniques provide a powerful *look-ahead ability* from concept to market, or any point in between. Such a capability reduces uncertainties enormously. It also provides the ability to iterate designs at the click of a computer button, hence testing out alternative assumptions, options, and applications.

Reducing Risks with Analytical Tools and Methods. Modern information technology provides the basis for many of today's analytical tools and methods of risk management. These tools cover a wide spectrum, ranging from very simple to highly complex, from highly quantitative to graphical and behavioral. Methods range from interviews, brainstorming, and focus groups to online databases for categorizing and sorting risks, to sophisticated Monte Carlo analysis for determining the probability of outcomes at specific project lifecycle points. These tools are especially helpful in processing complex arrays of data searching for risks, analyzing their impact, or supporting risk management decisions. These tools and methods also provide the engines for electronic testing, rapid prototyping, and other look-ahead techniques discussed in the previous paragraph.

Reducing Risks with People. People, through their work, behavior, organizational interaction, and imperfections, are one of the greatest sources of uncertainty and risk in any product development process. They are also one of the most important resources for reducing risk. The quality of communication, level of trust, respect, credibility, job security, skill sets, and maintainence of a minimal level of conflict all influence the collective ability of the organization to identify, process, and deal with risk factors. Based on our exploratory study, many of the favorable influences on people that allow them to deal with risk effectively are derived from the work itself, including personal interest, pride, and satisfaction with the work; having a professional challenge; achieving accomplishments; and receiving recognition. Other important influences include effective communication among team members and support units across organizational lines, good team spirit, mutual trust, respect, low interpersonal conflict, personal pride, and ownership, plus opportunities for career development, advancement, and, to some degree, job security. All of these factors seem to help in building a unified project team that focuses on cross-functional cooperation and desired results. In this mission-oriented process, the high-performing team also isolates and minimizes risk factors.

12.4   CONCLUSION

Taken together, effective risk management of high-technology developments involves a complex set of variables, related to task, tools, people, and organizational environment. Going beyond the mechanics of applying risk management tools requires a thorough understanding of the project work, organizational processes, tools, and techniques. Many of the weaknesses of risk management systems have their root cause in a "defective" project management platform. The indicators, such as conflict and disagreements over personal or project issues, insufficient sponsor involvement, lack of subject expertise, weak change management plans or updates, performance measurement problems, stovepipe execution, unclear deliverables, unclear objectives, and underfunding, are not conducive to early risk detection and treatment. These indicators, summarized in Table 12.3, can also be used for *bench-marking* the effectiveness of existing risk management capabilities, and can serve as an early warning system of changing organizational environments that are no longer conducive to effective risk management. Another leading indicator of product success is the ability to detect potential risks early. Many of the best success stories of our field study point at the critical importance of identifying and dealing with risks early in the development cycle. This requires broad scanning across all segments of the project team and its environment. Managers must foster a work environment where people can deal with uncertainties, ambiguities, and organizational imperfections. No single set of broad guidelines exists that guarantees product success. However, the process is not random! A better understanding of the organizational dynamics that affect project performance and the issues that result in unexpected problems and risks in the new product venture will help you to see the potential conditions that may occur and be able to connect them into an overall risk scenario. An important lesson is the critical importance of reducing complexity in all aspects of the product design and its organizational processes. The ability to exploit information technology inexpensively and quickly assess feasibility at an early development stage seems to be an important key to reducing uncertainties and costly iterations at an advanced stage of the product development. To be effective as "risk managers," project team leaders must be capable of understanding more than the tools and techniques of enterprise risk management. They must also understand the infrastructure of their organization and deal with the complex social, technical, and economic issues that determine the culture and value system of the enterprise. One of the more striking findings is that many of the drivers of effective risk management are derived

Table 12.3   Indicators of Weak Risk Management Ability

- Conflict (personal or organizational)
- Disagreements with project plan
- Insufficient sponsor involvement
- Lacking subject expertise
- No change management plan
- No project plan updates
- Performance measurement problems
- Stovepipe execution
- Unclear deliverables
- Unclear objectives
- Underfunding

from the human side. Organizational components that satisfy personal and professional needs seem to have the strongest effect on the team members' perception of trust, respect, and credibility. People who find their assignments professionally stimulating and challenging, leading to accomplishments, recognition, and professional growth, also seem to enjoy a climate of active participation, minimal dysfunctional conflict, and effective communication. They also seem to handle risks more effectively. To foster such a favorable work environment requires carefully developed skills in leadership, administration, organization, and technical expertise and the ability to involve top management to ensure organizational visibility, resource availability, and overall project support.

12.5   SUMMARY OF KEY POINTS AND CONCLUSIONS

The key points that have been made in this chapter include:

- Risk occurs when uncertainties emerge with the potential of adversely affecting business activities and performance.
- Many of product development–related risks can be traced to the organizational dynamics and multidisciplinary nature of today's business environment.
- Project leaders often understand the sources of risks well but focus their attention on *monitoring* the "derivatives of the cause," such as schedules and budgets.
- *Class-I risks* are the original problems that "cause" performance deterioration. *Class-II risks* are the derivatives of the original problem that affect project performance.

- Project leaders blame performance problems and failures predominately on situations outside their sphere of control, while senior management points at project managers for insufficient planning, tracking and control, poor communication, and weak leadership.
- Analytical approaches are important tools for risk management and are predominately used for quantifying probabilities of risk.
- The ultimate usefulness of analytical methods depends on the assumption that risk factors and their underlying parameters, such as economic, social, political, and market factors, can actually be quantified and reliably forecast over the project life cycle.
- Many contemporary approaches to *risk management* go beyond identifying risk factors and try to *eliminate the cause of risk.*
- The five categories of risk management in product development are (1) identifying and managing risk factors, (2) simplifying the product and product design, (3) simplifying the development process, (4) reducing product development time, and (5) testing product feasibility early.
- The major influences for dealing with risks in product developments are derived from three sources within the enterprise: (1) work and its process, (2) analytical tools and methods, and (3) people.
- Successful risk managers are able to identify and deal with risks early in the development cycle.
- Effective risk managers are capable of understanding more than the tools and techniques of enterprise risk management. They also understand the dynamics of their organizations and can deal with the complex social, technical, and economic issues that determine the culture and value system of the enterprise.

12.6   CRITICAL THINKING: QUESTIONS FOR DISCUSSION

1. Discuss the type of risks that may occur in a high-tech product development project of your choice (e.g., computer, pharmaceutical, automotive, etc.).
2. What kind of similarities and differences do you see in the types of risks between hardware, software, and service product developments? Discuss.
3. What kind of an "early warning system" could you implement to detect the type of risks discussed in Question 1?
4. Discuss risk management approaches within the five categories of (1) identifying and managing risk factors, (2) simplifying the product and product design, (3) simplifying the development process, (4) reducing product development time, and (5) testing product feasibility early. Choose a project or mission similar to the one you used in Question 1 or select an industry group such as computers, pharmaceuticals, or automotive.
5. Discuss the type of risks that may originate with each of the three areas within the enterprise: (1) work and its process, (2) analytical tools and methods, and (3) people.

6. Discuss the managerial leadership styles most conducive to risk management.

7. How can top management facilitate effective risk management at the project team level?

12.7 REFERENCES AND ADDITIONAL READINGS

Barlett, J. (2002) "Risk concept mapping," *Proceedings,* Fifth European Project Management Conference, Cannes, France, June 19–20.

Danneels, E. and Kleinschmidt, E. J. (2001) "Product innovativeness from the firm's perspective," *Journal of Product Innovation Management*, Vol. 18, No. 6, pp. 357–374.

Dey, P. K. (2002) "Project risk management," *Cost Engineering*, Vol. 44, No. 3 (March), pp. 13–27.

Elliott, M. (2001) "The emerging field of enterprise risk," *Viewpoint*, Vol. 2, No. 2 (March 2001).

Foti, R. (2002) "Priority decisions," *PM Network*, Vol. 16, No. 4 (April), pp. 24–29.

Gordon, P. (1999) "To err is human, to estimate, divine," *Information Week*, No. 711, pp. 65–72.

Hillson, D. (1999), "Business uncertainty: risk or opportunity," *ETHOS Magazine*, Vol. 13 (June/July), pp. 14–17.

Iansiti, M. and MacCormack, A. (1997) "Developing products on Internet time," *Harvard Business Review* (September/October), pp. 108–117.

MacCormack, A., Verganti, R., and Iansiti, M. (2001) "Developing products on internet time," *Management Science*, Vol. 47, No. 1, pp. 22–35; *Engineering Management Review*, Vol. 29, No. 2, pp. 90–104.

Nellore, R. and Balachandra, R. (2001) "Factors influencing success in integrated product development (ipd) projects," *IEEE Transactions on Engineering Management*, Vol. 48, No. 2, pp. 164–173

Pappas, D. (2002) "Determining appropriate levels of contingency," *Proceedings*, Fifth European Project Management Conference, Cannes, France, June 19–20.

Piney, C (2002) "Selecting appropriate risk management strategies," *Proceedings*, Fifth European Project Management Conference, Cannes, France, June 19–20.

Skelton, T. M. (2002), "Managing the development of information products: an experiential learning strategy for product developers," *Technical Communication*, Vol. 49, No. 1 (February), pp. 61–79.

Skelton, T. and Thamhain, H. (1993), "Concurrent project management: a tool for technology transfer, R&D-to-market" (coauthored with Terrance M.), *Project Management Journal*, Vol. 14, No. 4, pp. 41–48.

Standish Group International (2004), *Third Quarter Research Report*, West Yarmouth, MA: The Standish Group International.

Thamhain, H. J. (1994), "Designing project management systems for a radically changing world," *Project Management Journal*, Vol. 25, No. 4 (December), pp. 6–7.

Thamhain, H. J. (2001) "The changing role of project management," Chapter 5 in *Research in Management Consulting* (A. Buono, ed.), Greenwich, CT: Information Age Publishing.

Thamhain, H. J. (2002) "Criteria for effective leadership in technology-oriented project teams," Chapter 16 in *The Frontiers of Project Management Research* (Slevin, Cleland, and Pinto, eds.), Newton Square, PA: Project Management Institute, pp. 259–270.

Thamhain, H. J. and Wilemon, D. L. (1999) "Building effective teams for complex project environments," *Technology Management*, Vol. 5, No. 2 (May), pp. 203–212.

Thamhain, H. J. and Wilemon, D. L. (2000) "Team building in project management," in *Managing the Implementation of Development Projects*, Washington, DC: The World Bank.

Thieme, R., Song, M., and Shin, G. (2003) "Project management characteristics and new product survival," *Journal of Product Innovation Management*, Vol. 20, No. 2, pp. 104–111.

Wideman, R. M. (1992) *Project and Program Risk Management: A Guide to Managing Project Risks and Opportunities*, The PMBOK Handbook Series, Vol. 6, Newton Square, PA: PMI Publications.

DEVELOPING NEW
BUSINESS

MAKE YOUR CASE AS THE BASIS FOR COST-BENEFITS AND STRATEGIC VALUE

Not all new business opportunities are created equal. Next time you're wondering whether you should invest the time and resources developing a bid proposal for a potentially lucrative contract, think about the odds of winning. Not all proposal solicitations fit all enterprises. New contract opportunities must align with core competencies to be credible. They also have to align with organizational objectives, mapped into the strategic vision, according to Scott Fass, director of Global Risk Management Solutions at Pricewaterhouse Coopers LLP. Careful up-front analysis is crucial to avoid the costly mistake of bidding on the wrong project. "Pay now or pay later," says Fass. "Executives who make bid decisions by their gut feelings find themselves delivering proposals that do not meet customer expectations or do not support organizational objectives and lack senior management commitment." A well-developed *business case* can help in evaluating the new business opportunity and assessing the cost-benefits for the proposed program over its entire life cycle. "The real benefit is the process," claims Brian Coutanche, the principal of Brian Coutanche Consulting of Jersey in the United Kingdom. "Brainstorming the options, coming to the best decision as a team with vested interests is the ultimate benefit to the organization. When used effectively, the business case provides the

---

context for the executives to test out and confirm their views and to communicate their thinking to those who will be responsible for making it happen and those who will be affected." The discipline of thinking through the benefits and resource implications will sometimes show that your priority is not necessarily the organization's priority. Organizations are always faced with limited resources and have to judge the relative merits against the available alternatives, and the linkage of tactical and strategic drivers against a changing business background.

If done right, the business case specifies the constraints and boundaries of the new business opportunity. "It's not just the scope or acquisition plan," says Donna Fitzgerald, partner of Knowth Consulting in Kirkland, Washington. "It solely focuses on the business decision. How much we're willing to pay to win the proposed business and what we expect to get back from our efforts." According to Fitzgerald, a proper business case contains three essential components: (1) a statement of the opportunity, (2) a conceptual approach to a winning bid proposal, and (3) a justification for deploying the resources. Yet another benefit of the business case process is that once you agree on the key issues, you begin conceptualizing solutions and mapping out approaches to a winning proposal. You also start expanding your strategic vision beyond the obvious. "This requires system thinking and collective reasoning," says Fitzgerald," because you must examine the impact of the new business acquisition across the enterprise."

Data source: "Make Your Case," by Ross Foti, *pmNetwork*, Vol. 17, No. 11 (November 2003), pp. 36–43. For additional information on business case processes seePricewaterhouseCoopers' website and report "Project Advisory Services Case Study: What We Did for a Leading Telecommunication Company" by Scott Fass, at *http://www.pwc.com/Extweb/NewCoAtWork.nsf/docid/CB5087429860F4C385256D3D 005058DC* and Brian Coutanche's website at *http://www.coutanche.com/*.

13.1 THE IMPORTANCE OF NEW BUSINESS DEVELOPMENT

Many high-tech companies derive a substantial share of their income from contracts that must be won competitively. The bid proposal process provides the vehicle for gaining access to the annual multi-trillion-dollar market of contract business. Resulting contracts span all types of businesses and industries, ranging from simple services and supplies to complex multicompany defense contracts, from professional activities to business-to-business and business-to-government ventures. However, winning new business with bid proposals is a complex process. As discussed in the lead-in scenario, it is also expensive, exhausting, and highly uncertain. Among the top bidders, the field is usually very close. Beating most of the competition is not good enough. In most cases, there is only one winner. Yet, companies have no choice. For project-intensive enterprises in particular, new contracts are the lifeblood of the enterprise and must be pursued.

While the techniques for developing and winning contracts are highly specialized and differ for each market segment, they have some common dimensions, as summarized in Table 13.1. To score high, bid proposals require intense, disciplined team effort among all supporting functions and partner organizations, resulting in the following four characteristics:

- A well-defined and articulated solution, showing responsiveness to the customer's needs and requirements
- Credibility and trust that the bidding contractor will perform according to the proposal
- Competitive advantage, such as innovative solution, cost savings, or a licensing agreement
- Competitive pricing and cost credibility

Producing such a document is both a science and an art. It requires not just writing skills but also significant homework, customer contact, and specialized efforts to be successful. As was discussed, a clear understanding of the contract acquisitions process and its tools and techniques is critically important to organizing and managing the complexities of a proposal development effectively and predictability.

The key components and activities that come into play during the life cycle of a proposal-based business development, such as the ones in the following list, will be discussed in this chapter with the focus on the procedures and actions necessary for winning new contract business:

- Proposal solicitation
- Proposal types and formats
- Identification of new business opportunities
- Assessment of new business opportunities
- Writing a winning proposal
- Contract negotiation and closure

Table 13.1   Characteristics of Bid Proposal Development

The following dimensions characterize bid proposals:

1. Systematic effort.   A systematic effort is usually required to develop a new project lead into an actual contract. The project acquisition effort is often highly integrated with ongoing programs and involves key personnel from both the potential customer and the performing organization.

2. Custom design.   Although traditional businesses provide standard products and revisions for a variety of applications and customers, projects are custom-designed items that fit specific requirements for a single customer community.

3. Project life cycle.   Project-oriented businesses have a beginning and an end and are not self-perpetuating. Business must be generated on a project-by-project basis rather than by creating demand for a standard product or service.

4. Market phase.   Long lead times often exist between project definition, startup, and completion.

5. Risks.   Especially for technology-based projects, substantial risks are present. The contractor must not only manage and integrate the project within budget and schedule constraints, but also manage innovations, technology, and the associated risks.

6. Technical capability to perform.   This capability is a critical prerequisite for the successful pursuit and acquisition of a new project or program.

7. Customer requirements.   Projects are often unique regarding specific operational requirements. Applications in the specific customer environment must be properly understood and addressed in the bid proposal document.

8. Follow-on potential.   Winning one contract often provides opportunities for follow-on business such as spare parts, maintenance, training, or volume production.

9. Complex bidding process.   The acquisition process is often very complex and subtle, especially for larger proposal efforts. They often start a long time before the proposal writing phase.

10. Contract negotiations.   Although the proposal serves as a very important vehicle for narrowing the selection of potential contractors, the winning bidder is most likely selected—and its contract finalized—by negotiations. Often these negotiations involve intricate and subtle processes.

13.2   PROPOSAL SOLICITATION

Bid proposals come in many different types, shapes, sizes, and formats. They can be *solicited* or *unsolicited*. Most proposals are provided in response to a formal request for proposal (RFP), request for quotation (RFQ), or request for information (RFI). However, they can also be based on a less formal inquiry by letter or personal discussion. Regardless of their type or format, proposals are sales instruments that try to persuade potential customers to buy goods and services. More specifically,

> *bid proposals offer suggestions for filling a specific customer need or solving a particular problem.*

Depending on the scope and complexity of the customer requirements, a solicitation, such as an RFP, can range from a simple note to highly complex, multivolume documents. For the more complex programs, solicitations often stipulate not only the specific deliverables, but also the conditions under which the work is to be done, delivered, and procured.

13.3 PROPOSAL TYPES AND FORMATS

The responses to these solicitations or client inquires are termed *bid proposals*. They are classified in two major categories:

- Qualification proposal
- Commercial bid proposals

Qualification Proposals. The qualification proposal provides general information about the company, its organization and management, qualifications to perform the work, and procedures, methods, and technologies that would be appropriate for the type of work under consideration. Qualification proposals make no specific offer to perform services or deliver goods, nor do they make any commitments for contracting with the client. These documents are also called informational proposals if the contents relate only to company organization, general qualifications, and procedures. Furthermore, qualifying proposals are often presented under the label of *white papers* or *technical presentations*. Yet another special form of the qualification proposal is the *oral presentation*.

Commercial Bid Proposal. The commercial bid proposal offers a definite commitment by the company to provide specific work, goods, or services in accordance with explicit contract terms and conditions. In addition to the specific performance commitment, commercial bid proposals usually contain the same type of information found in qualification proposals, but in more detail.

13.3.1 Proposal Forms

Both qualification and commercial proposals may be presented to the client in various forms under a wide variety of titles, depending on the situation, the client's requirements, and the firm's willingness to commit its resources. No sharp distinctions exist among these proposals on the basis of content. The difference is mainly in the format and extent of preparation effort. The most common forms are the letter proposals, preliminary proposals, detailed proposals, and presentations.

Letter Propsals. These are either qualification or commercial proposals. They are brief enough to be issued in letter form rather than as bound volumes.

Preliminary Proposals. These are either qualification or commercial proposals, usually large enough to be issued as bound volumes. They are sent to the client for the purpose of dialogue, eventually leading to a detailed proposal development, rather than an immediate proposal evaluation.

Detailed Proposals. These are most often commercial bid proposals, which aside from the technical part include a detailed cost and time estimate. They are the most complex and inclusive proposals. Because of the high cost of preparation and the bid commitments offered, the organization and contents of these documents are defined and detailed to a much greater degree than for other kinds of proposals.

Presentations. These are generally in the format of *oral proposals*. Selected personnel, specialized in certain areas, discuss their proposed offerings verbally with client representatives. Typical presentation time periods vary from an hour to an entire day. While oral presentations have been common in business-to-business biddings for a long time, they have become a new and very important element in the federal government's procurement process. Most oral presentations are conducted after the written proposal has been evaluated. Sophisticated use of information technology with audiovisual support is very common and necessary for optimizing presentation effectiveness.

13.4 IDENTIFYING NEW BUSINESS OPPORTUNITIES

Identifying high-quality business opportunities is the first step toward any new business acquisition. New business opportunities do not just happen, but are the result of sophisticated, systematic customer relation efforts, supported by effective market research. Much can be done to drive and lead market activities and to increase the number of qualified target opportunities consistent with the company's business objectives. Managers who find the process of identifying new business opportunities subtle and unfairly biased toward "insiders" often don't utilize effectively the wealth of information available in the market and within their own customer community. Customer meetings on current programs, professional meetings, conventions, trade shows, trade journals, customer service, competitors' announcements, and personal contacts represent just a few of the many sources for identifying new business opportunities, as shown in the listing of "Reference Material and Contract Information Sources" at the end of this chapter.

Effective customer relations management (CRM), systematic data mining of the business environment, and contemporary tools, such as joint ventures, professional networking, online data services, consulting services, and the Freedom of Information Act, can result in identifying more timely and better-qualified opportunities. All of these front-end efforts must be well orchestrated as part of a new business development plan that is fully integrated with the overall business mission.

Ongoing Process. Identifying new bid opportunities is an ongoing activity. The primary responsibility falls on the marketing or sales department, but personnel at all levels throughout the company can help significantly in identifying new business leads. For most businesses, ongoing program activities are the best source of new business leads. Not only are the lines of customer communication better than for new markets, but equally important, the image of an experienced, reliable contractor helps in creating a favorable environment for open communication, and often results in sharing of privileged information, clearly a desirable competitive advantage!

Acquisition Life Cycle. Developing a new opportunity into a contract takes considerable time and resources. For large programs, this could take several years and millions of dollars. Few companies rush into a major proposal development without carefully evaluating the new opportunity or having a clear win strategy. The formal bid proposal process provides the toolset for pursuing new business opportunities and for systematically developing them into contract awards. Realizing both the complexities and the significance of new business acquisition, many companies have established an *internal proposal development group* or are seeking consulting help from the outside. To make the process more manageable and to break up its complexities, new business acquisitions are typically broken into six phases:

1. Identifying new business opportunities
2. Assessing new contract opportunities
3. Planning the business acquisition
4. Developing the new contract opportunity
5. Writing a winning proposal
6. Negotiating and closing the contract

All phases have strong interdependencies and time overlaps, as well as opportunities for selective concurrent execution.

13.5 ASSESSING NEW CONTRACT OPPORTUNITIES

Pursuing new contracts is a highly intricate process, involving technical complexities, functional interdependencies, evolving solutions, high levels of uncertainty, and highly complex forms of work integration. It is a risky business that requires significant resources and specialized skills. Yet, the win probability is often low. Furthermore, investment in acquisition activities alone does not guarantee success. In fact, many less successful companies find themselves in the quandary of bidding on too many opportunities, without realizing the amount of resources and skills necessary for seriously competing for any one contract. To have a realistic chance of winning, new bid opportunities must be carefully analyzed and assessed. The result of this analysis is a preliminary acquisition plan that provides the basis for a bid decision and the starting point for the final acquisition plan. Table 13.2 describes the bid decision process and includes a checklist for determining the win potential for a new contract. Since the components for organizing a winning proposal effort do not add up linearly, it is often better to consider the bid opportunity in perspective with the overall strength and weaknesses of the enterprise, rather that to try to quantify a narrow set of selected variables. Table 13.2 suggests a broadly defined framework of questions for gaining collective insight into the basic viability of the new opportunity. Brainstorming, focus teams, Delphi groups, and other expert group assessment techniques can be useful in determining the chances of winning the new business and justifying the commitment of further resources for developing a detailed win strategy and acquisition plan, as characterized in Table 13.3.

Table 13.2 The Bid Decision

Few decisions are more fundamental to new business development than the *bid decision*. Resources for pursuing new business come from operating profits. These resources set aside for new business development must be carefully allocated to opportunities with payoff potential. *Bid boards* serve as management gates for the release and control of these resources. *Bid boards* are expert panels that analyze the new business opportunity relative to its importance to the company's mission and competitive strengths, to determine the readiness of the company to invest the necessary resources for a winning proposal effort. Four major dimensions must be considered in a bid decision: (1) desire for and value of acquiring the new business, (2) cost of the acquisition effort, (3) relative strength of the company versus its competition, and (4) readiness of the company to execute the contract. The *new business acquisition plan* provides a framework for the bid board deliberation and ultimate decision.

Major acquisitions usually require a series of bid board decisions, ranging from preliminary to final. Some preliminary bid decisions are being made as early as 18 months before the RFP. Subsequent bid boards reaffirm the bid decision and help in updating the acquisition plan. They may also redirect or terminate the acquisition effort. It is the responsibility of the *proposal manager* to gather and present pertinent information in a manner useful to the bid board for analysis and decision making. The following checklist provides a simple tool for stimulating critical thinking toward an integrated bid evaluation and decision making.

Checklist for Evaluating Bid Decision

Conditions	Status
(Evaluate on a 5-Point Scale: 1 = Strongly Unfavorable . . . 5 = Strongly Favorable)	
1. We have sufficient resources and capabilities to perform	[1] [2] [3] [4] [5]
2. We can meet the client's schedule.	[1] [2] [3] [4] [5]
3. We are in a strong technical position to perform.	[1] [2] [3] [4] [5]
4. We have a unique technical solution for client.	[1] [2] [3] [4] [5]
5. We have a unique approach to project execution.	[1] [2] [3] [4] [5]
6. We have a unique resources for project/contract execution.	[1] [2] [3] [4] [5]
7. We have a competitive cost advantage.	[1] [2] [3] [4] [5]
8. We have a favorable reputation in this type of work.	[1] [2] [3] [4] [5]
9. The client is ready to start the project (incl. budget).	[1] [2] [3] [4] [5]
10. We are on the preferred contractor list.	[1] [2] [3] [4] [5]
11. We have established strong client relations on this bid.	[1] [2] [3] [4] [5]
12. We have a competitive pricing strategy.	[1] [2] [3] [4] [5]
13. The contract has significant follow-on potential.	[1] [2] [3] [4] [5]
14. The contract is consistent with the enterprise's mission and plans.	[1] [2] [3] [4] [5]
15. The contract will enhance our future technical capabilities.	[1] [2] [3] [4] [5]
16. The contract will enhance our future market position.	[1] [2] [3] [4] [5]
17. The contract will result in significant economic gain.	[1] [2] [3] [4] [5]
18. We understand the competition.	[1] [2] [3] [4] [5]
19. Number of qualified bidders	[1] [2] [3] [4] [5]
20. We are very familiar with this bidding process.	[1] [2] [3] [4] [5]
21. We have unique advantage over the competition.	[1] [2] [3] [4] [5]
22. We have a realistic chance of winning the contract.	[1] [2] [3] [4] [5]

Table 13.3   The New Business Acquisition Plan

The *new business acquisition plan* is an important management tool for supporting the bid decision and provides a roadmap for guiding the contract acquisition process. The plan also provides the basis for acquiring the resources required to pursue the new contract acquisition and the roadmap for organizing and executing the bid proposal development. Typically, the *new business acquisition plan* should include the following components:

Brief description of the new business opportunity. A statement of the customer requirements, including specifications, schedules, budgets, and key decision makers.

Rationale for bidding. A statement discussing the reason for bidding on the new contract opportunity, including perspectives on established business plans and desirable results such as profits, markets, and technology.

Competitive assessment. A description of each competing firm with regard to relevant past activities, related experiences, current contracts, customer relations, strengths and weaknesses, and potential baseline of approach.

Critical win factors. A listing of specific factors important to winning the new contract and their rationale.

Ability to write a winning proposal. Discussion of the specific resources and timing required for preparing a winning bid proposal. Factors to be considered should include available personnel, understanding of customer problem, competitive advantage, ability to meet customer budget constraints, willingness to bid competitively, special factors such as licensing, joint ventures, and long-range investment.

Win strategy. A statement describing the actions to be taken for positioning the enterprise uniquely in the competitive field, including a chronological listing of critical actions and milestones necessary to guide the acquisition effort from its current position to contract award. Responsible personnel and timing should be defined for each milestone.

Capture plan. A detailed action plan that supports the win strategy, integrated with the overall business plan. All actions should include timing, budgets, and responsible personnel. The capture plan is a working document that serves as a roadmap in a dynamically changing competitive landscape. It should be updated regularly.

Ability to perform under contract. This is often a separate document. However, a summary should be included in the acquisition plan, including the ability to meet technical requirements, staffing , facilities, program schedules, and subcontracting.

Problems and risks. A list of problems, challenges, and risks regarding the capture plan implementation should be presented.

Resource plan. A budget summary, including the key personnel, support services, and other resources needed to capture the new contract.

Analyzing a new business opportunity and preparing the acquisition plan is a highly interactive effort among the various resource groups of the enterprise, its partners, and the customer community. Often, many meetings are needed between the customer and the performing organization before a clear picture emerges of both customer requirements and matching contractor capabilities. A valuable side benefit of such customer involvement is the potential for building confidence, trust, and credibility with the customer community. These meetings provide a platform for communicating the understanding of customer requirements and the capacity to perform, both important prerequisites for winning the contract. The acquisition plan, as outlined in Table 13.3, provides the foundation and framework for winning the new contract, providing a roadmap for favorably positioning the enterprise. Four dimensions are crucial for positioning a winning proposal:

- Significant customer contact
- Relevant experience
- Technical readiness to perform
- Organizational readiness to perform

13.5.1 Significant Customer Contact

Establishing a liaison with the customer is vital to learning the specific customer requirements and needs. It is necessary to define the project's baseline, potential problem areas, and risks. Having a liaison with the customer also allows participation in customer problem solving and building a favorable image as a competent, credible contractor. Today's complex customer organizations involve many people in the bid decision-making process. Confusing requirements and customer biases are realities and must be dealt with. Multinational involvement at various levels of both contractor and customer organizations is often necessary to reach all decision-making parties. The *new business acquisition plan* is a good source of information and a roadmap for the development effort.

13.5.2 Prior Relevant Experience

Nothing is more convincing to a potential customer than demonstrated prior performance in the area of the proposed program. It reduces the perceived technical risks, as well as the associated budget and schedule uncertainties. This image of an experienced contractor can be communicated in many ways: (1) field demonstrations of working systems and equipment; (2) listing of previous or current customers, their equipment, and applications; (3) model demonstrations; (4) technical status presentations; (5) promotional brochures: (6) technical papers and articles; (7) trade show demonstrations and exhibits; (8) audiovisual presentation of equipment in operation; (9) simulation of the systems, equipment, or services; (10) specifications, photos, or models of the proposed equipment; and (11) media advertisements. Demonstrating prior experience should be integrated with the customer liaison activities.

13.5.3   Technical Readiness to Perform

Once the basic requirements and specifications for the new program are known, it is often necessary to mount a substantial technical preproposal effort to advance the baseline design to a point that permits a clear definition of the new program. These efforts may be funded by the customer or absorbed by the contractor. Typical efforts include (1) feasibility studies, (2) system design, (3) simulation, (4) design and testing of certain critical elements in the new system or the new process, (5) prototype models, (6) developments necessary to bid the new job within the desired scope, and (7) developments necessary to minimize technical and financial risks. Although these precontract efforts can be expensive, they are often essential for winning new business. These early developments reduce the implementation risks and enhance the contractor's reputation that it can perform under contract.

13.5.4   Organizational Readiness to Perform

Another element of credibility is the readiness of contractor organization to perform under contract. This includes facilities, key personnel, support groups, and management structure. Credibility in this area is particularly critical when bidding on a large program relative to the contractor. Organizational readiness does not necessarily mean reorganization prior to the contract award, but it requires a clearly defined organization plan, detailed procedures that can be followed after contract award. The following checklist defines typical organizational components that might be required and should be clearly defined in the proposal and discussed with the customer. If possible, such a customer dialogue should be conducted prior to submitting the formal proposal:

- Organizational structure
- Authority and responsibility relationships
- Project charter
- Company policy, procedures, and management guidelines
- Staffing plan
- Job descriptions of key contract personnel
- Type and number of laboratories, offices, and facilities
- Floor plans
- Milestone schedule and budget for reorganizing under contract

13.6 WRITING A WINNING PROPOSAL

Bid proposals are payoff vehicles. They are the final deliverable in the bid proposal cycle (of course, contract negotiations and closure are yet another phase of the overall acquisition process). Regardless of the type or nature of the work, whether bidding on a service or hardware contract, a government or commercial program, the basic process is the same.

The bid proposal is *the* marketing tool, often the only one, for formally communicating the contract offer. The program requirements, soundness of proposed approach, possible alternatives, the company's credibility, and so forth, hopefully have been established during the preproposal phase of the contract development. Yet, a superior proposal is still necessary for winning a new contract in a competitive environment. Your competition is most likely working with great intensity toward the same goal of winning this program. They, too, may have sold the customer on their approaches and capabilities. Usually, only one company will emerge as the winner. Therefore, writing a superior proposal is crucial to winning. It is a serious business by itself.

13.6.1 Organizing for Group Writing

Proposal development requires hard work and long hours, often in a work environment filled with tension and constant pressure to perform against deadlines. As with any project, proposal development requires a multifunctional effort, well orchestrated for disciplined execution. Special tools are available to help large programs in particular to integrate the many activities needed for developing a high-scoring proposal. Smaller proposals often can be managed with less formality. Yet, any proposal plan should include at least the following components:

- Proposal team organization
- Proposal schedule
- Win strategy
- Categorical outline
- Writing assignments and page allocation
- Synopsis of approach for each topic
- RFP analysis
- Technical baseline review
- Proposal draft writing
- Development of illustration
- Reviews
- Cost estimating and pricing
- Proposal production
- Final management review

13.6.2 Storyboarding Facilitates Group Writing

Most bid proposals are group writing efforts. Organizing, coordinating, and integrating these team efforts can add significantly to the complexities and difficulties of managing the proposal development. Especially for larger efforts, storyboarding is a useful technique that facilitates the group writing process. It helps in breaking down the complexities and facilitates incremental integration of the proposal document.

Storyboarding is based on the idea of (1) splitting up the proposal writing into modules, assigned to various contributors and (2) developing the text incrementally via a series of writing, editing, and review phases. The development sequence for a typical proposal effort is given in this list along with the percentage of effort required relative to the overall development:

1. Categorical outline 3% . . . . .Completion at Day 01
2. Synopsis of approach 6% . . . . .Completion at Day 03
3. Roundtable review 4% . . . . .Completion at Day 04
4. Topical outline 5% . . . . .Completion at Day 05
5. Storyboard preparation 20% . . . . .Completion at Day 10
6. Storyboard review 4% . . . . .Completion at Day 11
7. Storyboard expansion 25% . . . . .Completion at Day 22
8. Staff review 3% .Completion at Day 24
9. Final proposal draft 15% . . . . .Completion at Day 26
10. Final edit 10% .Completion at Day 28
11. Publication and delivery 5% . . . . .Completion at Day 30

The number and type of phases, and the relative effort, might be typical for a major bid proposal development with a 30-day response cycle. In addition, this listing can serve as a guide for smaller or larger proposals. For smaller proposals, the effort can be scaled back to include fewer phases, possibly eliminating the first three, and require fewer iterations. For very large proposal efforts, more formal project management systems and additional stages and iterations are being used. In recent years, *Integrated Product Development (IPD) concepts, including concurrent engineering* and *Stage-Gate concepts,* have gained wide acceptance for managing more complex, large proposal efforts with the primary objective of reducing project cycle time. Each of the 10 phases is briefly described next.

Categorical Outline. Whether managed by storyboarding or conventional methods, the first step in the proposal process is the development of a categorical outline. This is a *listing of the major topics or chapters to be covered* in the proposal. The outline should also show for each category the following information: responsible author, page estimate, and references to related documents. The categorical outline can often be developed before the receipt of the RFP, and should be finalized at the time of proposal-writing kickoff. A sample categorical outline is shown in Table 13.4 for a typical technology system proposal, subdivided into three major sections or volumes: (I) Technical, (II) Management, and (III) Cost.

Synopsis of Approach. A synopsis is developed for each *proposal category* by each responsible author. As an alternative, the proposal manager can complete these forms and issue them as writing guidelines to each of the responsible authors. This approach works especially well for development projects that have a professional proposal support group. These synopses can further be used as a basis for technical brainstorming and searching for innovative solutions. The *synopsis is a top-level outline of the proposed approach to be articulated in each category.* At the minimum, the synopsis should address three questions for each of the categories:

1. What does the customer require?
2. How are we planning to respond?
3. How is the approach unique and effective?

The typical *synopsis of approach format is an 8 1/2" x 11"* sheet of paper, subdivided into six sections:

1. Proposal category and responsible writer
2. Objective to be communicated within this proposal category
3. Understanding of customer requirements
4. Proposed approach and compliance
5. Soundness of approach and effectiveness
6. Risks, alternatives, and options

In preparation for the review, the categorical outline and completed synopsis forms are posted on a wall, in sequential order. This method of display facilitates effective open group reviews, analyses, and integrated proposal development.

Roundtable Review. During this phase, all synopsis forms are analyzed, critiqued, augmented, and approved by the proposal team and its manager. This is the first time that the proposed approach is displayed in a complete and continuous summary form. In addition to the proposal team, key members of functional support groups, such as technical resource managers, marketing managers, contract specialists, and upper management, should participate in this review. The review typically starts four days after the proposal kickoff.

Topical Outline. Concurrent with the review and revision of the synopsis, or shortly thereafter, the *categorical outline is expanded into the specific topics* to be addressed in the proposal. This topical outline becomes the *table of contents* for the bid proposal. The number of pages needs to be estimated for each topic and references to other documents should be made. Similar to the categorical outline, a responsible individual should be assigned for each topic.

Storyboard Preparation. Storyboards are expansions of each synopsis according to the topical outline. All storyboards put together represent the complete bid proposal in summary form. Preparation is straightforward. Typically, a one-page storyboard is prepared for each topic by the responsible writer. As shown in the following list, the storyboard represents an outline and content summary of the author's approach to the write-up for a topical module. Often the storyboard template (A-sized form) is divided into the following four parts:

1. **Writing Guidelines (given by proposal manager).** Proposal category, topic, objective, and proposal address; responsible writer and due date
2. **Theme Section (given by proposal manager).** Tone and emphasis of this proposal topic
3. **Text Summary (to be prepared by responsible writer).** To be developed in blank space on *left side* of form
4. **Illustration Summary (to be prepared by responsible writer).** To be developed in blank space on *right side* of form

The storyboard takes a first cut at articulating the key issues and proposed solutions for each topic. The lead-in statement and conclusion should be written in detailed draft format, as they are intended for the final text. Storyboard text must be relevant, responsive, logical, and emphatic to be useful for final proposal text development. Sophistication of expression is important.

If done properly, the completed storyboards can be given to a professional writing team for storyboard expansion, the final composition, and the editing effort. Storyboards are one of the most important elements in the proposal development process. They should be typed for clarity and easy comprehension, hence ensuring effective review sessions.

Storyboard Review. The completed storyboard forms are pasted on the walls of the review room in a logical sequence, together with the earlier displays of outlines and synopses. The set of storyboards is, in essence, the bid proposal. It presents the complete project plan, that is, the total story that the contractor wants to tell the customer.

Typically, storyboard reviews should start within seven working days after the proposal writing kickoff. The reviews should be held in the same room that was established as a control and display room for the ongoing proposal development. All storyboards should be displayed on the walls. Reviews should be attended by the entire proposal team, including the acquisition management group, authors, and key members of the resource support functions. The storyboard review permits a dialogue between the author, the proposal team, and its management. For very large proposals, it may be impractical to bring the whole proposal team together in one meeting, but it might be necessary to review storyboards in categorical modules. This increases the challenge for the proposal manager of developing a fully integrated, seamless document. Professional proposal development teams, such as proposal specialists, professional writers, and consultants can provide useful resources in these more complex situations.

The storyboard review provides the team with the single most important opportunity to change approaches or direction in the proposed bid. It provides the team with an integrated overview of the proposed work, and a forum for collectively deciding what material to insert, modify, or eliminate. Similarly to the synopsis review, storyboarding is an interactive process. During the reviews, a copy of the latest storyboard of the entire proposal should be on display in the control room.

Storyboard Expansion. After the storyboard review, each author prepares a storyboard expansion. Storyboard expansion is the development of each topic from the original storyboard into a narrative of approximately 500 words. As part of the storyboard expansion, all authors finalize their art work and give it to the publications specialist for processing. This is the first draft of the final proposal. The material is given to the technical editor who will edit the draft for clarity. Each responsible author should review and approve the final draft, which might cycle through the editing process several times.

This final text generation is the major activity in the proposal development process. All prior activities are preparatory yet incrementally cumulative to this final writing assignment. If preparations are done properly, writing the final text will be a logical and straightforward task without the need for additional technical clarification and worries about integration with other authors' work.

As a guideline, ten working days out of a 30-day proposal development cycle may be a reasonable time for this final text generation. Because of its relatively long duration, it is particularly important to set up specific milestones for measuring intermediate progress. The process of final text generation should be carefully controlled. The proposal specialist, copyeditor, and other internal consultants, if available, will play a key role in the integration, coordination, and controlling of this final text and its publication. The final text should be submitted incrementally to the publication department for editing, processing, art preparation, and final integration.

Staff Review. The final proposal review is conducted by the proposal team, its management group, and selected functional managers. In addition, a specialty review committee may be organized to fine-tune the final draft for feasibility, rationale, and responsiveness to the RFP. Typically, this staff review is completed in less than a day. The comments are reviewed by the original authors for approval. The staff review can be repeated if necessary.

Final Proposal Draft. Each author finalizes his or her section of expanded storyboards, incorporating the staff review, comments, and recommendations.

Final Edit. After the final revision, the entire proposal is turned over to the publication department for copyediting and final layout. The authors should be given a last opportunity to look at the completed proposal in its final form. Any major flaws or technical errors that may delay *publication and delivery.* The proposal is now ready for printing, binding, and delivery to the customer.

Table 13.4   Categorical Outline for Technology System Proposal, Subdivided into Three Sections

Section I Technical Proposal	Section II Management Proposal	Section III Cost Proposal
1. Executive summary	1. Executive summary	1. Executive summary
2. Problem statement and analysis	2. Management commitment	2. Scope of work and cost model
3. Recommended solution	3. Recommended solution	3. Contract type
4. Alternative solutions	4. Statement of work	4. Cost summary by workgroups
5. Scope of work and limitations	5. Work breakdowns	5. Cost escalation
6. Method of approach	6. List of deliverables	6. Taxes
7. Detailed solutions	7. Project organization	7. Subcontracting
Subsystem I	8. Task responsibilities	8. Progress payments
Subsystem II	9. Project management process	9. Options
Subsystem III	10. Project tracking and reporting	10. Basis of c-estimate, and assumptions
8. Prototyping	11. Project control	11. Liabilities
9. Field installation, testing	12. Make-buy analysis	12. Overhead rates
10. Specifications	13. Subcontracting	13. Support facilities
11. Reliability assessment	14. Quality control	14. Assurances for cost effective contract work
12. Maintenance	15. Qualifications of key personnel	15. Detailed cost schedules
13. Training	16. Contractor qualifications	16. Appendix
14. Risk analysis	17. Appendix	17. Index
15. Related experiences	18. Index	
16. Appendix		
17. Index		

13.7   NEGOTIATING AND CLOSING THE CONTRACT

Sending off the bid proposal signals the start of the postsubmission phase. Regardless of the type of customer or the formalities involved, even for an oral proposal, the procurement will proceed through the following principal steps:

- Bid proposals delivery and verification
- Proposal evaluation (by customer)
- Proposal values competitively compared (by customer)
- Alternatives assessed (by customer)
- Clarifications solicited from bidders
- Proposal ranking by value
- Negotiations
- Source selection and contract award

Although bidders have no direct influence on the proposal evaluation or source selection process, they can prepare for upcoming opportunities for customer inquiries and negotiations. Depending on the type of procurement, opportunities for improving the competitive position come in many forms, such as:

- Follow-up calls and visits
- Responses to customer requests for additional information
- Fact finding requested by customer
- Oral presentations
- Invitations to field visits
- Samples or prototypes
- Supportive advertising
- Contact via related contract work
- Plant or office visits
- Press releases
- Negotiations

Postsubmission activities can significantly improve the bidder's competitive position. Any opportunity for customer contact should be used. Follow-up calls and visits are effective in less formal procurements, whereas fact finding and related contract work are often used by bidders in formal procurements. Many companies use the bid evaluation period to conduct postproposal reviews, trying to emulate the customer's evaluation process. Although the bidding company mounted an outstanding effort and prepared the best proposal document possible within the given time and resources, this review can provide valuable additional insight into the strength and weaknesses of the submitted proposal. This insight provides the basis for clarification, corrections, enhancements, and image building during the upcoming postsubmission period.

The proposal evaluation period is highly dynamic in terms of changing scores, particularly among the top contenders. Only through active customer contact is it possible to assess realistically the competitive situation and to improve the emerging proposal score. The bidder who is well organized and prepared for interacting with the customer community stands the best chance of being called first for final contract negotiations.

13.8   RECOMMENDATIONS TO MANAGEMENT

Winning contracts involves more than just price, market position, or luck. Winning a piece of new business via bid proposals depends on many factors that can be controlled by management, at least to some degree, during the acquisition process. Successful contract acquisitions start with a keen assessment of the bid opportunity and a sound bid decision, followed by significant homework during the pre-RFP period, intense efforts to develop a responsive and unique bid proposal, and postsubmission customer interactions.

While aggressive pricing is important and can win certain contracts, research shows that for most solicitations, a low-price bid is advantageous primarily for contracts of low complexity and modest technical risk. In most other situations, price is a significant factor in winning, but only within the context of many other competitive components, including compliance with customer requirements, having a unique best-fitting solution, past experience, soundness of approach, cost credibility, delivery, and after-sale support. The better a firm understands its customer, the better it will be able to communicate the value of the proposed solution and the strength of its organization in performing under contract. The following recommendations can help business managers and bid proposal professionals in preparing their organizations for effectively competing for new contract acquisitions:

- **Develop a Detailed Business Acquisition Plan.** This should include a realistic assessment of the new opportunity with specific milestones.
- **Form a Committee of Senior Personnel.** Ensure that the right people become involved early in the acquisition cycle.
- **Maintain Close Contact with the Customer Community.** Try to understand the customer's requirements well and to develop credibility regarding your company's ability to perform.
- **Select Bid Opportunities Carefully.** Submitting more proposals does not necessarily improve your win ratio, but most certainly will drain your resources.
- **Be Sure That You Have the Resources to Go the Full Distance.** Up front, develop a detailed cost estimate for the entire proposal effort. Decide what to do in case the customer extends the bid submission deadline, which will cost additional money for the extended proposal effort.
- **Obtain Commitment from Senior Management.** Make the necessary resources and facilities available when needed.
- **Gain Competitive Perspective.** Before starting the proposal writing, ensure that you have a clear picture of strengths, weaknesses, and limitations of all competing firms relative to your position. Gather marketing intelligence from trade shows, bidder briefings, customer meetings, professional conferences, competitors' literature, and special market service firms.
- **Take a Project-Oriented Approach.** Plan, organize, and manage your proposal development as a *project.*
- **Use a Proposal Specialist.** Enhance the effectiveness of the proposal effort with a professional proposal specialist.
- **Cultivate Your "Unfair Advantage."** Define your market niche by understanding and exploiting your company's strengths relative to your competition, and focus your win strategy on this "unfair advantage."
- **Use a Storyboard Process.** Use this method to develop your proposal text incrementally.
- **Don't Make Exceptions to Customer Requirements**. Don't do this unless it's absolutely unavoidable.

- **Demonstrate Your Ability to Perform.** Focus on past related experiences, which will score the highest points. Showing that (1) your company performed well on similar programs, (2) you have experienced personnel, and (3) you have thoroughly analyzed the requirements will score favorably with the customer and enhance the value of other advantages such as an innovative solution, streamlined schedules, or competitive pricing.

- **Review Proposal Effectiveness.** As part of the incremental proposal development, ensure that there effective reviews, checking compliance with customer requirements, soundness of approach, effective communication, and proper integration of topics into one proposal.

- **Use Red-Team Reviews.** Set up a special review team, especially for "must-win" proposals. This review team evaluates and scores the proposal, emulating the evaluation process used by the customer. Deficiencies that may otherwise remain hidden can often be identified and dealt with during the proposal development. Such a review can be conducted at various stages of the proposal development. It is important to budget the time and money needed for revising the proposal after a red-team review.

- **Use Editorial Support.** Have a professional editor work side by side with the technical proposal writers.

- **Price Competitively.** For most proposals, a competitively priced bid has the winning edge. Pricing should be considered at the time of the bid decision.

- **Prepare for Customer Inquiries and Negotiations.** Prepare these to occur immediately after proposal submission.

- **Conduct Postbid Analysis.** Review the proposal effort regardless of the final outcome. The lessons learned should be documented for the benefit of *future proposal efforts.*

Taken together, winning new contract business competitively is a highly complex and resource-intensive undertaking. To be successful, it requires special management skills, tools, and techniques that range from identifying new bid opportunities and making bid decisions to developing proposals and negotiating the final contract. Companies that win their share of new business usually have a well-disciplined process that is being fine-tuned and improved continuously. They also have experienced personnel who can manage the intricate process and lead the multidisciplinary team through the complex effort of developing a winning bid proposal. Successful companies target specific bid opportunities very selectively, using careful judgment in the bid decision making. They position their enterprise uniquely in the competitive field by building a strong image of a contractor well qualified to perform the required work and by exploiting their strengths. Finally, winning proposals are fully responsive to the customer's requirements and are competitively priced. Winning new contracts in today's continuously changing word of business is not an easy feat. No single set of broad guidelines exists that guarantees success. However, the bid proposal process is not random! A better understanding of the customer criteria and market dynamics that drive contract awards can help managers in fine-turning their acquisition processes and organizational support systems, and will therefore enhance the chances of winning new business via bid proposals.

13.9 SUMMARY OF KEY POINTS AND CONCLUSIONS

The key points that have been made in this chapter include:

- Bid proposals are payoff vehicles.
- Winning new contract business competitively is a highly complex and resource-intensive undertaking.
- To score high, bid proposals require intense, disciplined team effort among all supporting functions and partner organizations.
- Key components and activities of a proposal-based business development include (1) proposal solicitation, (2) proposal types and formats, (3) identification of new business opportunities, (4) assessment of new business opportunities, (5) writing a winning proposal, and (6) contract negotiation and closure.
- Bid proposals must offer suggestions for solving a particular problem.
- *Bid proposals can be* classified in two major categories: (1) qualification proposal and (2) commercial bid proposals.
- New business acquisitions are typically broken into six phases to break up complexities and make the process more manageable: (1) identifying new business opportunities, (2) assessing new contract opportunities, (3) planning the business acquisition, (4) developing the new contract opportunity, (5) writing a winning proposal, and (6) negotiating and closing the contract.
- Four dimensions are crucial for winning bid proposals: (1) significant customer contact, (2) relevant experience, (3) technical readiness to perform, and (5) organizational readiness to perform.
- Storyboarding is a useful technique for facilitating the group writing process, especially for larger acquisition efforts.
- Winning new contracts involves more than just price, market position, or luck; it depends on many factors that can be controlled by management.
- Successful contract acquisitions start with a keen assessment of the bid opportunity and a sound bid decision, followed by significant homework during the pre-RFP period.
- Winning proposals are fully responsive to the customer's requirements and are competitively priced. Price is a significant factor toward winning, *but only within the context of many other competitive components*.
- Aggressive pricing is advantageous primarily for contracts of low complexity and modest technical risk.

13.10 CRITICAL THINKING: QUESTIONS FOR DISCUSSION

1. Discuss the importance of assessing new contract opportunities systematically and early in the acquisition cycle.
2. Under what circumstances can a qualifying proposal be a better vehicle for new business development than a conventional bid proposal? Discuss.
3. Considering the six phases of a typical new business acquisition cycle, develop a list of activities in support of each phase.

4. Find a new business acquisition procedure that is actually being used in a high-tech company. Compare this procedure (benchmark) against the concept of "identifying and assessing new business opportunities" discussed in this chapter.

5. Try to determine the win strategy of a bid proposal development that is known to you within your organization or was published in the media. Evaluate and critique the strategy. Discuss your findings in a group.

6. Develop a win strategy for an upcoming new business development of your choice (select a bid proposal opportunity known to you within your company, or select a suitable opportunity published in the media).

7. How can you influence the proposal evaluation of a "closed bidding process"? Select a bid proposal opportunity known to you within your company, or select a suitable opportunity published in the media.

8. Develop a post-proposal-submission plan in preparation for upcoming proposal discussions with the customer and subsequent contract negotiations.

13.11 REFERENCE MATERIAL AND CONTRACT INFORMATION SOURCES

HANDBOOKS AND PROFESSIONAL REFERENCE BOOKS

Barakat, Robert A (1991) "Developing winning proposal strategies," *IEEE Transactions on Professional Communication*, Vol. 34, No. 3, pp. 130–140.

Battaglia, Robert (1997) "The capture plan—the foundation for preparing winning proposals," *Journal of Technical Writing and Communication*, Vol. 27, No. 3, pp. 291–302.

Fisher Chan, Janis and Lutovich, Diane (1997) *Professional Writing Skills*, San Anselmo, CA: Advanced Communication Designs.

Freed, Richard C. and Freed, Shervin (1995) *Writing Winning Business Proposals: Your Guide to Landing the Client, Making the Sale, Persuading the Boss*, New York: McGraw-Hill.

Hamper, Robert J. and Baugh, L. Sue (1995) *Handbook for Proposal Writing*, New York: McGraw-Hill.

Holtz, Herman (1998) *The Consultant's Guide to Proprosal Writing: How to Satisfy Your Clients and Double Your Income,* New York: Wiley & Sons.

Kampmeier, Curt (1996) "Writing winning business proposals," *Consulting to Management*, Vol. 9, No. 1, pp. 60–61.

Kantin, Robert F. (1999) *Strategic Proposals: Closing the Big Deal*, New York: Vantage Press.

Kantin, Robert F. (2001) *Sales Proposals Kit for Dummies*, New York: Wiley & Sons.

Lester, Mel (1998) "Pulling ahead," *Civil Engineering*, Vol. 68, No. 9, pp. 71–73.

Porter-Roth, Bud (1995) *Proposal Development: How to Respond & Win the Bid*, Waterloo, Ontario, Canada: Entrepreneur Press.

Porter-Roth, Bud R. (2001) *Request for Proposal: A Guide to Effective RFP Development*, New York: Pearson Education.

Roberts, J. and Storms, J. (1996) *The Art of Winning Contracts: Proposal Development for Government Contractors* (2nd edition), Las Vegas, NV: J Melvin Storm Co.

Sant, Tom (1992) *Persuasive Business Proposals: Writing to Win Customers, Clients and Contracts*, New York: AMACOM.

Theriault, Karen (2002) *Handbook for Writing Bids and RFP's*, Florida Atlantic University: Academics Press.

PERIODICALS AND NEWSPAPERS

Acquisition Review Quarterly, Defense Acquisition University, Fort Belvoir, VA, *www.dau.mil*.

Acquisition Today, quarterly, electronic newsletter published by Defense Acquisition University, *http://aitoday.dau.mil*.

Commerce Business Daily, U.S. Department of Commerce Office of Field Services, U.S. Government Printing Office, Washington, DC 20402, *http://cbdnet.access.gpo.gov*.

Contract Management Magazine, National Contract Management Association, 8260 Greensboro Drive, Suite 200, McLean, VA 22102, *www.ncmahq.org*.

Defense Daily International, Newsletter, PBI Media, *www.mindbranch.com*.

Federal Grants and Contracts Weekly (Project Opportunities in Research, Training and Services), Capitol Publications, 1101 King St. Alexandria VA 22314, *www.uri.edu/research/tro/library.htm*.

Federal Publications' Briefing Papers, bimonthly, 1725 K Street NW, Washington, DC 20006.

Federal Register, National Archives and Records Administration, Washington DC, 20408, *http://fr.cos.com*.

Government Procurement Journal, monthly, The Penton Government Media Group, Cleveland, OH, *www.govpro.com*.

GovExec.Com, electronic journal, monthly, *www.govexec.com*.

NCMA Newsletter, National Contract Management Association, 8260 Greensboro Drive, Suite 200, McLean, VA 22102, *www.ncmahq.org*.

Program Manager Magazine, Defense Acquisition University, Fort Belvoir, VA; *www.dau.mil*.

DIRECTORIES

Catalog of Federal Domestic Assistance, General Service Administration, Washington, DC 20402, *www.cfda.gov/public/cat-writing.htm*.

Directory of Government Production Prime Contracts, Government Data Publication, Washington, DC 20402, *www.loc.gov/rr/news/govtcon.html*.

Directory of Research Grants, The Oryx Press, 4041 N. Central Avenue, Phoenix, AZ 85012, *www.uri.edu/research/tro/library.htm*.

Federal Acquisition Regulations, Published by GSA's FAR Secretariat on Behalf of DoD, GSA, and NASA; FAR Secretariat (202) 501-4755.

Federal Register, National Archives and Records Administration, Washington, DC 20408, *http://fr.cos.com*.

Government Contracts Directory, Government Data Publications, Washington, DC 20402, *www.govcon.com*.

The Government Contract Law Report, *www.attny.com/kwkgcin.html*.

ONLINE DATA BASES

Acquisition Support Center, Defense Acquisition University, Fort Belvoir, VA, *www.dau.mil*.

Dialog on Disc, The Grants Database, Dialog Information Services, Inc, 3450 Hillview Avenue, Palo Alto CA 94304, *http://web.utk.edu/~hoemann/disciplines.html*.

Markets.frost.com, Industry portal for contract information, Frost and Sullivan Company, *www.frost.com*.

Newspaper and Current Periodical Reading Room, On-Line Sources of Information on Federal Government Contracts, http://www.loc.gov/rr/news/govtcon.html

ASSOCIATIONS AND SOCIETIES

American Association for Cost Engineering (AACEI International), 209 Prairie Avenue, Morgantown, WV 26501, (304) 296-8444, *www.aacei.org*.

Association for Federal Information Resources Management, Washington, DC.

Association of Proposal Management Professionals, 300 Smelter Avenue NE #1, Great Falls, MT 59404, Phone/Fax: (406) 454-0090, *www.apmp.org*.

International Association of Contract and Commercial Managers (IACCM), 90 Grove Street, Suite 01, Ridgefield, CT 06877, (203) 431-8741, *http://www.iaccm.com*.

National Contract Management Association, 8260 Greensboro Drive, Suite 200, McLean, VA 22102, *www.ncmahq.org*.

Project Management Institute (PMI), Newtown Square PA. 19073, (610) 356-4600, *http://pmi.org*.

CONSULTING IN TECHNOLOGY MANAGEMENT

DEMONSTRATE THE VALUE OF YOUR KNOWLEDGE

The stakes are getting higher for companies that need project management to drive business performance in areas such as new product development, new business acquisition, organization development, and R&D. Senior managers realize that the combination of technology and best practices, skillfully applied, can generate huge competitive benefits ranging from improved time to market to reduced cost and superior product performance. In fact, William Smillie, a partner in the Learning and Knowledge Group of IBM's Business Consulting Services, found that "clients won't invest in projects unless there is a clear and compelling business case to do so." In the past, more traditional projects typically focused on implementations within existing operations. However, the recent trend toward business transformation challenges project managers to maximize the benefits of the project and to wring out the highest level of effectiveness from the implementation process. That's where consultants can help with their specialized management skills, knowledge of best practices, and latest project management support technology. They clearly have an advantage over front-line managers who also oversee project activities within their companies.

However, gaining the trust, respect, and credibility of these managers is a different matter. Credentials and technology alone won't do it. This will merely raise client expectations. "Creating trust and loyalty requires human interactions and diligence that many consultants find hard to deliver," says Jeff Pappin, a former director of KPMG, now with BearingPoint in McLean, Virginia. More to the point, Smillie of IBM Business Consulting explains that "effective communications with key personnel is critical to stakeholder commitment and effective consulting. It will affect the timeliness and quality of decisions our projects need to stay on track. And, it's not always happiness that we bring to our sponsors. Our job is to tell them what they need to know, not necessarily what they want to hear."

How, then, can one build competencies in project management consulting? One important action that IBM Consulting Services has taken is to adopt a simple but powerful language to discuss critical issues. Called the *seven keys to success*, it provides a universal framework for understanding critical *project health issues* and for taking the necessary corrective actions. The first two of the seven keys, "stakeholders are committed" and "business benefits will be realized," immediately focus senior management attention. This language is highly effective with business executives across a variety of companies and industries, say Smillie.

Sources: "IBM Combines Technology with Best Practices" by William G. Smillie, *pmNetwork*, Vol. 17, No. 9 (September 2003), p. 18, and "Stay True" by William Atkinson, *pmNetwork*, Vol. 17, No. 11 (November 2003), pp. 46–51.

14.1. THE NEED FOR TECHNOLOGY MANAGEMENT CONSULTANTS

To survive and prosper in our hypercompetitive world, each firm must have the best business processes and practices. This is especially true for project management, an organizational subsystem that connects with many critical enterprise functions, such as new product development, R&D, field operations, and bid proposal activities. At the same time, project management tools, techniques, and processes have become very sophisticated and complex. Business managers and project leaders often do not have the specialized knowledge and skills needed to optimize their project operations. In fact, for many managers, the advances of modern project management have outpaced their understanding of how and why project-oriented interventions work and why they create value in some situations while leading to disappointment or even outright disaster in others. Therefore, it is not surprising that many managers seek expert help from inside or outside the organization.

14.1.1 How They Add Value

As mentioned, the emergence of contemporary management techniques has outpaced the understanding of how and why complex business processes work, creating value in some situations while leading to disappointment or even outright disaster in others. There is little argument among business leaders that systems such as integrated product development, concurrent engineering, project management, spiral developments, and Stage-Gate concepts provide important toolsets for implementing multidisciplinary ventures, ranging from new product, service, and process development initiatives to acquisitions and foreign assistance programs. Yet, there is also a growing sense of disappointment and frustration that not all techniques work equally well or are equally applicable to all business ventures (see Iansiti and MacCormack 1997, Rasiel 1999, Thamhain 1994). Realizing both the complexities and the significance for business performance, many technology companies have established an *in-house* management consulting group or sought consulting help from the *outside*. As emphasized in the lead-in scenario, merely engaging consultants—whether internal or external—however, does not automatically guarantee business success. Both William Smillie of IBM Consulting Services and Jeff Pappin of KPMG agree that to be effective as change agents, these consultants must be capable of understanding more than the tools and techniques of modern management. They must also understand the infrastructure of their client organization and deal with the complex social, technical, and economic issues that determine the culture and value system of the enterprise, as summarized in Table 14.1. A traditional focus on management tools alone will seldom be sufficient to compete effectively in today's dynamic world of business.

Few companies rush into consulting services, which are often viewed as costly and disruptive, at times even in conflict with deeply held organizational beliefs about the way things should be done. Given the increasing complexities associated with the myriad projects organizations are dealing with today, however, a growing number of

Table 14.1   High-Technology Consulting: Complexities and Challenges

- Dealing with technology-related complexities, challenges, uncertainties, and risks
- Understanding client needs, communicating a vision
- Aligning tools and techniques with existing business process, organizational culture, and values
- Dealing effectively with organizational conflict, power, and politics
- Dealing effectively with the human side of the organization and management
- Obtaining senior management buy-in and commitment
- Aiming at long-term improvements while maintaining existing operating efficiency
- Dealing with process improvement challenges that often require total organizational involvement and broad commitment to change
- Developing effective team leaders and project managers
- Creating sustainable improvements
- Improving the business process and operations with minimal interference in the business

companies literally have no choice. They have exhausted their expertise in dealing with these challenges and see consulting services as a possible lifeline for finding solutions. Others take a more strategic view toward continuing improvement, trying to push forward the frontier of their project environment. These managers realize the complexities of their organizations and the limitations of general management. These companies, in turn, often hire specialists who are experts in particular business areas and have effective diagnostic skills (see Block 1999, Reimus 2000). Achieving benefits in this arena, however, involves complex organizational issues, administrative tools, and human factors. Companies that make the process work are able to attract and engage consulting services that can help the enterprise to:

- Augment expertise and skills in specialty areas
- Benchmark internal and external capabilities
- Challenge conventional wisdom and stimulate innovative thinking
- Develop and evaluate options and alternatives
- Diagnose and develop operation and management systems
- Expedite organizational developments
- Facilitate cross-functional collaboration and alliances
- Function as a change agent
- Lower organizational barriers and conflict
- Make the organization more aware of changes and more receptive to innovation
- Move the organizational development process forward
- Provide an extension of the company management
- Transfer pretested management techniques/models
- Transfer in new capabilities

14.2 CRITERIA FOR EFFECTIVE CONSULTING SERVICES

Companies do not go lightly into reorganizing their established business processes. At best, it is inconvenient, painful, disruptive, and costly. At worst, it can destroy any existing operational effectiveness and the ability to compete successfully in the marketplace. Achieving benefits from technology management consulting involves complex organizational issues, administrative tools, and human factors. Companies and consultants that make the process work see consulting as an intervention tool for building and fine-tuning the project system as part of the overall business process (see Thomas, Delisle, Jugdev, and Buckle 2001). Yet, engaging the organization in an evaluation of the current system and aiming for improvement can be a threatening process. Such threats increase with people's perception that the new environment will deviate significantly from the current one.

What senior consultants and managers find consistently is that for the consulting processes to work effectively, they must blend with the culture and value system of the organization. Project management consultants cannot ignore the human side of the enterprise. That is, consultants must work to establish an environment conducive to mutual trust, respect, candor, and risk sharing. Equally important, consultants must foster effective two-way communication and cross-functional linkages capable of connecting people, processes, and support functions and resolving inevitable conflict.

14.3   A MODEL FOR HIGH PERFORMANCE

Based on the preceding discussion, the issues affecting consulting effectiveness and high-tech project performance can be grouped into seven categories consistent with the concepts presented in Chapter 6. A graphical summary of these performance influences are shown in Figure 14.1: The first three are (1) the *people* of the project team, (2) managerial *leadership* and guidance from the enterprise to the project organization and *leadership* within the project organization, and (3) *project tools, techniques, and business processes* that power and support the project activities. These three categories are overlapping and also affected by (4) *the organizational infrastructure and support systems,* (5) *managerial support,* (6) *project complexity,* and (7) the *overall business environment.* These categories not only determine project performance but also hold the DNA for the type of consulting services that are best suited for improving specific project management situations. The model focuses on the core influences toward successful consulting interventions and project success, areas that have been consistently emphasized by business managers and consultants alike: (a) *people,* (b) *leadership,* and (c) *tools and techniques and business process.*

The model emphasizes the human side as critically important to successful consulting. The project team's attitude, effort, and commitment all are influenced by managerial leadership and the work environment and have a strong effect on consulting effectiveness. Research (see Thamhain 2001, 2002, 2004ab) shows consistently and measurably that the strongest drivers toward cooperation with

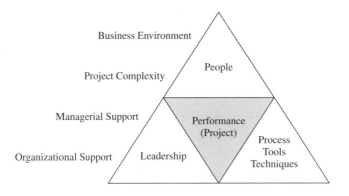

Figure 14.1   Influences on project performance.

management are *derived from the work itself,* including *personal interest, pride and satisfaction with the work, professional work challenge, and accomplishments and recognition.* Other important influences include effective communication among team members and support units across organizational lines, good team spirit, mutual trust and respect, low interpersonal conflict, opportunities for career development and advancement, and, to some lesser degree, job security. All of these factors help in building a unified project team that can leverage organizational strengths and competencies, and will support the organizational objectives.

Other crucial factors relate to the organizational structure and business process, including the technology transfer system, which, by and large, relies on the tools and techniques of modern project management. This has implications for top management. Many of the variables of the organizational process, management tools, and technology have their locus outside the project team organization. These variables are often a derivative of the company's business strategy, developed and controlled by senior management. It is important for management to recognize that these variables can directly affect the quality of the work environment, including the perception of organizational stability, availability of resources, management involvement and support, personal rewards, stability of organizational goals, objectives, and priorities. Since all of these influences are images reflecting personal perception, it is important for management to understand the personal and professional needs of the team members and to foster an *organizational environment* conducive to these needs. *Proper communication* of the organizational vision and perspective is especially important. The relationship of managers to consultants and the project team, including mutual trust, respect, and credibility, is critical in building an effective partnership among all the stakeholders of the project organization.

14.4   RECOMMENDATIONS TO CONSULTING PROFESSIONALS

A number of suggestions have been derived from various field studies in project management consulting (see Block 1999, Bower 1997, Deschamps and Nayak 1995, Rasiel 1999, Sarvary 1999, Thamhain 2004). These suggestions are aimed at helping both managers and consultants to understand the complex interaction of organizational and behavioral variables involved in creating change and improvements in today's intricate project management systems. The findings also help to increase our awareness of what is and is not effective in terms of building high-performance project management systems.

> *Align Project System Improvements with Company Goals.* It is critical to identify value from the start. Organizational members are more likely to engage in a development initiative if they perceive it to be clearly related to the goals and objectives of the company. Clear linkages between the intervention and the company's mission, goals, and objectives create enthusiasm and the desire to participate as well as lowering anxieties and helping to unify the people behind the intervention. Thus, it is important for the consultant to both understand and

communicate the significance of the project to the company's mission. Explicitly securing senior management's support and endorsement further enhances these benefits.

Take a Systems Approach. Many project management improvement efforts fail due to a poor understanding of the need to integrate project management with the total business process and its management systems. Subsystems, such as specific project management initiatives, are easier and quicker to develop than is the overall organizational system to which they belong. Yet, they function suboptimal at best unless they are designed as an integrated part of the total enterprise. System thinking, as described by Senge (1990), provides a useful tool for front-end analysis and organizational design.

Define Clear Objectives and Direction. Consulting leaders must ensure that the people in the organization perceive the objectives of the intervention to be attainable, with a clear sense of direction toward reaching those ends. Maintaining this perception can be influenced by both the consultant and the organization's managers, and should be continuously reinforced during the implementation cycle. More than any other tool, well-articulated business performance metrics—ranging from product plans to mission statements, technology objectives, and financial goals—can cut through the complexities and fog of "management performance," further connecting the project to the mission and strategic direction of the enterprise.

Build Trust, Respect, and Credibility with People Affected by the Consulting Intervention. While top-down expert-based consulting has dominated the high-technology management field, it is critical to engage also in process-oriented consulting (O'Connell 1990), ensuring that relevant organizational members are part of the intervention. Consultants should act as facilitators of the organizational development, encouraging clients to help themselves. Operational procedures, quality standards, and management practices must be generated from within the organization as a prerequisite for becoming institutionalized. Key project personnel and managers from all functions and levels of the organization should be involved in assessing the situation, searching for solutions, and evaluating new tools and techniques. Consultants need to ensure organizational acceptance of any changes, including the introduction of new tools and methods, especially in self-directed team environments. While direct participation in decision making is the most effective way to obtain buy-in toward a new system, it is not always possible, especially in large organizations. Critical factor analysis, focus groups, and process action teams are also good vehicles for team involvement and collective decision making, leading to ownership, greater acceptance, and willingness to work toward continuous improvement.

Assess Benefits and Challenges. Before introducing a new project management tool or technique to an organization, it is important to conduct a thorough assessment of the potential benefits, problems, and challenges that can be anticipated. A focus group—selected from the project population at large or specific project team—can provide valuable insight into the challenges of getting the new tool accepted and working at the team level. This assessment can also provide many of the necessary inputs for customizing the new tool to the organization, identifying the support systems and interfaces needed for effective implementation.

Build on Existing Management Systems. This is similar to suggestions made in Chapter 4 in support of new project management process implementation. Brand-new methods are greeted with suspicion. The more congruent the new order is with already existing practices, procedures, and distributed knowledge of the organization, the more self-sustaining and successful the change intervention will be. The highest level of acceptance and success is found in areas where new tools are *added incrementally* to already existing management systems. These situations should be identified and addressed first. Consultants should make an effort to integrate already established and proven procedures for project definition, documentation, status reports, reviews, and sign-offs into the new process. Building upon an existing project management system also facilitates incremental enhancement, testing, and fine-tuning of the new managerial tools or process.

Define Checks and Balances. Consultants should set up a system, enabling the people in the organization to monitor, validate, and audit the implementation of the new system, and to assess the impact on organizational performance. Such checks and balances also provide in-process feedback, early warnings of potential problems, and the opportunity for in-process correction and fine-tuning of the new management system.

Foster a Culture of Continuous Support and Improvement. A successful project intervention centers on people's behavior and their role within the project itself. Companies that are effective in creating and integrating new management processes into their organizations have cultures and support systems that demand broad participation in their organizational developments. Ensuring that organizational members are more proactive and aggressive toward change is not an easy task, yet it must be facilitated systematically by both consultants and management. Any project management tool or technique, for example, must be integrated into the continuously changing business process. Provisions should be made for updating and fine-tuning these tools on an ongoing basis to ensure relevancy to current project management challenges. Therefore, it is important to establish support systems—such as discussion groups, action teams, and suggestion systems—to capture and leverage the lessons learned and to identify problems as part of a continuous improvement process. Consultants should work with senior management to establish incentives, norms, and practices for going beyond encouraging meetings and dialogue, and encourage organizational members to take proactive approaches toward continuous organizational improvement.

Ensure Senior Management's Leadership and Support. Consulting interventions require top-down support to succeed. Team members will be more likely to work within new guidelines and processes and cooperate with necessary organizational requirements if senior management clearly articulates the criticality and relevancy to business performance and the benefits to the enterprise. Senior management involvement and encouragement are often seen as recognition of team competence and a validation of effective team work. Throughout the implementation phase, both senior management and consultants can influence the attitude and commitment of their people toward a new concept or process by their own actions. Concern for team members, assistance with the work, and enthusiasm for the project all foster a climate of high motivation, involvement, open communication, and willingness to cooperate with the consultant-guided project management process.

14.5 CONCLUSIONS

In today's dynamic and hypercompetitive world of business, project management consultants must understand all components of the project environment. This includes not only the latest tools and techniques of project management, but also the people, business processes, cultures, and value systems of the organization. Being able to leverage the tools and organizational resources, and to provide tangible value added to the client company, requires more than just writing a new procedure, delivering a best-practice workshop, or installing new information technology. It requires the ability to engage the client organization in a systematic evaluation of its competencies, assessing opportunities for improvement and linking existing competencies to the overall enterprise system and its strategy. Effective consultants understand the complex interaction of organizational and behavioral variables. They can work with the client organization, creating win-win situations between the project team and senior management. By involving the firm's most experienced and respected project personnel and business leaders, and by analyzing the critical functions that drive project performance, consultants can help to shake up conventional thinking and create a vision for their clients. Through the interaction of all stakeholders, management can gain insight into the critical functions and cultures that drive project performance and identify those critical components that could be further optimized. It is important to keep in mind that proactive participation and commitment of the key stakeholders is critical to success. In addition, congruency of the intervention with the overall business process and the management system greatly facilitates and expedites the implementation of any new or changed operational method.

Too many clients end up disappointed that the latest management technique did not produce the desired result. Regardless of their conceptual sophistication, any management tool, such as tracking software, concurrent engineering, design-build, Stage-Gate, voice of the customer, and theory of constraints, is *just a framework* for processing project data and for aligning organizational strategy, structure, and people. To produce benefits to the firm, these tools must be fully customized to the business process and congruent with the organizational system and its culture. As a result, few firms hire consultants based just on their knowledge of management tools or techniques. Such knowledge is considered a threshold competency and is expected. True value is added to the client firm by helping managers to identify the areas that hold potential for improvement and by designing the new tool or system as an integrated part of the business process, fully leveraged with all other organizational functions.

Most importantly, project management consultants must pay attention to the human side. While traditionally such an emphasis has not been a strong point of specialized area experts, it is critically important for enhancing cooperation among project stakeholders. Such cooperation is greatly enhanced when consultants foster a work environment where people see the purpose and significance of the intervention for the enterprise and its specific mission. One of the strongest catalysts to change and cooperation is the professional pride and excitement of organizational members, fueled by visibility and recognition. Such a professionally stimulating environment seems to lower anxieties over organizational change, reduce communication barriers and conflict, and enhance the desire of those involved to cooperate and succeed. Effective consultants are social architects who can foster a climate of active participation by involving people at all organizational levels in the assessment of the current project management system and in the planning and implementation of the change process. They also can build alliances with support organizations and upper management to ensure organizational visibility, priority, resource availability, and overall support for sustaining the organizational improvement beyond its implementation phase. A better understanding of the existing project management system, and the criteria and organizational dynamics that drive the system, can help consultants, together with the project team and senior management, to diagnose and implement potential system improvements.

14.6 SUMMARY OF KEY POINTS AND CONCLUSIONS

- There is a growing sense of disappointment and frustration that not all project management techniques work equally well or are equally applicable to all business ventures.

- The complexities of project management and its significance for business performance have convinced many technology companies to establish an *in-house* management consulting group or seek consulting help from the outside.

- Merely hiring consultants, whether internal or external, does not automatically guarantee business success.

- Consultants must be capable of understanding more than the tools and techniques of modern management. They must also understand the infrastructure of their client organization and deal with the complex social, technical, and economic issues that determine the culture and value system of the enterprise.

- Few companies rush into consulting services, which are often viewed as costly and disruptive, at times even in conflict with deeply held organizational beliefs about the way things should be done.

- Companies that make the consulting process work are able to attract and engage services that can help the enterprise to augment expertise and skills in specialty areas, benchmark internal and external capabilities, and challenge conventional wisdom.
- Engaging the organization in the evaluation of a current system and aiming for improvements can be a threatening process.
- The strongest drivers toward cooperation with management are derived from the work itself, including personal interest, pride, and satisfaction with the work; professionally challenging work; a feeling of accomplishment; and recognition.
- Effective consultants understand the complex interaction of organizational and behavioral variables. They can work with the client organization, creating win-win situations between the project team and senior management.

14.7   CRITICAL THINKING: QUESTIONS FOR DISCUSSION

1. What type of help are client companies typically looking for from consultants?
2. Why are client organizations often not fully supportive and cooperative with consulting services?
3. What type of characteristics are client organizations looking for in a management consultant? Discuss. How can you win the trust, respect and credibility of your clients at (1) the senior management level and (2) the project team level?
4. Can consulting on work process improvements be separated from consulting on project implementation? Discuss.
5. Why is senior management critical to successful consulting at the project team level? Discuss.
6. Develop an action plan of homework in preparation for an initial client visit.
7. How would you assess your effectiveness as a consultant (1) at the early stages of your assignment, (2) at the midpoint of your assignment, and (3) at the conclusion of your assignment? Discuss.

14.8   REFERENCES AND ADDITIONAL READINGS

Bahrami, H. (1992) "The emerging flexible organization: perspectives from Silicon Valley," *California Management Review*, Vol. 34, No. 4, pp. 33–52.

Barner, R. (1997) "The new millennium workplace," *Engineering Management Review (IEEE)*, Vol. 25, No. 3, pp. 114–119.

Bishop, S. K. (1999) "Cross-functional project teams in functionally aligned organizations," *Project Management Journal*, Vol. 30, No. 3, pp. 6–12.

Block, P. (1999) *Flawless Consulting*, San Francisco: Jossey-Bass.

Bower, M. (1997). *Will to Lead: Running a Business with a Network of Leaders*, Cambridge, MA: HBS Press.

Dawes, P. (2004) "A model of the effects of technical consultants on organizational learning in high-technology purchase situations," *Journal of High Technology Management Research*, Vol. 14, No. 1, pp. 1–20.

Deschamps, J. and Nayak, P. R. (1995) "Implementing world-class process," Chapter 5 in *Product Juggernauts*, Cambridge, MA: HBS Press, pp. 175–214.

Gupta, A. and Govindarajan, V. (2000) "Knowledge management's social dimension," *Sloan Management Review*, Vol. 42, No. 1, pp. 71–80.

Iansiti, M. and MacCormack, A. (1997) "Developing product on Internet time," *Harvard Business Review*, Vol. 75, No. 5, pp. 108–117.

Kruglianskas, I. and Thamhain, H. (2000) "Managing technology-based projects in multinational environments," *IEEE Transactions on Engineering Management*, Vol. 47, No. 1, pp. 55–64.

O'Connell, J. J. (1990) "Process consulting in a context field: Socrates on strategy," *Consultation*, Vol. 9, No. 3, pp. 199–208.

Rasiel, E. (1999) *The McKinsey Way*, New York: McGraw-Hill.

Reimus, B. (2000) "Knowledge sharing within management consulting firms," *Report*, Fitzwilliam, NH: Kennedy Information, Inc., *www.kennedyinfo.com/mc/gware.html*.

Sarvary, M. (1999) "Knowledge management and competition in the consulting industry," *California Management Review*, Vol. 41, No. 2, pp. 95–107.

Senge, P. M. (1990) *The Fifth Discipline: The Art and Practice of the Learning Organization*, New York: Doubleday/Currency.

Scott, B. (200) *Consulting from the Inside*. New York: American Society for Training and Development.

Thamhain, H. J. (1994) "Designing project management systems for a radically changing world," *Project Management Journal*, Vol. 25 No. 4, pp. 6–7.

Thamhain, H. (2001) "Team management," Chapter 19 in *Project Management Handbook* (J. Knutson, ed.), New York: Wiley & Sons.

Thamhain, H. (2002) "Criteria for effective leadership in technology-oriented project teams," Chapter 16 in *The Frontiers of Project Management Research* (Slevin, Cleland, and Pinto, eds.). Newton Square, PA: Project Management Institute, pp. 259–270.

Thamhain, H. (2003) "Managing innovative R&D teams," *R&D Management*, Vol. 33, No. 3 (June), pp. 297–312.

Thamhain, H. (2004) "15 rules for effective project management consulting," *Consulting to Management Journal*, Vol. 15, No. 2 (June), pp. 42-46.

Thamhain, H. (2004). "Leading technology-based project teams," *Engineering Management Journal*, Vol. 16, No. 2, pp. 42–51.

Thamhain, H. J. and Wilemon, D. L. (1999) "Building effective teams for complex project environments," *Technology Management*, Vol. 5 No. 2, pp. 203–212.

Thomas, J., Delisle, C., Jugdev, K., and Buckle, P. (2001) "Selling project management to senior executives," *pmNetwork*, Vol. 15, No. 1, pp. 59–62.

Appendix 1

POLICY AND PROCEDURE EXAMPLES FOR MANAGEMENT OF TECHNOLOGY

Four sample documents are shown in this Appendix to illustrate the type of tools that should be developed in support of the organizational design for the enterprise, and to provide the basis for managerial direction, communication, and control:

Appendix 1.1
Policy Directive: Engineering Management

Alpha Technology Company
ABC Division

Policy Directive No. 01.2345.05
Engineering Management   Effective Date: May 1, 2005

1. Purpose

To provide the basis for the management of engineering activities within the ABC Division. Specifically, the policy establishes the relationship between program and functional operations and defines the responsibilities, authorities, and accountabilities necessary to ensure engineering quality and operational performance consistent with established business plans and company policies.

2. Organizations Affected

All engineering departments of the ABC Division.

3. Definitions

3.1 *Unit Manager.* Individual responsible for managing an engineering group. The unit manager reports to a department manager.

3.2 *Department Manager.* Individual responsible for managing an engineering department, usually consisting of various engineering groups, each headed by a unit manager. The department manager reports to the director of engineering.

3.3 *Lead Engineer or Task Manager.* Individual jointly appointed by the program manager and the functional superior. The lead engineer or task manager is responsible for the cross-functional integration of the assigned task.

3.4 *Task Authorization.* A one-page document summarizing the program requirements, including tasks, results, schedules, budgets, and key personnel.

3.5 *Annual Engineering Plan.* Document summarizing the specific goals and objectives of the engineering function within the ABC Division, as well as specific programs to be funded for the next year.

3.6 *Department Budget.* Document establishing the funding level for each engineering department based on activities and other organizational needs.

4. Applicable Documents

4.1 Supervisor's Handbook.

4.2 Program Management Guidelines.

4.3 Procedures on Engineering Management, No. 04.11.03.

5. Policy

5.1 All engineering departments are organized along functional lines with each individual reporting to one functional superior. The chain of command is established as: unit manager, department manager, director of engineering. The director of engineering is appointed by the divisional general managers.

5.2 Specific engineering departments are chartered by the director of engineering, who is also responsible for making the managerial appointments.

5.3 The engineering organization, through its various departments, is chartered to accomplish the following functions:

1. Execute engineering programs according to company-external or -internal requests.

2. Support the business divisions in the identification pursuit, and acquisition of new contract business.

3. Provide the necessary program direction and leadership within engineering for planning, organizing, developing, and integrating technical efforts. This includes the establishment of program objectives, requirements, schedules, and budgets as defined by contract and customer requirements.

4. Maintain close liaison with customer, vendor, and educational resources, keeping abreast of state-of-the-art and technological trends.

5. Plan, acquire, and direct outside and company-funded research, which enhances the company's competitive position.

6. Plan, utilize, and develop all engineering resources to meet current and future requirements for quality, cost-effectiveness, and state-of-the-art technology.

5.4 The activity level is defined by department budgets, which are based on company-internal or -external requests for specific services or programs.

5.5 Resource requirements for new engineering programs can be introduced at any point of the engineering organization. However, these requests must be reported via the established chain of command to the director of engineering, who coordinates and assigns the specific activities or programs to organizational units.

5.6 The basis for resource planning is the annual engineering plan, which is to be coordinated with the division's annual business plan.

5.7 The program manager is responsible for defining a preliminary program plan as a basis for negotiating workforce resources with functional managers.

5.8 Unit managers appoint lead engineers or task managers as needed for each program executed within the unit manager's organization.

5.9 No engineering program will be started without proper program definition, which includes at the minimum (1) the specific tasks to be performed, (2) the overall performance specifications, (3) a definition of the overall task configuration, (4) the deliverable items or results, (5) a master schedule, and (6) the program budget. Five percent of the total program budget is a reasonable expenditure for the program definition phase.

5.10 Responsibility for defining an engineering program rests with the program manager, who will coordinate the requirements and capabilities among the sponsor and resource organizations, including engineering.

5.11 The principle instrument for summarizing contractual requirements of (1) technical requirements, (2) deliverable items, (3) schedules, (4) budgets, and (5) responsible personnel is the *task authorization*.

5.12 The lead engineer or task manager is responsible for the cross-functional integration of the assigned engineering task. He or she has dual accountability to (1) his/her functional superior for the most cost-effective technical implementation, including technical excellence and quality of workmanship, and (2) the program manager for the most efficient implementation of the assigned task, according to the established specifications, schedules, and budgets.

5.13 Functional management, including their designated task leaders, will be responsive to the program manager regarding the contractual requirements, including performance, schedules, budgets, and changes, as well as reporting requirements and customer presentations.

5.14 The responsibility for (1) quality of workmanship and (2) method of task implementation rests explicitly with the engineering organization.

5.15 Conflicts over administrative or technical issues between the functional and program organization must be dealt with in good faith and on an timely basis. If no resolution can be found, the problem must be reported to the responsible manager at the next level (if necessary up to the divisional general manager) who is responsible for resolving the conflict.

Appendix 1.2
Position Charter for Product Design Manager

POSITION TITLE: PRODUCT DESIGN MANAGER

Authority. The manager of design engineering has the delegated authority from the director of engineering to establish, develop, and deploy organizational resources according to established budgets and business needs, both long and short range. This authority includes the hiring, training, maintaining, and terminating of personnel; the development and maintenance of the physical plant, facilities, and equipment; and the direction of the engineering personnel regarding the execution and implementation of specific engineering programs.

Responsibility. The manager of design engineering is accountable to the director of engineering for providing the necessary direction for planning, organizing, and executing the engineering efforts in his or her cognizant area. This includes (1) establishment of objectives, policies, technical requirements, schedules, and budgets as required for the effective execution of engineering programs; (2) assurance of engineering design quality and workmanship; and (3) cost-effective implementation of established engineering plans. The product design manager further seeks out and develops new methods and technology to prepare the company for future business opportunities.

Appendix 1.3
Position Charter for Team Leader, Lead Engineer, or Task Manager in a Matrix Organization

POSITION TITLE: TEAM LEADER, SOLAR POWER CONTROL PANEL (SPCP) DEVELOPMENT

Authority. Appointed jointly by the director of engineering and the program manager of the *Portable Solar Power Generator System (PSPGS)*, the lead engineer has the authority to direct the implementation of the SPCP development within the functional organization according to the established plans and requirements.

Responsibility. Reporting to both the director of engineering and the PSPGS program manager, the lead engineer is responsible: (1) to the functional resource manager(s) for the most cost-effective technical implementation of the assigned program task, including technical excellence and quality of workmanship; and (2) to the program manager for the most efficient implementation of the assigned program task according to the program plan, including cost, schedule, and specified performance. Specifically, the team leader:

1. Represents the program manager within the functional organization.
2. Develops and implements detailed work package plans from the overall program plan.
3. Acts for the functional manager in directing personnel and other functional resources toward work package implementation.
4. Plans the functional resource requirements for his or her work package over the life cycle of the program and advises functional management accordingly.
5. Directs subfunctional task integration according to the work package plan.
6. Reports to program office and functional manager on the project status.

Appendix 1.4
Job Description for R&D Project Manager

JOB DESCRIPTION NO. 07.1285.05: R&D PROJECT MANAGER

The project manager is the key individual charged with the responsibility for the total project and is delegated the authority by management. He or she is also assigned a supporting staff, which may include individuals directly responsible to him or her (that is, administratively and technically) as well as others technically responsible to the project manager though administratively located in other parts of the organization. The responsibilities and authorities of the project manager can be summarized as follows:

1. Establish and effectively operate an integration effort that is based on rapid feedback from each element of project activity.

2. Implement an appropriate philosophy of how project work will be done, avoiding cumbersome formality without sacrificing traceability and project intercommunications.

3. Plan the work and break it down into understandable elements with well-defined technical, schedular, and financial aspects. Maintain an up-to-date plan that integrates and reflects the actual work status of these work elements.

4. Analyze the total project work against allocated resources and establish an optimum project organization to achieve the project objectives. Reorganize this project organization as required to meet the changing needs of the project.

5. Assign responsibility for discrete elements of project work (contract and in-house) to members of his or her project organization and redelegate sufficient of his or her own authority to these members to enable them to contribute effectively to the successful completion of the project.

6. Serve as the focal point of responsibility for all actions relating to accomplishment of the total project, including both in-house and contract effort.

7. Ensure, by direct participation and review of the work of others, that each contract effort in support of the project is well planned and fully defined during the preparation of the technical package, thus supporting the customer solicitation (that is, RFPs) with a technically responsive proposal.

8. Ensure by direct participation in negotiations, that the sponsor organization and our company have a common understanding of the goals and requirements of each contract before contract execution.

9. Maintain a continuous surveillance and evaluation of all aspects of the work being conducted by each contractor, including technical, schedular, and financial status.

10. Identify, devise, and execute effective solutions to management and/or technical problems that arise during the course of the work.

11. Ensure the timely detection and correction of oversights in any aspect of each contract to minimize cost overruns, schedule delays, and technical failures.

12. Make final decisions within the scope of work of each contract.

13. Give guidance and technical direction for all elements of the project work, or concur in contractor actions in accordance with provisions of each contract.

14. Continually evaluate the quality of the work performed by each contractor and verify that the requirements are properly carried out by the contractor.

15. Ensure that configuration management requirements are adhered to by each contractor in accordance with the requirements of the contract.

16. Help to indoctrinate and regularly monitor contractor performance to ensure full reporting of all new technology evolved under each project contract.

17. Participate in contract closeout or in termination proceedings to ensure that the best interests of the company are safeguarded.

Appendix 2

PROFESSIONAL SOCIETIES IN SCIENCE, ENGINEERING, AND TECHNOLOGY MANAGEMENT

The listing that follows represents professional societies with members interested in the practice and advancement of MOT knowledge. The listing focuses on U.S. organizations, but also includes others from the international field with connections to U.S. societies and professional joint activities with the U.S. The listing is for reference only, and is strictly based on the subjective visibility of these organizations, and their MOT-related activities, as they appear to the author. No attempt is being made to rank or exclude any particular organization.

Academy of Applied Science (AAS); 24 Warren Street, Concord, NH 03301; Ph (603) 228-4530, Fx (603) 228-4730; *info@aas-world.org*; *www.aas-world.org*.

Academy of Management (AOM); P.O. Box 3020, Briarcliff Manor, NY 10510-8020; Ph (914) 923-2607, Fx (914) 923-2615; *www.aomonline.org*.

American Association for the Advancement of Science (AAAS); 1515 Massachusetts Avenue, NW, Washington, DC 20005; Ph (202) 467-4400.

American Association of Industrial Management (AAIM); Stearns Building, Suite 206, 293 Bridge Street, Springfield, MA 01103; Ph (413) 737-8766, (413) 737-9724; *aaimn-mta@aol.com, www.aaimnmta.com*.

American Chemical Society (ACS); 1155 Sixteenth Street, NW, Washington, DC 20036; Ph (800) 227-5558, Fx (202) 872-6067; *help@acs.org, www.chemistry.org*.

American Institute of Aeronautics and Astronautics (AIAA); 1801 Alexander Bell Drive, Suite 500, Reston, VA 20191-4344; Ph (703) 264-7500, Fx (703) 264-7551; *custserv@aiaa.org, www.aiaa.org*.

American Institute of Chemical Engineers (AIChE); 3 Park Ave, New York, N.Y., 10016-5991; Ph (212) 591-7338, Fx (212) 591-8897; *xpress@aiche.org*, *www.aiche.org*.

American Institute of Mining, Metallurgical, and Petroleum Engineers (AIME); 8307 Shaffer Parkway, Littleton, CO 80127-4012; Ph (303) 948-4255, Fx (303) 948-4260; *aime@aimehq.org*, *www.aimeny.org*.

American Institute of Physics (AIP); One Physics Ellipse, College Park, Maryland 20740-3843; Ph (301) 209-3100; *www.aip.org/index.html*.

American Management Association (AMA); 1601 Broadway New York, NY 10019; Ph (212) 586-8100, Fx (212) 903-8168; *www.amanet.org/service/index.htm*.

American Physical Society (APS); One Physics Ellipse, College Park, MD 20740-3844; Ph (301) 209-3200, Fx (301) 209-0865; *opa@aps.org*, *www.aps.org*.

American Society for Engineering Education (ASEE); 1818 N Street, N.W., Suite 600, Washington, DC 20036-2479; Ph (202) 331-3500, Fx (202) 265-8504; *aseeexec@asee.org*, *www.asee.org*.

American Society for Engineering Management (ASEM); Stevens Institute of Technology, Castle Point on Hudson Hoboken, NJ 07030; Ph (201) 216-8103/8025, Fx (201) 216-5541; *jfarr@stevens-tech.edu*, *www.asem.org/home.html*.

American Society for Quality (ASQ); 600 North Plankinton Avenue, Milwaukee, WI 53203; Ph (414) 272-8575, Fx (414) 272-1734; *help@asq.org*, *www.asq.org*.

American Society for Training and Development (ASTD); 1640 King Street, Box 1443, Alexandria, Virginia 22313-2043; Ph (703) 683-8100, Fx (703) 683-8103; *www.astd.org/astd*.

American Society of Civil Engineers (ASCE); 1801 Alexander Bell Drive, Reston, Virginia 20191-4400; Ph (800) 548-2723, Fx (703) 295-6222; *www.asce.org*.

American Society of Mechanical Engineers (ASME); Three Park Avenue, New York, NY 10016-5990; Ph. (800) 843-2763, Fx (973) 882-1717; *infocentral@asme.org*, *www.asme.org*.

American Society of Metals (ASM); 9639 Kinsman Road, Materials Park, OH 44073-0002; Ph (440) 338-5151, Fx (440) 338-4634; *cust-srv@asminternational.org*, *www.asminternational.org*.

Association for Computing Machinery (ACM); 1515 Broadway, New York, NY 10036; Ph (212) 626-0500, Fx (212) 944-1318; *support@acm.org*, *www.acm.org*.

Association for Consulting Chemists and Chemical Engineers (ACC&CE); P.O. Box 297, Sparta, New Jersey 07871; Ph (973) 729-6671, Fx (973) 729-7088; *info@chemsonsult.org*, *www.chemconsult.org*.

Association for Facilities Engineering (AFE); 8160 Corporate Park Drive, Suite 125, Cincinnati, OH 45242; Ph (513) 489-2473, Fx (513) 247-7422; *mail@afe.org*, *www.afe.org*.

Association for Federal Information Resources Management (AFFIRM); P.O. Box 2851, Washington, DC 20013; Ph (202) 208-2780; *www.affirm.org*.

Association for Project Management Professionals (APMP); 300 Smelter Avenue NE #1, Great Falls, MT 59404; Ph (406) 454-0090; *www.apmp.org*.

Association for Science, Technology, and Innovation (ASTI); P.O. Box 1242, Arlington, VA 22210; (703) 241-2850; *www.washacadsci.org/asti*.

Association for the Advancement of Cost Engineering International (AACE); 308 Monongahela Building, Morgantown, WV 26505; Ph (304) 296-8444; *info@aacei.org*; *www.aacei.org*.

Association for Women in Science (AWIS); 1200 New York Ave., Suite 650, Washington, DC 20005; Ph (202) 326-8940, Fx (202) 326-8960; *awis@awis.org*, *www.awis.org*.

Decision Sciences Institute (DSI); 35 Broad Street, Atlanta, GA 30303; Ph (404) 651-4073, Fx (404) 651-2804; *dsi@gsu.edu, www.decisionsciences.org*.

Electro-Chemical Society (ECS); 65 South Main Street, Building D, Pennington, NJ; Ph (609) 737-1902, Fx (609) 737-2743; *ecs@electrochem.org, www.electrochem.org*.

Engineering Management Society (EMS), Division of IEEE; Institute of Electrical and Electronics Engineers (IEEE); 3 Park Avenue, 17th Floor, New York, New York, 10016-5997; Ph (212) 419-7900, Fx (212) 752-4929; *www.ewh.ieee.org/soc/ems*.

Industrial Relations Research Association (IRRA); 121 Labor and Industrial Relations, University of Illinois, 504 E. Armory, MC-504 , Champaign, IL 61820; Ph (217) 333-0072, Fx (217) 265-5130; *irra@uiuc.edu, www.irra.uiuc.edu*.

Industrial Research Institute (IRI); 2200 Clarendon Boulevard, Suite 1102 Arlington, VA 22201; Ph (703) 647-2580, Fx (703) 647-2581; *www.iriinc.org*.

Institute for Operations Research and the Management Sciences (INFORMS); 901 Elkridge Landing Road, Suite 400, Linthicum, MD 21090-2909; Ph (800) 4INFORMS, Fx (410) 684-2963; *informs@mail.informs.org, www.informs.org*.

Institute for the Advancement of Engineering (IAE); Box 26241, Los Angeles, CA 90026; Ph (213) 413-4036.

Institute of Electrical and Electronics Engineers (IEEE); 3 Park Avenue, 17th Floor, New York, New York 10016-5997; Ph (212) 419-7900, Fx (212) 752-4929; *webmaster@ieee.org, www.ieee.org*.

Institute of Industrial Engineers (IIE); 3577 Parkway Lane Suite 200 Norcross, GA 30092; Ph (800) 494-0460, (770) 441-3295; *cs@iienet.org, www.iienet.org*.

Institute of Industrial Engineers (IIE); 3577 Parkway Lane Suite 200 Norcross, GA 30092; Ph (800) 494-0460, Fx (770) 441-3295; *cs@iienet.org, www.iienet.org*.

Institute of Management Consultants (IMC); 2025 M Street N. W. Suite 800, Washington, DC 20036-3309; Ph (202) 367-1134, Fx (202) 367-2134; *gaylen@imcusa.org, www.imcusa.org*.

Institute of Management Sciences (IMS); 146 Westminster Street, Providence, NJ 02903; Ph (401) 274-2525.

Instrumentation, Systems, and Automation Society (ISA); 67 Alexander Drive, Research Triangle Park, NC 27709; Ph (919) 549-8411, Fx (919) 549-8288; *info@isa.org, www.isa.org*.

Insulated Cable Engineers Association (ICEA); P.O. Box 1568 Carrollton, Georgia 30112; www.icea.net.

International Association for Management of Technology (IAMOT); University of Miami, College of Engineering, 248294 Coral Gables, FL 33124-0623; Ph (305) 284-2344, Fx (305) 284-4040; *iamot@miami.edu, www.iamot.org*.

International Association of Contract and Commercial Managers (IACCM); 90 Grove Street, Suite 01, Ridgefield, CT 06877; Ph (203) 431-8741, Fx (203) 431-9305; *info@iaccm.com, www.iaccm.com*.

International Consultants Foundation (ICF); 5605 Lamar Road, Bethesda, MD 20816; Ph (301) 320-4409.

International Project Management Association (IPMA); P.O. Box 1167, 3860 BD, Nijkerk, The Netherlands; *www.ipma.ch/intro*.

National Academy of Engineering (NAE); 500 Fifth Street, NW, Washington, DC 20001; Ph (202) 334-3200, Fx (202) 334-2290; *www.nae.edu*.

National Contract Management Association (NCMA); 8260 Greensboro Drive, Suite 200, McLean, VA 22102; Ph (571) 382-0082, Fx (703) 448-0939; *www.ncmahq.org*.

National Mining Association (NMA); 101 Constitution Avenue, NW, Suite 500 East, Washington, DC 20001-2133; Ph (202) 463-2600, Fx (202) 463-2666; *craulston@nma.org*, *www.nma.org*.

National Society of Professional Engineers (NSPE); 1420 King Street, Alexandria, VA 22314-2794; Ph (703) 684-2800, Fx (703) 836-4875; *pr@nspe.org*, *www.nspe.org*.

Optical Society of America (OSA); 2010 Massachusetts Ave., N.W., Washington, DC 20036-1023; Ph (202) 223-8130, (202) 223-1096; *info@osa.org*, *www.osa.org*.

Product Development and Management Association (PDMA); 17000 Commerce Parkway, Suite C, Mount Laurel, NJ 08054; Ph (800) 232-5241, Fx (856) 439-0525; pdma@pdma.org, *www.pdma.org*.

Project Management Institute (PMI); Four Campus Boulevard, Newtown Square, PA 19073-3299; Ph (610) 356-4600, Fx (610) 356-4647; *pmihq@pmi.org*, *www.pmi.org*.

Society for the Advancement of Management (SAM); SAM International Office, Texas A&M University, 6300 Ocean Drive - FC 111, Corpus Christi, TX 78412; Ph (888) 827-6077, Fx (361) 825-2725; *moustafa@cob.tamucc.edu*, *www.cob.tamucc.edu/sam*.

Society of Automotive Engineers (SAE); 400 Commonwealth Drive, Warrendale, PA 15096-0001; Ph (724) 776-4841, Fx (724) 776-0790; *customerservice@sae.org*, *www.sae.org*.

Society of Concurrent Product Development (SCPD); P.O. Box 68, Dedham, MA 02027-0068; Ph (949) 643-5046; *webmaster@soce.org*, *www.scpdnet.org*.

Society of Research Administrators International (SRA); 1901 North Moore Street, Suite 1004, Arlington, VA 22209; Ph (703) 741-0140, Fx (703) 741-0142; *Info@srainternational.org*, *www.srainternational.org*.

Society of Women Engineers (SWE); 230 East Ohio St., Suite 400 Chicago, IL 60611; Ph (312) 596-5223; *hq@swe.org*, *www.swe.org/stellent/idcplg?IdcService=SS_GET_PAGE&nodeId=5*.

Technology Transfer Society (TTS); 435 North Michigan Avenue, Suite 1717, Chicago, IL 60611-4067; Ph (352) 955-0066, Fx (312) 644-8557; *t2s@t2s.org*.

Appendix 3

PROFESSIONAL JOURNALS IN ENGINEERING AND TECHNOLOGY MANAGEMENT

The listing that follows represents professional journals dedicated, at least in part, to the advancement of knowledge in the field of management of technology, MOT. The listing focuses on U.S.-based publications, but also includes publications from other countries with readership in the United States. Because of the similarities in managerial issues, challenges and support systems, many of these journals are being read worldwide, regardless of their origin. The listing is for reference only, and is based on the visibility of these publications, and their MOT-related content, as perceived by the author. No attempt is being made to rank or exclude any particular publication.

Academy of Management Executive. Published by the Academy of Management, Briarcliff Manor, NY; *http://aom.pace.edu/AME.*

Academy of Management Journal. Published by the Academy of Management, Briarcliff Manor, NY; *http://aom.pace.edu/amjnew.*

Academy of Management Review. Published by the Academy of Management, Briarcliff Manor, NY; *http://aom.pace.edu/AMR.*

Aerospace America. Published by the American Institute of Aeronautics and Astronautics, Reston, VA; *www.aiaa.org, www.aiaa.org/content.cfm?pageid=167.*

AIAA Journal. Published by the American Institute of Aeronautics and Astronautics, Reston, VA; *www.aiaa.org, www.aiaa.org/content.cfm?pageid=167.*

California Management Review. Published by the University of California at Berkley; *www.haas.berkeley.edu/news/cmr.*

Cost Engineering Journal. Published by the Association for the Advancement of Cost Engineering International, Morgantown, WV; *www.aacei.org/resources/costengineering.shtml.*

Creativity Research Journal. Published by Lawrence Erlbaum Associates; *www.erlbaum. com/ shop/tek9.asp?pg=products&specific=1040-0419.*

Decision Analysis. Published by the Institute for Operations Research and the Management Science (INFORMS), Linthicum, MD; *http://da.pubs.informs.org.*

Decision Sciences Journal of Innovative Education. Published by the Decision Sciences Institute, Atlanta, GA; *www.decisionsciences.org.*

Decision Sciences Journal. Published by the Decision Sciences Institute, Atlanta, GA; *www.decisionsciences.org, www.decisionsciences.org/dsj/index.htm.*

Engineering Management Journal. Published by the American Society for Engineering Management; *www.asem.org/publications/index.html, www.asem.org/home.html.*

Environmental Engineering and Policy. Published by Springer Publishing, Heidelberg, Germany; *http://springerlink.metapress.com/app/home/journal.asp?wasp=g275yjxurh7juq64 qwtm&referrer=parent&backto=linkingpublicationresults,1:101787,1.*

Harvard Business Review. Published by Harvard University, Cambridge, MA; *http://harvard businessonline.hbsp.harvard.edu/b02/en/hbr/hbr_home.jhtml.*

IBM Systems Journal. Published by IBM, White Plains, NY; *www.research.ibm.com/ journal/sj.*

IEEE Engineering Management Review. Published by the IEEE Engineering Management Society, New York, NY; *www.ieee.org/organizations/pubs/magazines/emr.htm.*

IEEE Transactions on Engineering Management. Published by the IEEE Engineering Management Society, New York, N Y; *www.andromeda.rutgers.edu/~ieeetem.*

Information Systems Research. Published by the Institute for Operations Research and the Management Science (INFORMS), Linthicum, MD; *http://isr.katz.pitt.edu.*

International Journal of Innovation Management. Published by World Scientific Publishing, Hackensack, NJ; *www.worldscinet.com/ijim/ijim.shtml.*

Journal of Engineering and Technology Management. Published by Elsevier, New York, NY; *http://authors.elsevier.com/JournalDetail.html?PubID=505648&Precis=DESC.*

Journal of Management in Engineering. Published by the American Society of Civil Engineers. Reston, VA; *www.pubs.asce.org/journals/me.html.*

Journal of Management Systems. Published by Bentley College, Waltham, MA; *http://jmis. bentley.edu.*

Journal of Manufacturing Science and Engineering. Published by the American Society of Mechanical Engineers, Fairfield, NJ; *www.asme.org/pubs/journals.*

Journal of Product Innovation Management. Published by the Product Development & Management Association, Mount Laurel, NJ; *www.blackwellpublishing.com/journal. asp?ref=0737-6782.*

Journal of Professional Issues in Engineering Education and Practice. Published by the American Society of Civil Engineers, Reston, VA; *www.asce.org, www.pubs.asce.org/ journals/jrns.html.*

Leadership and Management in Engineering. Published by the American Society of Civil Engineers. Reston, VA; *www.pubs.asce.org/journals/lenews.html.*

Management and Organization Review. Published by the International Association for Chinese Management Research, Hong Kong University of Science and Technology and Peking University; *www.blackwellpublishing.com/journal.asp?ref=1740-8776.*

Management Science. Published by the Institute for Operations Research and the Management Science (INFORMS), Linthicum, MD; *www.decisionsciences.org, http://man-sci.pubs.informs.org.*

Operations Research. Published by the Institute for Operations Research and the Management Science (INFORMS), Linthicum, MD; *www.jstor.org/journals/0030364X.html.*

Organizational Dynamics. Published by Emerald/Ingenta plc, Bath, UK; *www.ingentaconnect. com/content/mcb/502;jsessionid=9arr8xqnaybk.victoria?.*

Plant/Operations Progress, Process Safety Progress (Quarterly). Published by the American Institute of Chemical Engineers, New York, NY; *www.aiche.org/safetyprogress.*

Project Management Journal. Published by the Project Management Institute, Newtown Square, PA; *www.pmi.org/info/PIR_PMJournal.asp.*

Proposal Management. Published by the Association of Proposal Management Professionals, *www.apmp.org, www.apmp.org/PubsAPMP.asp.*

Quality Management Journal. Published by the American Society for Quality, Milwaukee, WI; *www.asq.org/pub.*

Quality Progress. Published by the American Society for Quality, Milwaukee, WI; www.asq.org/pub/qualityprogress.

R&D Management. Published by Blackwell, Oxford, UK; *www.blackwellpublishing. com/journal.asp?ref=0033-6807.*

Research-Technology Management. Published by the Industrial Research Institute, Arlington, VA.

Six Sigma Forum Magazine. Published by the American Society for Quality, Milwaukee, WI; *www.asq.org/pub.*

Sloan Management Review. Published by MIT, Cambridge, MA; *www.sloanreview. mit.edu/smr.*

Technology Review. Published by MIT, Cambridge, MA; *www.techreview.com.*

The Futurist. Published by the World Future Society, Bethesda, MD; *www.wfs.org/futurist.htm.*

Appendix 4

PROFESSIONAL CONFERENCES IN ENGINEERING AND TECHNOLOGY MANAGEMENT

The listing that follows represents major professional conferences at national or international level, focused on the advancement of MOT knowledge. Most of the conferences listed are managed by U.S.-based organizations, but are often held in countries outside the United States. The listing also includes some conferences organized and managed by institutions outside the United States. Because of the similarities of MOT issues worldwide, many of these conferences draw delegates internationally, regardless of their venue. The listing is for reference only, and is based on the visibility of these conferences, and their MOT-related content, as perceived by the author. No attempt is being made to rank or exclude any particular conference.

Annual Meeting of AACE International. Organized by the Association for the Advancement of Cost Engineering International (AACE), Morgantown, WV; *www.aacei.org*.

Annual Meeting of the AAAS. Organized by the American Association for the Advancement of Science (AAAS), Washington, DC; *www.aaas.org/meetings/Annual_Meeting*.

Annual Meeting of the Academy of Management, AOM. Organized by the Academy of Management (AOM), Briarcliff Manor, NY; *www.aomonline.org*.

Annual Meeting of the Decision Sciences Institute (DSI). Organized by the Decision Sciences Institute (DSI), Atlanta, GA; *www.decisionsciences.org*.

ASCE Civil Engineering Conference and Exposition. Organized by American Society of Civil Engineers (ASCE); Reston, VA; *www.asce.org.*

ASEE Annual Conference and Exposition. Organized by the American Society for Engineering Education (ASEE), Washington, DC; *www.asee.org.*

ASEE/AAAE Global Colloquium on Engineering Education. Organized by the American Society for Engineering Education (ASEE), Washington, DC; *www.asee.org.*

ASEM National Conference. Organized by the American Society for Engineering Management (ASEM); Hoboken, NJ; *www.asem.org/home.html.*

ASTD International Conference and Exposition. Organized by American Society for Training and Development (ASTD), Alexandria, VA; *www.astd.org/ASTD/conferences/ice/ice05/ice05_home.*

Hawaii International Conference of Systems Science, HICSS. Organized by the University of Hawaii; *www.hicss.hawaii.edu.*

IAMOT 14th International Conference on Management of Technology. Organized by International Association for Management of Technology (IAMOT); Coral Gables, FL; *iamot@miami.edu, www.iamot.org.*

IIE Annual Conference & Exhibition. Organized by Institute of Industrial Engineers (IIE); Norcross, GA; *www.iienet.org.*

INFORMS Annual Meeting. Organized by Institute for Operations Research and the Management Sciences (INFORMS), Linthicum, MD; *www.informs.org.*

International Engineering Management Conference, IEMC (IEEE). Organized by the Engineering Management Society (EMS), Division of IEEE; Institute of Electrical and Electronics Engineers (IEEE), New York, NY; *www.ewh.ieee.org/soc/ems.*

Management Roundtable. Organized by The Management Roundtable (Resource for Product & Technology Development), Waltham, MA; *www.roundtable.com.*

PDMA International Conference. Organized by Product Development and Management Association (PDMA), Mount Laurel, NJ; *www.pdma.org.*

PMI Global Congress. Multiple venues: North America, Europe, Asia, Latin America. Organized by the Project Management Institute (PMI), Newtown Square, PA; *www.pmi.org.*

Portland International Conference for Management of Engineering and Technology, PICMET. Organized by the International Center for Management of Engineering and Technology at Portland State University, Portland, OR; *www.picmet.org/main.*

SAM International Business Conference. Organized by Society for the Advancement of Management (SAM), Corpus Christi, TX; *www.cob.tamucc.edu/sam.*

Appendix 5

CENTERS OF TECHNOLOGY MANAGEMENT AND OTHER RESOURCES

The listing that follows represents just a small sample of Internet-accessible centers with technology and innovation management. These sites often provide links to other related resources, and hence provide a starting point for further research. The listing is for reference only and is not intended to be comprehensive.

Centers of Technology and Innovation Management

Boeing Center for Technology, Information & Manufacturing:
http://bctim.wustl.edu

Cambridge Centre for Technology Management at the University of Cambridge:
http://www.ifm.eng.cam.ac.uk/common/az.html

Center for Applied Information Technology at the University of Dallas: *www.caitud.org*

Center for Innovation in Product Development at MIT: *http://mitsloan.mit.edu/faculty/c-innovation.php*

Centre for Management of Technology and Entrepreneurship at the University of Toronto: *http://cmte. chem-eng.utoronto.ca.*

Centre for Management of Technology at the National University of Singapore: *www.fba.nus.sg/rsearch/cmt/index.htm*

Management of Innovation and New Technology (MINT) Research Centre at MacMaster University, Hamilton, Ontario, Canada: *http://mint. mcmaster.ca.*

Management of Technology Program at the University of Minnesota, Minneapolis, Minnesota: *www.cdtl.umn.edu/mot/mot.asp*

Management of Technology Programs at University of Texas, San Antonio

Technology and Innovation Management Division of the Academy of Management: *www.aom.pace.edu/tim.*

US-Asia Technology Management Center at Stamford University: *http://asia. stanford.edu*

Websites on MOT Acronyms, Terms and Definitions

Source: Fiatech	*www.fiatech.org/projects/roadmap /keywords.html*
Source: John Grout	*www.campbell.berry.edu/faculty/ jgrout/acronym.html*
Source: Juran Institute	*www.isixsigma.com/dictionary*
Source: NASA	*http://appl.nasa.gov/resources/ lexicon/terms.html*
Source: National Institute of Standards and Technology	*www.nist.gov/admin/cams_external/ pmp_ver7/pmpacroabbr.html*
Source: National Institute of Standards and Technology	*www.encyclopedia.com/html/N/ NatlI1nstS1T1.asp*

INDEX